LLANELLI PUBLIC LIBRARY LOCAL HISTORY RESEARCH GROUP SERIES:

No. 1 Coal mining in the Llanelli area: Vol. 1. 16th Century to 1829 by M. V. Symons.

Frontispiece — The Llanelly Copperworks and Dock 1818. Artist unknown. A view from high ground at Machynis. (By kind permission of the National Museum of Wales).

COAL MINING IN THE LLANELLI AREA

Volume One: 16th Century to 1829

M. V. SYMONS

1015.

Llanelli Borough Council

SOUTH WALES
MINERS LIBRARY

Copyright © Llanelli Borough Council 1979

Published by
Llanelli Borough Council
Public Library
Llanelli

All rights reserved. No part of this publication may be reproduced, stored in a retrieval system, or transmitted, in any form or by any means, electronic, mechanical, photocopying, recording or otherwise, without the prior permission of Llanelli Borough Council.

ISBN 0 906821 00 2

Printed in Wales by
Salesbury Press Ltd
Llandybïe, Dyfed

To the memory of
John Clive Griffiths

PREFACE

The Llanelli Public Library Local History Research Group was formed in 1970 following an approach to me, as Borough Librarian, by a small number of researchers who were interested in various aspects of Llanelli's history, for special facilities regarding access to the extensive local collection material at the public library. Their original intention was to produce a single volume history of Llanelli to supplement, if not replace, the very readable but in many ways outmoded "Old Llanelly" by John Innes, which had been published 65 years before. After several meetings and a study of the local material available, it was realised that the subject could not be dealt with satisfactorily in a single volume and it was decided to form a Research Group, based on the public library, with the Borough Librarian acting as co-ordinator of the Group's activities, which would be directed towards each member producing a monograph or monographs on his or her particular area of study. The Public Library Committee and the Council of the then Borough endorsed the proposal and resolved to sponsor the Group and to be responsible for the publication of any material produced by members, undertakings which were readily taken over by the new Recreation Committee and Borough Council of Llanelli in 1974.

It is believed that this Research Group, officially backed and supported financially by the local authority, is the first of its kind. Whether this is true or not, I wish to record my appreciation, on behalf of the members of the Group, of the Committee's and the Council's enlightened and progressive attitude towards the publication of local history material, exemplified in the publication of this volume, the first of a series which it is intended to publish under the Borough Council imprint in the immediate future. By their actions the Borough Council members live up to the Borough motto "Ymlaen Llanelli".

It is fortuitous — in that it was the first manuscript ready for the press — but perhaps fitting that the first volume published by the Council should deal with the history of coalmining in Llanelli to 1829. The exploitation of its coal resources laid the foundation on which modern industrial Llanelli was built. Important initially as an industry in its own right, the presence of suitable coal which could be fairly easily won led to the setting up of metal industries, which provided the vast majority of Llanelli's jobs in the second half of the Nineteenth Century and which, in its turn, laid the foundation of an industrial tradition, particularly in metal working and engineering, which has attracted many new industries to Llanelli in the second half of the Twentieth Century.

The history of coal mining in Llanelli will be completed with the publication of a second volume taking the story to nationalisation in 1947 and a separate publication on the anthracite industry in the Gwendraeth Valley is also planned. Further volumes will deal with the Geology and Geomorphology of the Llanelli Area (to be published during 1980), the metal industries, the port and shipbuilding, the growth of land settlements within the Borough and a demographic study. It is hoped, eventually, to produce volumes on religion, education, sport and the development of industry generally in Llanelli in comparatively recent times.

This programme of research and publication should result in the Llanelli area being studied more widely and in greater depth than any other comparable area, with the results of the studies being conveniently available, between the covers of a series of volumes, not only to the people of Llanelli, who have an intense interest in their origins and heritage, but also to researchers and students working on wider national canvasses, whose conclusions can only be valid if they are based on a study of the parts which make up the whole. If such studies in depth are carried out in other important industrial centres in Wales, and the results are written up with such meticulous regard for accuracy and attention to detail as in this volume by Dr Symons, future historians will have at their disposal reliable, accurate background material on which to base their general conclusions on industrial or social development, regionally or nationally. These conclusions could well call into question judgements by past writers on regional or national progress and development in the same way as Dr Symons, using sources not easily available to researchers in the past, has been obliged to question judgements by earlier researchers into Llanelli's coal industry.

H. A. PRESCOTT

ACKNOWLEDGEMENTS

My gratitude is expressed to the Recreation Committee of Llanelli Borough Council for providing the publishing costs of this book, and to the Department of Civil Engineering, University of Wales Institute of Science and Technology, for allowing me the necessary time to carry out the research and writing.

Acknowledgement is given to Col. W. H. Buckley, Rt. Hon. The Earl of Cawdor and Mr M. S. Murray Threipland for kindly allowing family documents, deposited at Carmarthen Record Office, to be photographed and included as plates in this book, and to the Director General of the Ordnance Survey and the Director of the Institute of Geological Sciences for allowing information to be abstracted from Ordnance and Geological Survey plans of the Llanelli area and included in the figures accompanying the text.

The following people supplied information, advice and encouragement at different stages in the preparation of the work and I remain indebted to them:-

the staff at Llanelli Public Library, particularly Miss Ann Thomas who typed the edited manuscript and Mr David Griffiths, the reference librarian; Mr G. Dart (Librarian) and the reference staff at Cardiff Central Library; Mrs Gillian Howells, formerly Librarian of the South Wales Division of the National Coal Board; Miss Cerrys Evans, University of Wales Institute of Science and Technology Library; the late Mr L. M. Rees, formerly Librarian of Swansea Central Library; Miss Maureen Patch (Dyfed Archivist), Miss Susan Beckley and the staff at Carmarthen Record Office; Messrs. D. Bowen (former Area Surveyor), G. Lewis (Area Surveyor), V. Jones and C. Hopkins, all of the Area Survey Department of the South Wales Division of the National Coal Board; Messrs. W. A. M. Jones and D. Pennington of the Valuation Office Inland Revenue, Mineral Valuer (Wales); Dr Douglas Bassett (Director) and Dr Gerwyn Thomas, National Museum of Wales; Mr R. O. Roberts, University College Swansea; Dr Huw Owen, University College, Cardiff; the late Mr J. P. Burnell, Port Talbot; Messrs. Stuart Cole, T. A. James, A. B. Thomas, John Williams, Brynmor Voyle, all of Llanelli; Mr M. C. S. Evans,

Carmarthen; Mr W. H. Morris, Kidwelly; and the late Messrs. Robert Thomas and David Lewis of Llanelli who both sadly died before this work could be completed.

My particular thanks are expressed to Mr Harold Prescott, Llanelli Borough Librarian and co-ordinator of the Local History Research Group at the Library, for his unfailing help, encouragement and friendship over the past nine years, for the production of the photographs for the majority of the plates in this book and for his hard-working and valued editorship over the past year.

Finally, I wish to thank my wife, Menna, who has helped to bring this work to completion by filling the several roles of draft typist, Welsh language translator, proof reader and site visit companion.

CONTENTS

LIST OF PLATES

LIST OF FIGURES

EXPLANATORY NOTES ON THE REFERENCE SOURCES

1. The reference authorities for all statements made and opinions expressed have been given at the end of each chapter to enable subsequent researchers to build on this work and correct any mistakes or misconceptions contained within it. Abbreviations have been employed where the location of a reference needs to be given:-

(CCL) — Cardiff Central Library
(CRO) — Carmarthen Record Office
(GRO) — Glamorgan Record Office
(IGS) — Institute of Geological Sciences
(LPL) — Llanelli Public Library
(NCB) — National Coal Board
(NLW) — National Library of Wales
(SCL) — Swansea Central Library

2. Important document collections relating to Llanelli's coal industry are referred to throughout the book and, to save repeated reference to their locations, the following summary is provided:-

Abandonment plans (NCB); Accessions (CRO); Alltycadno (and Gwylodymaes) Estate (GRO); Brodie Collection (LPL); Castell Gorfod (CRO); Cawdor (CRO); Cawdor II (CRO); Cawdor (Vaughan) (CRO); Cilymaenllwyd (NLW); Cwrt Mawr (NLW); Derwydd (CRO); Derwydd (NLW); Edwinsford (NLW); Local Collection (LPL); Mansel Lewis (CRO); Mansel Lewis London Collection (CRO); Museum Collection (CRO); Nevill (CRO); Nevill (NLW); Penller'gaer A (NLW); Penller'gaer B (NLW); Stepney Estate (LPL); Stepney Estate Office (CRO).

Both the National Library of Wales and Carmarthen Record Office hold document collections entitled Derwydd and Nevill. Unless (CRO) is appended after the reference the collection at the National Library of

Wales is being referred to. Both Llanelli Public Library and Carmarthen Record Office hold document collections entitled Stepney Estate. Unless (CRO) is appended after the reference the collection at Llanelli Public Library is being referred to.

3. Newspapers have been consulted as source references: *The Times* (through Palmer's Index), 1790-1804 (CCL); *The Cambrian,* 1804-1863 (CCL and SCL); and *The Llanelly Guardian,* 1863 onwards (LPL). *The Times* yielded no direct information on Llanelli's coal industry; *The Cambrian* yielded important contemporary knowledge, not obtainable from other sources, for 1803 to 1829; *The Llanelly Guardian* provided valuable retrospective knowledge from its publication in 1863.

4. Published and unpublished works on Llanelli's history, written between 1856 and 1904, have been adopted as source references in the absence of manuscript evidence. They are: "*Hanes Llanelli*" by David Bowen (1856) (LPL); "*History of Llanelly*" by James Lane Bowen (1886) (LPL); "*Llanelly Parish Church*" by Arthur Mee (1888); "*Thomas Mainwaring's Commonplace Book*" (probably compiled between 1870 and 1899) (LPL); and "*Old Llanelly*" by John Innes (1904). These writers did not have the free access to manuscript collections enjoyed by present-day researchers and many of their comments on Llanelli's coal industry were based on incomplete and hearsay evidence. Nevertheless, they did have access to some documents and to oral traditions which have since been lost, and the reliability of their works needs to be assessed. "*Thomas Mainwaring's Commonplace Book*" has proved to be consistently accurate and is confidently quoted as a source reference. Virtually all the coal mining information in John Innes's "*Old Llanelly*" was taken from Mainwaring's work. Arthur Mee's "*Llanelly Parish Church*", based mainly on study of the Llanelly Parish Registers, is also an accurate source reference. "*Hanes Llanelli*" and "*History of Llanelly*" are generally inaccurate in coal mining matters and the statements contained within them must be treated with caution.

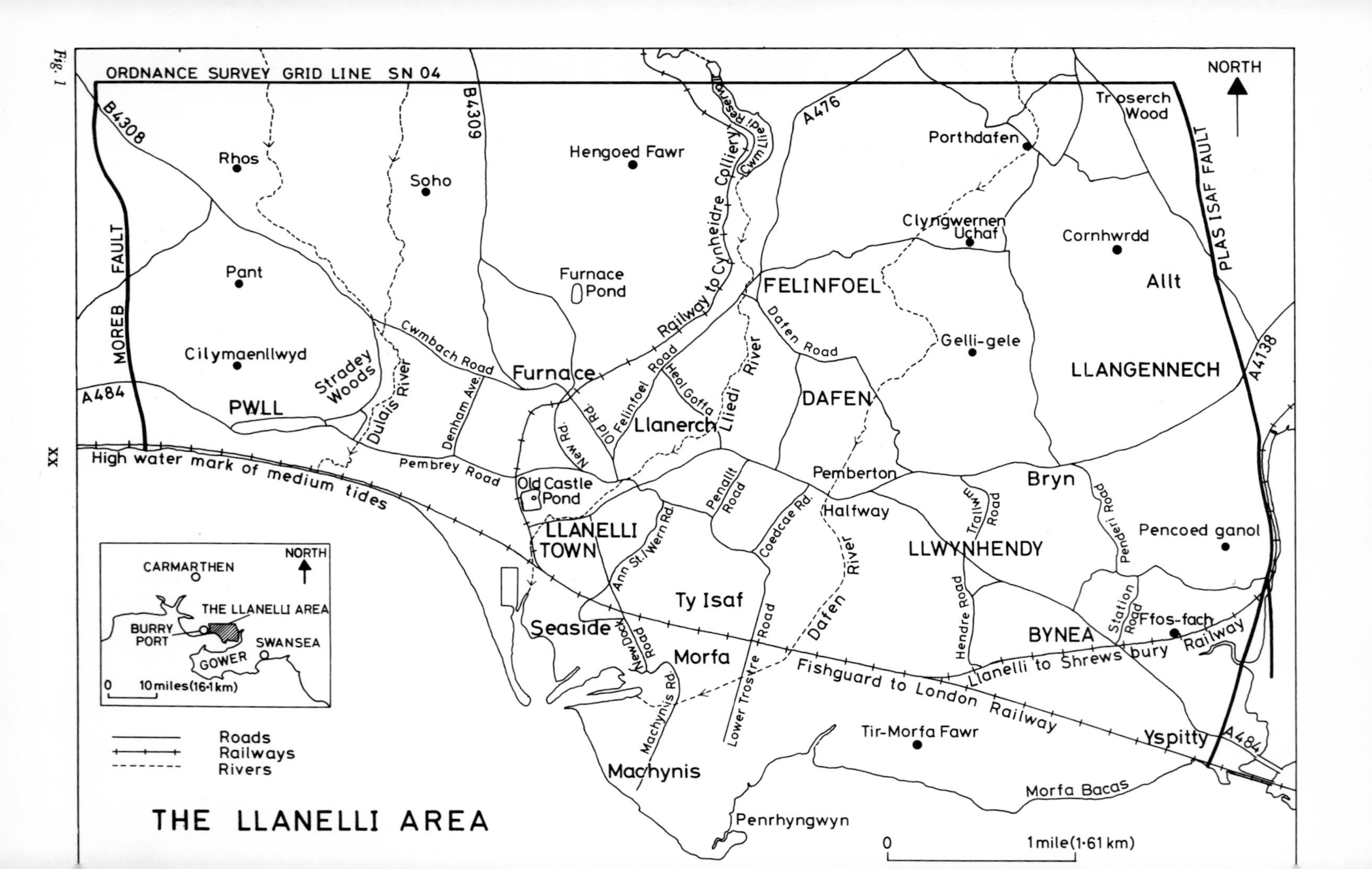
ORDNANCE SURVEY GRID LINE SN 04
NORTH
Troserch Wood
PLAS ISAF FAULT
MOREB FAULT
B4308
B4309
A476
A4138
A484
Rhos
Soho
Hengoed Fawr
Porthdafen
Clyngwernen Uchaf
Cornhwrdd
Allt
Pant
Furnace Pond
Railway to Cynheidre Colliery
Cwm Lliedi Reservoir
FELINFOEL
Dafen Road
Gelli-gele
LLANGENNECH
Cilymaenllwyd
Stradey Woods
Cwmbach Road
Furnace
Felinfoel Road
Heol Goffa
Lliedi River
DAFEN
PWLL
Dulais River
Denham Ave.
Old Rd.
New Rd.
Llanerch
High water mark of medium tides
Pembrey Road
Old Castle Pond
Pemberton
Bryn
Penallt Road
Coedcae Rd.
Halfway
Trallwm Road
Penderi Road
Pencoed ganol
LLANELLI TOWN
Ann St./Wern Rd.
LLWYNHENDY
Dafen River
Ty Isaf
Hendre Road
Station Road
Ffos-fach
Seaside
New Dock Road
BYNEA
Llanelli to Shrewsbury Railway
Morfa
Lower Trostre Road
Fishguard to London Railway
Machynis Rd.
Tir-Morfa Fawr
Yspitty
Machynis
Morfa Bacas
Penrhyngwyn
CARMARTHEN
NORTH
THE LLANELLI AREA
BURRY PORT
SWANSEA
GOWER
0 10 miles (16·1 km)
Roads
Railways
Rivers
THE LLANELLI AREA
0 1 mile (1·61 km)

Fig. 1

INTRODUCTION

Llanelli was one of Britain's early coalfields. The bituminous coals that occur in and around the present Town area were mined from the 16th century through to the 20th century, with peak output occurring in the late 1860s. The industry fell into steady decline after that time, as the easily available seams became worked out and as the port facilities failed to expand to meet the increasing size of the coal export trade ships. At Nationalisation, in 1947, just one small colliery was taken over by the National Coal Board and this was abandoned five years later. The only vestiges of this once major activity that survive today are the occasional small, private mines, working under licence from the National Coal Board, which extract minor quantities of coal on a scale similar to that of the 16th century. Llanelli's coal mining industry has thus completed a full life-cycle of events. This book recounts the history of that life-cycle over its first 300 years, from its first documented origins in the early 16th century through to the year 1829, which was a time of important change in its ownership and organisation, as the Llanelly Copperworks Company took control of most of the coalfield, and as the Llangennech Coal Company's plans for the large-scale development of the steam coal resources of the Llangennech area were initiated.

Scope of work — It had long been realised that, to understand and appreciate Llanelli's post 16th century industrial and social history, a factual reference work of Llanelli's coal industry was needed. The decision was, therefore, taken at Llanelli Public Library to make this study a mainly factual appraisal of the growth of the industry, with greater emphasis being placed on detailed physical considerations, such as the location of collieries, canals, railways and shipping places, than would normally be the case for more general works covering larger areas. Additionally, it was decided that the motives, activities and fortunes of the people known to have been involved, together with the inter-relationships and rivalries which existed between them, should be studied to obtain a full picture of the industry's development.

The Llanelli area examined is shown in Figure 1. It covers approximately 20 square miles (5238 hectares) and mainly practical criteria have been adopted in defining its boundaries. The area extends from the

Ordnance Survey National Grid line SN04 in the north (delineating a barren inland zone between bituminous and anthracite surface coal where exploitation has been minimal) down to the shore-line of the Burry Estuary in the south. The Plas Isaf Fault, which represented a natural barrier to eastwards advancing workings from within the area and to westwards advancing workings from the Loughor area, has been taken as the eastern boundary. Similarly, the Moreb Fault, which provided a natural discontinuity between westwards advancing workings from Pwll and eastwards advancing workings from Burry Port, has been taken as the western boundary. The Burry Port/Pembrey area has been excluded from the study because of the need to keep the work to a manageable size, and also because its coal mining industry developed independently of that at Llanelli.

Virtually no reference is made in this study to Llanelli's work-force, or to the working conditions that existed prior to 1830, because records covering the period contained such scant reference to these aspects of the industry that no positive statement can be made, and general comparisons with known conditions in other areas would not necessarily be valid. Nevertheless, it can be concluded from post-1830 evidence,* that Llanelli's coal miners and their families fared as badly as their counterparts in other coalfields. Children, sometimes only 5 or 6 years old, were employed in Llanelli's collieries and often suffered crippling injuries because they were far too young to cope with the dangers to which they were exposed. The safety standards of Llanelli's collieries were also poor and were strongly criticised by the first Inspector of Mines, as late as the middle of the 19th century. There is little doubt that a steady toll of life and health took place in Llanelli's mines prior to 1830, although its extent will never be known because official records were not kept.

The limitations of knowledge and concept in this work will be evident to all specialist readers but it is hoped that it will be accepted as a skeletal, but factually correct, account of the critical years of the industry, which provided the social and economic foundations to the industrial township of Llanelli. Statements and opinions have been based on the evidence of manuscript sources whenever possible and care has been taken not to mislead subsequent researchers, by expressing certainty where doubt exists. In this respect, it is hoped that the liberal use of qualifying words, defining the degree of uncertainty in the writer's mind, will be acceptable to the reader.

---

* Children's Employment Commission, Appendix to First Report of Commissioners, Part II (1842); H.M. Inspector of Mines Report for the South Western District, H. Mackworth (31 Jan 1854); The Cambrian (21 Jul 1854).

CHAPTER 1

THE LLANELLI COALFIELD*

INTRODUCTION

Physical factors are of paramount importance in all coal mining enterprises and any account of Llanelli's coal mining history must be based on an appreciation of the geological structure of the exploited coalfield. The historical pattern of mining in the area was strongly influenced by, and often governed by, the prevailing geological features.

Llanelli's early miners knew little of the science of geology, however, and, even as late as 1800, their knowledge of the area's geological structure was rudimentary. For most of the 300 years, they were just following the coal from easily-locatable surface or shallow occurrences down to the limited depths which their mining techniques and financial resources allowed. They knew little about the structure of the coalfield and even less about the correlation of the coal seams. Because of this, their decisions and actions cannot be judged solely on the basis of our knowledge of the coalfield today. To give a satisfactory account of Llanelli's coal mining history it is, therefore, necessary to consider:-

(I) The Llanelli Coalfield and its exploited coal seams in terms of relevant present knowledge.

(II) The probable state of knowledge of the Coalfield that existed at different times during the development of Llanelli's coal industry prior to 1830.

(III) The many different names known to have been allocated to the coal seams at Llanelli prior to 1830. This is an important consideration in the context of coal mining history because manuscript sources contained the seam names in contemporary usage. Correct interpretation of coal mining activities can only be achieved if the old names are correctly correlated with the present known seams.

---

* The term Llanelli Coalfield has been used in this work to describe the exploited coalfield lying within the boundaries specified in Figure 1. This is a smaller geographical area than that normally associated with the term in geological texts.

It is assumed that the reader is not familiar with geological terminology and all technical terms which have to be used will be fully explained.

(I) THE LLANELLI COALFIELD IN TERMS OF PRESENT KNOWLEDGE

A detailed account of the geology of the Llanelli area will be given in a separate Llanelli Public Library publication by D. Q. Bowen[1] and the Llanelli Coalfield will be fully evaluated in his work. This section merely attempts to describe, in as non-technical a manner as possible, the relevant structural features, with a particular emphasis on coal seam occurrences, of the upper part of the Coalfield within which all past mining has taken place.

(1) *The exploited coalfield*

Coal has been mined in the Llanelli area to a depth of approximately 1300 feet (396 metres)* below ground surface level and consideration of the structure of its Coalfield and coal seams will be made over this depth range.

The coal seams:- In geological terms, the rocks of the Llanelli area belong to the Carboniferous System** and, in particular, those underlying the Llanelli area within the depth range under consideration, belong to a sub-division of younger rocks within the Carboniferous System known as the Pennant Measures or Upper Coal Measures. These Measures are further sub-divided into various beds. In the Llanelli area coal has been extracted from the Grovesend, Swansea and Hughes Beds of the Upper Pennant Measures and from the Brithdir and Rhondda Beds of the Lower Pennant Measures. Down to the depth of 1300 feet (396m) the Pennant Measures consist principally of a succession of sandstones and shales, with more than 20 coal seams occurring at different horizons within these rocks. The coal seams vary in thickness between a few inches and 9 feet (a few centimetres and 2.7m) and the majority of them have been exploited at some time in the past. Most of the seams have been allocated names and the sequence of known coal seams in the Llanelli area, together with the range of thickness of the beds within which they

---

* The Llanelly Copperworks Company mined to this depth after 1850.

** The word Carboniferous, derived from the Latin words "*carbo*" (charcoal) and "*fero*" (bear), literally means coal-bearing and most of the true coal seams of the world are found in the upper or younger part of the Carboniferous System. These seams would have been originally deposited in the form of vegetation between 270 and 330 million years ago with subsequent pressure causing them to change in character to form the soft rock known as coal. In a similar manner the sands, silts and clays deposited at the same time as the vegetation have been changed into the relatively hard sedimentary rocks known as sandstone and shale.

TABLE I — COAL SEAM SEQUENCE IN THE LLANELLI AREA

System	Series	sub-division	Beds and their thickness ranges.	Coal seam name(s)	Known thickness range*	Notes
CARBONIFEROUS	Pennant Measures (or Upper Coal Measures)	Upper Pennant Measures	Grovesend Beds 800 - 840 ft.	Gelli Group of coal seams.	thin to 2ft. 6in.	5 or more seams, mainly thin, in the Bryn/Llangennech region. The Gelli Vein (2ft 6in) and the Gelli Rider have been the most exploited.
				Un-named seam(s)	thin	1 or 2 thin, undesignated seams.
				Penyscallen	2ft. approx	
				Un-named seam(s)	thin	Up to 2 thin, undesignated seams.
			Swansea Beds 780 - 850 ft.	Swansea Four Feet (or Llanelly Six Feet)	2 to 9 ft.	
				Un-named seam(s)	thin	The existence of this seam is not proven.
				Swansea Five Feet (or Llanelly Four Feet)	3 to 9 ft.	
				Upper Swansea Six Feet (or Rosy Vein)	1 to 3 ft.	Although the Rosy Vein was generally 40 to 70 ft. above the Fiery Vein the two came together at certain locations in the Llanelli area.
				Lower Swansea Six Feet (or Fiery Vein)	3 to 4 ft.	
				Swansea Three Feet (or Golden Vein)	2 to 3ft.	
			Hughes Beds 875 - 900 ft.	Swansea Two Feet (or Bushy Vein)	1 to 2 ft.	
				Un-named seam	thin	
				Cille No. 1	thin	Three seams, all less than 2 ft. in thickness, with at least one more thin seam present at certain locations. The Cille No. 2 has been the most exploited.
				Cille No. 2	thin	
				Cille No. 3	thin	
				Pwll Little	1 to 2 ft.	
				Hughes (or Pwll Big)	2 to 4 ft.	
		Lower Pennant Measures	Brithdir Beds 900 ft.	Cilmaenllwyd	1 to 2 ft.	
				Cwmmawr	1 to 3 ft.	
				Un-named seam(s)	thin	Up to 2 thin, undesignated seams.
			Rhondda Beds 260-380 ft. to Gwscwm seam.	Goodig	thin	
				Gwscwm	thin to 3 ft.	

* All thicknesses in this Table are given in feet and inches. To convert to metric units take 1 foot = 0.305 metres and 1 inch = 0.025 metres.

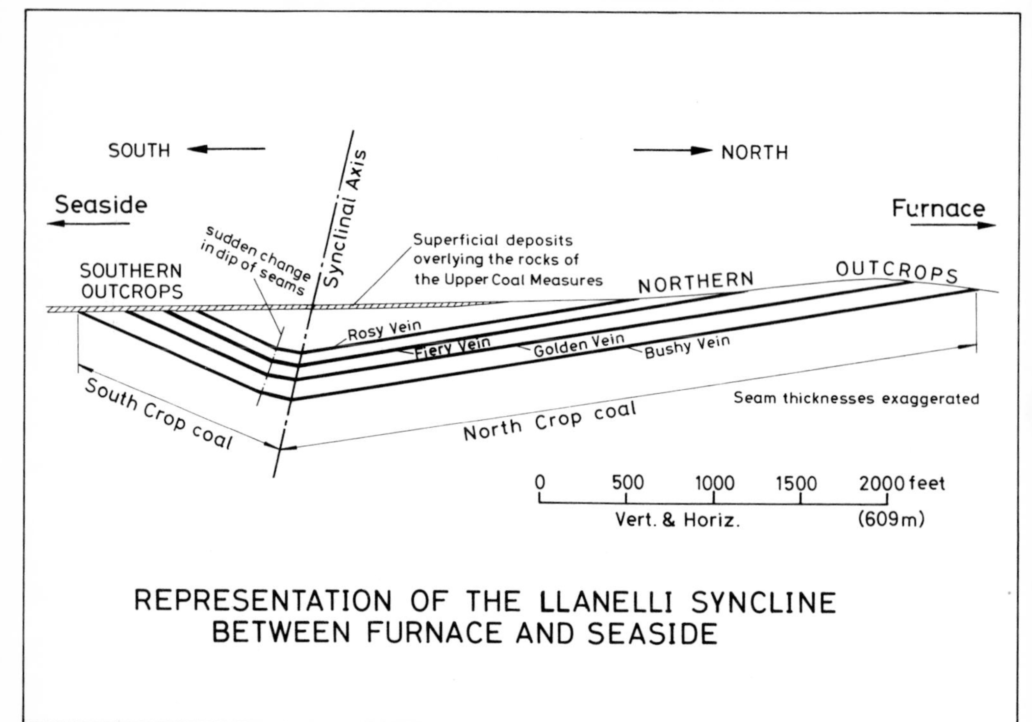

Fig. 2

occur, is given in Table I.[2] Only those seam names* in accepted present usage are given in this Table and the numerous historical names, known to have been allocated in the past, will be considered in the third section of this chapter.

The Structure of the Coalfield:- Although the sandstones, shales and coal seams of the Upper Coal Measures were originally deposited as flat or gently sloping beds, subsequent pressures and ground movements have led to widespread folding and fracturing of the strata. As a result, the Llanelli Coalfield has been formed into a trough-shaped structure, in which the strata generally slope downwards from the northern and southern sides towards the centre. This particular geological structure is known as a "*syncline*" and Figure 2 shows a representation of Llanelli's synclinal structure between Furnace to the north and Seaside to the south,** a region where four seams — the Rosy, Fiery, Golden and Bushy Veins — have been worked down to a maximum depth of about 570 feet (174m). It can be seen that the synclinal structure causes each seam to come to the rock surface (or "*outcrop*") at two locations, termed its

* It is now common practice to refer to a stratum of coal as a "*seam*" but at Llanelli and throughout South Wales, in the past, it was referred to as a "*vein*". Because this is an historical work, use will have to be made of both words and the reader should treat "*seam*" and "*vein*" as synonymous terms for the Llanelli Coalfield.

** Representations of the synclinal structure in other parts of the Llanelli Coalfield are given in many of the figures included in Chapters 3 and 4.

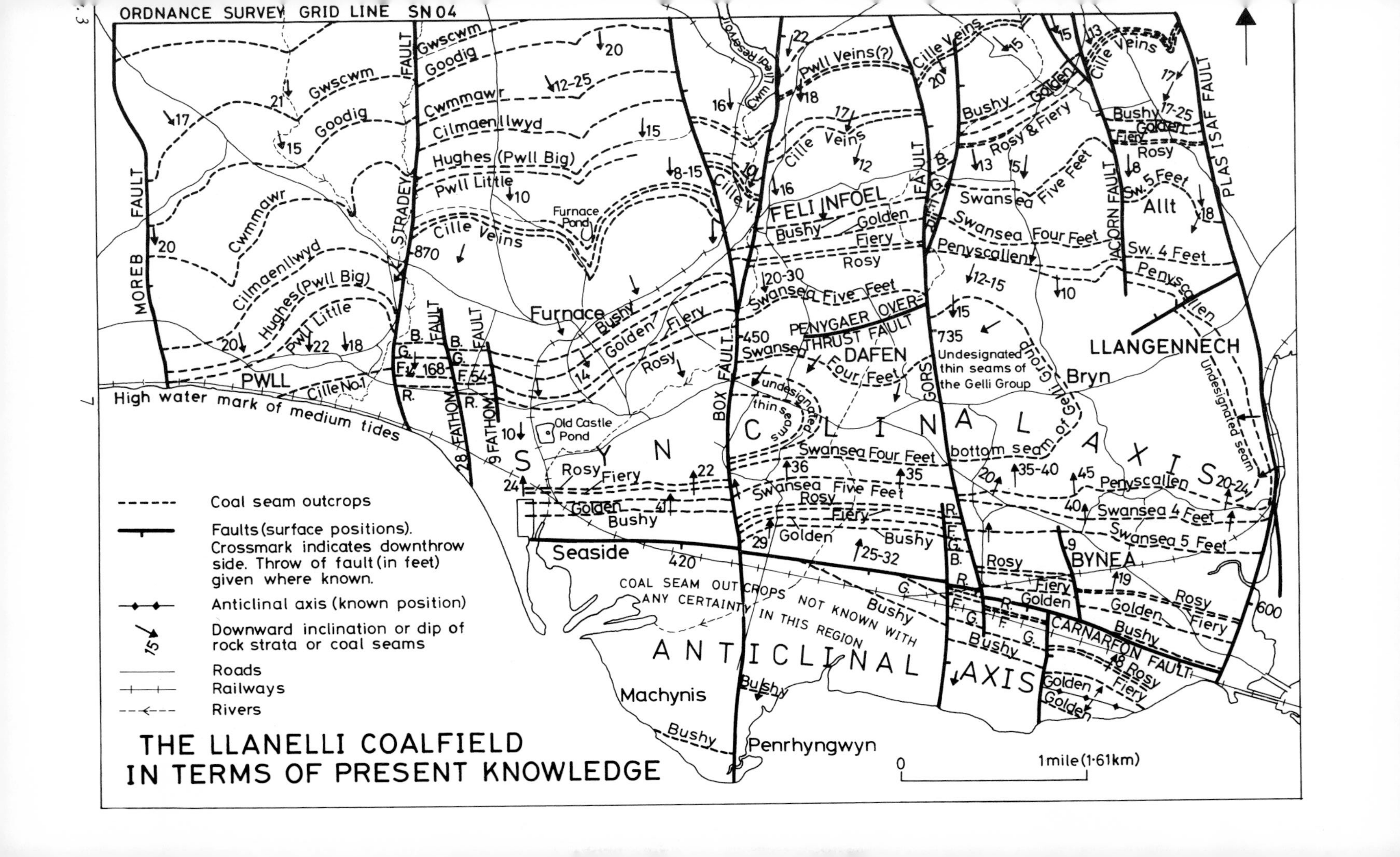
THE LLANELLI COALFIELD
IN TERMS OF PRESENT KNOWLEDGE
ORDNANCE SURVEY GRID LINE SN04
Coal seam outcrops
Faults (surface positions). Crossmark indicates downthrow side. Throw of fault (in feet) given where known.
Anticlinal axis (known position)
Downward inclination or dip of rock strata or coal seams
Roads
Railways
Rivers
0
1mile (1·61km)
MOREB FAULT
STRADEY FAULT
BOX FAULT
GORS FAULT
ACORN FAULT
PLAS ISAF FAULT
PENYGAER OVER-THRUST FAULT
CARNARFON FAULT
28 FATHOM
9 FATHOM
SYNCLINAL AXIS
ANTICLINAL AXIS
COAL SEAM OUTCROPS NOT KNOWN WITH ANY CERTAINTY IN THIS REGION
PWLL
FELINFOEL
DAFEN
LLANGENNECH
BYNEA
Bryn
Allt
Furnace
Furnace Pond
Old Castle Pond
Seaside
Machynis
Penrhyngwyn
Cwm Lliedi Reservoir
High water mark of medium tides
Gwscwm
Goodig
Cwmmawr
Cilmaenllwyd
Hughes (Pwll Big)
Pwll Little
Cille Veins
Cille No.1
Pwll Veins(?)
Cille V.
Bushy
Golden
Fiery
Rosy
Rosy & Fiery
Swansea Five Feet
Swansea Four Feet
Sw. 5 Feet
Sw. 4 Feet
Swansea 4 Feet
Swansea 5 Feet
Penyscallen
Undesignated thin seams of the Gelli Group
bottom seam of Gelli Group
undesignated thin seams
Undesignated seam
870
168
54
450
735
420
600

northern and southern outcrop respectively. Llanelli's syncline is also asymmetric in that the coal seams slope downwards (*"dip"*) more steeply from their southern outcrops than from their northern outcrops.

The Llanelli Coalfield considered in this work is shown in plan view in Figure 3[3] with the coal seam outcrops represented by heavy dashed lines. The downward slopes or dips of the strata are represented directionally by the arrows on the figure, and in magnitude (measured in degrees downwards from the horizontal) by the figures accompanying the arrows. The fractures in the strata, termed "*faults*", are shown at their rock surface positions* as heavy full lines and it can be seen that they run mainly in a north/south direction, with only the Carnarfon Fault, the Penygaer Overthrust Fault and an un-named fault, near Llangennech, running east to west. As relative movement has taken place between the rocks on each side of the faults, discontinuities in the coal seams have resulted in these regions. The major north/south faults in the Llanelli Coalfield all "*downthrow*" to the east, i.e. a given stratum or coal seam immediately to the east of each fault occurs at a lower level than it does to the west of each fault. The only exceptions to this pattern of easterly downthrows are the ancillary 28 Fathom, 9 Fathom and Acorn Faults, which all downthrow to the west. The direction of downthrow of each fault is shown by the crossmarks on the fault line, and the figures accompanying the crossmarks represent the approximate range of values (in feet) known for the downthrow of the faults. Coal seams may be displaced hundreds of feet vertically by faulting. In the context of coal mining, the five major north/south faults, viz., the Moreb, Stradey, Box, Gors and Plas Isaf Faults, acted as barriers to the continuity of working in any given seam and divided the Llanelli Coalfield into four subsidiary coalfields, each of which was developed independently of the other three parts. The Carnarfon Fault also acted as a barrier to coal workings and the area to the south of the fault can be regarded as a fifth subsidiary coalfield, which itself is further sub-divided by the major north/south faults. Although the detailed structure of most of this subsidiary coalfield is not known with any certainty to the present day, it can be concluded that it is more complex than that to the north of the Carnarfon Fault, being governed by an east/west anticline,** which has been proved near Yspitty, and which

---

* Because the faults do not follow a true vertical path their locations at the levels of the various coal seams will not correspond to the rock surface positions shown although, in most cases, they will be near to them. The five main north/south faults are all inclined downwards to the east and the Penygaer Overthrust Fault is steeply inclined downwards to the south. The Carnarfon Fault apparently consists of at least five closely situated faults downthrowing to the south but the direction of their inclination is not known.

** An anticline is the opposite form of structure to a syncline with the strata forming an arch-shaped fold.

probably runs in the direction approximately represented by the words ANTICLINAL AXIS, on Figure 3. This limitation of knowledge will not detract from an understanding of Llanelli's early coal industry, however, because little exploitation took place, or was even attempted, to the south of the Carnarfon Fault prior to 1830.

The axis of the Llanelli Syncline has a pronounced southwards inclination with depth and, therefore, occupies a different position in plan view for each individual coal seam. This, combined with a lack of certainty regarding its actual positions and inclinations to the east of the Box Fault, prevents its accurate representation on Figure 3, although its general path is approximately given by the words SYNCLINAL AXIS. It can be assumed, however, that the synclinal axis for any given coal seam is situated between the northern and southern outcrops of that seam and in closer proximity to the southern than to the northern outcrop.

(2) *The superficial deposits overlying the Pennant Measures*

Over much of the Llanelli area the solid rock of the Pennant Measures is covered by loose, superficial deposits, ranging in thickness from a few inches to over 100 feet. These deposits affected early coal mining by concealing the outcrop locations of the seams and, in areas where the deposits were water-bearing and unstable, by making the sinking of pits impossible. Knowledge of their distribution and thickness is, therefore, necessary to reach a proper interpretation.

The superficial deposits have been laid down over the Pennant Measures rock during the last one million years or so* and have not themselves been subjected to the pressures which caused the folding and faulting of the rocks in previous geological times. Although 13 separately identifiable superficial deposits are listed on the geological plans,** glacial deposits (mainly boulder clay but also some sand and gravel, left behind by melting glaciers in the Ice Age, when the area was covered by sheets of ice) and alluvial deposits (mainly sand, silt and clay laid down by rivers, streams and by estuarine waters), account for almost all of the superficial deposits in the Llanelli area. This is illustrated in Figure 4, which gives the distribution of glacial deposits, alluvial deposits and other superficial deposits, together with their thicknesses (in feet), at several

* The geological period during which the superficial deposits have been laid down is termed "*Recent and Pleistocene*" or alternatively "*Quaternary*".

** The following separately identifiable deposits are listed on the geological plans of the area:- blown sand, peat, alluvium, alluvial fan, 1st Terrace (gravel), 2nd Terrace (gravel), marine beach deposits and tidal flats, marine alluvium, storm beach head, fluvio-glacial gravel, sand and gravel (glacial), boulder clay.

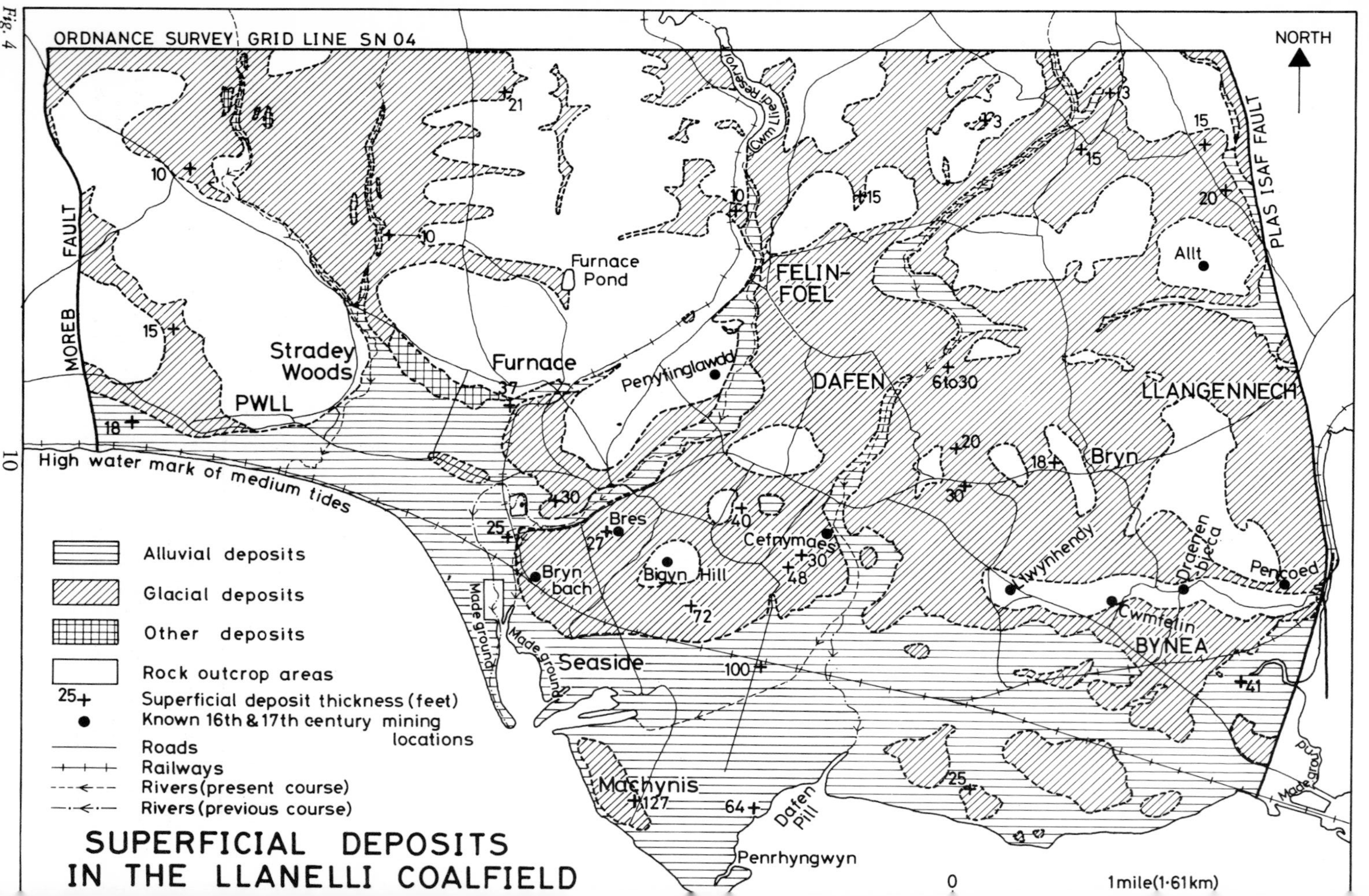

ORDNANCE SURVEY GRID LINE SN 04
NORTH
PLAS ISAF FAULT
MOREB FAULT
Furnace Pond
Furnace
Stradey Woods
PWLL
FELIN-FOEL
Cwm Lliedi Reservoir
Allt
LLANGENNECH
DAFEN
6 to 30
Penyfinglawdd
Bryn
High water mark of medium tides
Bres
Bryn bach
Bigyn Hill
Cefnymaes
Llwynhendy
Draenen-bieca
Pencoed
Cwmfelin
BYNEA
Seaside
Made ground
Machynis
Dafen Pill
Penrhyngwyn
Alluvial deposits
Glacial deposits
Other deposits
Rock outcrop areas
25+ Superficial deposit thickness (feet)
Known 16th & 17th century mining locations
Roads
Railways
Rivers (present course)
Rivers (previous course)
SUPERFICIAL DEPOSITS
IN THE LLANELLI COALFIELD
0
1 mile (1·61 km)

Fig. 4

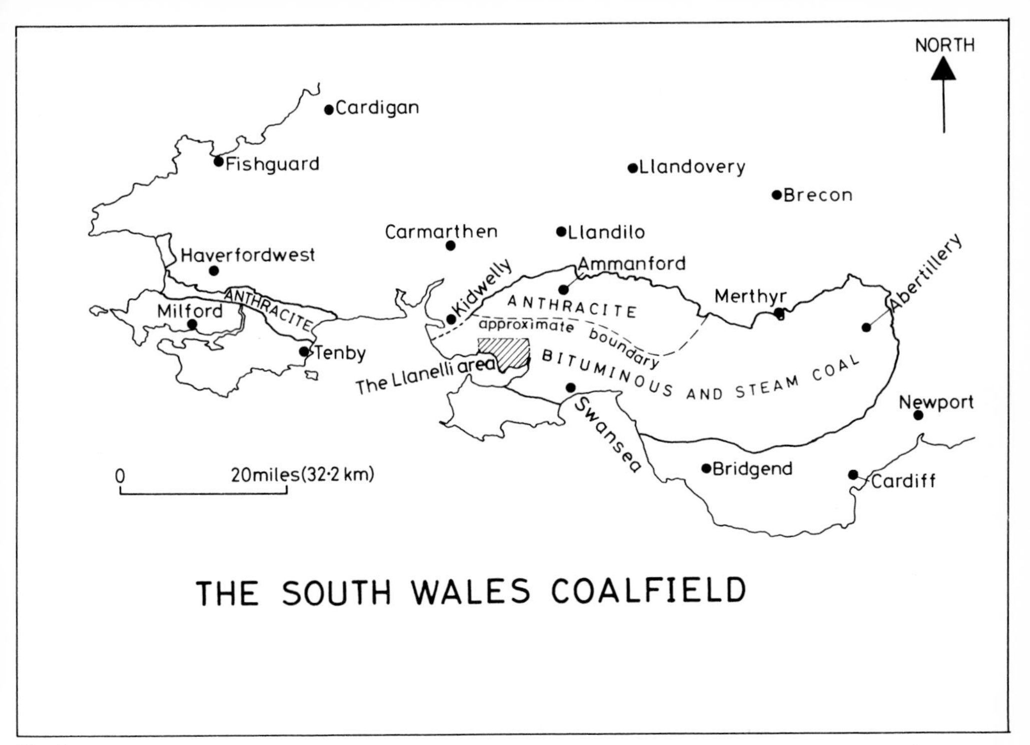

Fig. 5

locations where they are known with any certainty.* Alluvial deposits predominate near the shore, with the boulder clay lying mainly inland (apart from isolated but quite large patches of it south of Llwynhendy). A significant proportion of the area is shown as having a rock outcrop, i.e. no superficial deposits present, but weathering of the solid rock has produced an overlying, loose residual deposit, usually of minor thickness, at most localities.

(3) *The coal types in the Llanelli Coalfield*

The type of coal found in the Llanelli area was a significant factor in its emergence as one of Britain's first coalfields. In the early days of the British coal industry, when coal was replacing wood as the nation's main fuel source, the demand was for an easily-ignitable coal which could be kept burning without special draught arrangements, and which would not produce a fierce heat, injurious to the unsophisticated hearths and furnaces of the day. Bituminous coal, with a high percentage of volatile

---

* Many of the thicknesses shown have been taken from 19th and early 20th century sources which relate to ground surface levels that existed at those times. Present ground surface levels may have changed at a number of these locations due to subsequent filling or excavation.

matter quickly released as ignitable gases when heated, satisfied these requirements. It also gave off flames of pleasant appearance, reminiscent of the wood fuel it was replacing. Thus, both practically and psychologically, bituminous coal came into favour for most industrial and domestic purposes from the 16th century onwards. It held sway until the first half of the 19th century, when the industrial advantages of coals with a lower percentage of volatile matter and impurities were recognised, and steam and anthracite coals came into general use.*

The South Wales Coalfield is shown in terms of the main coal types occurring at shallow and easily worked depths in Figure 5, and it can be seen that the Llanelli area considered in this work lies in a designated bituminous and steam coal region.** Within the Llanelli area itself, bituminous coal probably predominates, with less-volatile steam coal occurring at certain locations. Because most of Llanelli's coals were last mined prior to the start of this century, little reliable information has survived to indicate which seams yielded bituminous and which steam coal, as we define these terms today. Nevertheless, all references consulted suggest that true steam coal occurred mainly in the north of the area, its excellent steam-raising properties being first recognised nationally in the 1820-30 period, when coal from the Llangennech region was used in the London brewers' steam engines and in the steamships of the East India Company and the Indian Navy.

Because there is no evidence that anthracite was mined in the area, it must be concluded that Llanelli's coals were in the bituminous to steam coal range. There is also insufficient evidence to allocate a specific type in this coal range to any particular seam and so "*coal*" and "*bituminous and steam coal*" should be regarded as synonymous terms throughout this book.

(II) KNOWLEDGE OF THE LLANELLI COALFIELD BEFORE 1830

From the earliest times, local inhabitants would have been aware of the existence of coal in the Llanelli area. They would have observed coal seams outcropping in those regions where the Pennant Measures were not overlain by superficial deposits, and would doubtless have used some

* Solid fuels are normally classified on the basis of their "*rank*" which is assessed mainly by the percentage of volatile matter present. In increasing order of rank, solid fuels fall into four general categories:- lignite, peat, bituminous coal, anthracite. Bituminous coal and anthracite are divided into a number of sub-groups with steam coals lying in a range which overlaps between high-rank bituminous and low-rank anthracite.

** The anthracite seams which outcrop to the north of the Llanelli Coalfield probably occur at great depth under the Llanelli area. These seams have not been proved at Llanelli, however, and all workings have taken place in the shallower bituminous and steam coal seams.

of the outcrop coal for their own purposes. There was no reason for their interest in this coal to extend beyond mere observation or localised surface extraction, and it is unlikely that any thought was devoted to the structure of the Coalfield. However, when coal began to replace wood as Britain's primary fuel source in the 16th century, and the necessary stimulus was provided for the establishment of an organised coal industry and export trade at Llanelli,[4] the early professional miners would have spent considerable time, and effort, locating and following the seams. Knowledge of the Llanelli Coalfield would have started to accumulate and a slow but growing awareness of its structure would have dawned in the minds of these intuitive geologists.

(1) *The 16th and 17th centuries*

No direct evidence is available to tell us what was known about the Llanelli Coalfield in the 16th and 17th centuries, but interpretation of surviving coal mining documents allows some tentative conclusions to be drawn. In particular, if the sites of early coal mining can be established, some assessment can be made of the knowledge which would have been gained as the various seams were worked. In this respect, evidence exists, mainly in the form of old leases, giving the approximate locations of many of the early coal ventures. To interpret this evidence correctly, it is first necessary to appreciate the main practical constraints upon early coal mining and the measures that could be taken to overcome them.

The early miners had to contend with three main practical problems, each of which influenced their choice of site:-

(a) the coal seams had to be located. Although prospecting techniques were coming into use in these early years,[5] they were probably insufficiently advanced or practiced to exploit those areas where the Pennant Measures were overlain by substantial thicknesses of superficial deposits which concealed the seams. It is, therefore likely that coal mining in the Llanelli area was concentrated into seam outcrop areas of little or no superficial deposition.

(b) the workings had to be de-watered. Normal percolation of surface water through rock, and the presence of standing water tables,both led to an accumulation of water in the spaces produced by mining activity and, unless this water was removed, the workings had to be abandoned. Primitive pumping systems were available in the 16th and 17th centuries[6] but coal winnings were normally restricted to shallow depths. The simplest and cheapest method of de-watering workings was to concentrate mining activity in areas of localised high ground, and use drainage adits or gutters to run the accumulating water away to lower ground. (See Appendix B).

(c) the extracted coal had to be transported to the vessels of the export trade. In the 16th and 17th centuries, land transport systems, in the form of canals or railroads, were non-existent and those roads which did exist were little better than cart tracks. The haulage of a high-bulk, low unit-value commodity such as coal posed both a practical and economic problem to the early miners, and the most effective way of overcoming it was to mine the coal as near to the seashore or navigable rivers and pills as possible.

An ideal site for early mining activity, therefore, would have been a coal seam outcrop area where little or no superficial deposits were present, situated on localised high ground, in close proximity to the shipping places of the export trade.

The available evidence relating to the siting of coal mining activity can now be examined, on the basis of these conclusions, in an attempt to discover what might have been known about the geology of the Llanelli Coalfield in the 16th and 17th centuries. Although a large number of coal leases and other references are extant for this period (see Chapter 2), only 10 individual coal mining sites or areas can actually be located. These are: Penyfinglawdd,[7] Bryn bach,[8] Allt,[9] Llwynhendy,[10] Pencoed,[11] Cwm-felin,[12] Draenenbicca,[13] Bigyn Hill,[14] Cefn y Maes[15] and Bres,[16] and their locations are given on Figure 4. With the exception of Bryn bach, Cefn y Maes and Bres, these sites are situated in, or very close to, areas where the Pennant Measures are not overlain by superficial deposits. This is particularly marked between Llwynhendy and Pencoed, where four separate sites have been located, lying along a narrow strip where superficial deposits are absent, and where the Penyscallen and Swansea Four Feet Veins and an un-named thin seam outcrop. Two of the located sites, Penyfinglawdd and the Allt on the northern limb of the Llanelli Syncline, were up to a mile away from the shipping places, but both were situated in areas where there was no superficial deposit above the Pennant Measures, where there was localised high ground suitable for drainage adits, and from which the haulage of coal to the export vessels would be entirely downhill. (Plate 1)

Surviving evidence of the location of pits worked in the 16th and 17th centuries, therefore, supports the conclusions that early coal mining was mainly limited to those areas where the coal seams were exposed, or covered only by minor thicknesses of superficial deposits; that proximity to a shipping place was an important consideration in the siting, because

Plate 1 — Sites of known 16th and 17th century mining. (Top) The Allt, Llangennech. (Bottom) Bigyn Hill. No superficial deposits, coal seams outcrop and dewatering of workings by drainage adits was possible.

8 out of 10 located sites lay along the southern outcrop; and that high ground, where shallow workings could be de-watered by simple drainage adits, was favoured, because 8 of these sites were on high ground.

By deduction from these conclusions, we can summarise the probable knowledge of the Llanelli Coalfield at the end of the 17th century as follows:-

In coal seam outcrop areas, where superficial deposits were of minor thickness or non-existent, the seam occurrences were probably known over much of their lengths. The opposing dips of the northern and southern outcrops must have been observed but it is unlikely that the synclinal structure had been recognised, or the fact that the same seams were occurring at both outcrops had been appreciated. References to the coal seams, in all the leases and documents of the period, were extremely vague and no seam was allocated a name. The discontinuities, represented by the Box and Gors Faults, may have been encountered as workings were extended, but their significance was not understood beyond the fact that they represented barriers to the working of the coal.

(2) *The 18th century*

Knowledge of the geology of the Llanelli Coalfield would have continued to accumulate, as workings were extended laterally along the seam outcrops and also in depth, but little direct evidence has survived to confirm this for the first half of the 18th century. Nevertheless, by 1729 at the latest, coal seams were being identified by name in legal documents[17] and significant gains in knowledge must have been made to allow seams to be distinguished one from the other. Substantial exploitation of the Rosy, Fiery, Golden and Bushy Veins on the southern limb of the Llanelli Syncline west of the Box Fault took place from 1733 onwards[18] and, by the mid 1750s, the synclinal structure of that part of the Coalfield had been recognised, leases of 1754 and 1758[19] making reference to the Fiery Vein at both its southern and northern outcrops respectively.

A rapid expansion of the local coal industry took place after 1750, as industrialists, with financial resources and deep mining experience, came to the area and, although this first phase of industrialisation faltered after 1770, knowledge of the Coalfield's geological structure increased. Exploitation of many of the southern outcrop seams, north of the Carnarfon Fault, took place, or was attempted, during this period, and the employment of highly-developed water engines and the introduction of steam engines (See Appendix B) eliminated the need to confine workings to shallow depths. Many pits were sunk in close proximity to suitable shipping places, where significant thicknesses of superficial deposits often overlaid the Pennant Measures rock, and the previous practice of confining workings only to high ground areas of minor superficial deposition was abandoned. Consequently, many of the coal seams along the southern outcrop, north of the Carnarfon Fault, became known, and awareness of the synclinal structure of the Coalfield must have led to attempts to correlate the northern and southern outcrops of known seams.

A full understanding of the structure of the Coalfield had not been gained, however, and the main reason for this was the fact that the significance of the main north/south faults, in relation to the coal seam occurrences, had not been appreciated. The 18th century miners knew only that the faults constituted a barrier to their workings, and evidently did not understand that seams could be displaced vertically by up to hundreds of feet at these fractures. Consequently, it was assumed that the same seams occurred, in the same sequence, from the ground surface downwards over the entire Coalfield. This misconception is illustrated by the fact that, in 1772, the Swansea Four Feet Vein, (which was the second known seam below the ground surface, along the southern outcrop between the Gors and Plas Isaf Faults) was considered to be the Fiery Vein[20] (which was the second seam along the southern outcrop between the Stradey and Box Faults). Similarly, it was thought, c1776, that the Swansea Four Feet Vein (which was the top known seam between the Box and Gors Faults) was the Rosy Vein (which was the top seam west of the Box Fault).[21] This lack of understanding lasted throughout the 18th century and persisted, even among the most knowledgeable local miners, until the 1840s.

At the end of the 18th century, therefore, the extent of geological knowledge can be summarised thus:-

The synclinal structure of the Coalfield was understood and the outcrop locations of many of the seams were known; the practice of allocating names to the coal seams was adopted in all areas; the lack of understanding of the effect of the north/south faulting upon the coal seam occurrences led to the assumption that the sequence of seams from the surface downwards was identical in all parts of the Coalfield; this resulted in considerable confusion which persisted for some time regarding the correlation of the seams.

(3) *Between 1800 and 1829*

A second, sustained phase of industrialisation took place in the Llanelli Coalfield from the mid 1790s onwards and deep mining was expanded on an unprecedented scale. Geological knowledge of the Coalfield increased rapidly at this time, although many of the old misconceptions regarding the correlation of seams were maintained right up to 1829. Detailed information is available relating to the knowledge of the Coalfield held in the mid 1830s, and this will be examined on the grounds that it indicates the extent of understanding that had been gained by 1830.

The development of mining in the 18th century had been largely speculative, with only one know instance of a commissioned report on the potential of a colliery.[22] Even when Alexander Raby became involved in large scale development, he had accepted hearsay evidence of the thickness of coal seams when obtaining leases, in 1797/98, of all coal under Stradey Estate lands to the west of the Lliedi River.[23] This practice changed from the beginning of the 19th century and reports were commissioned for most major colliery projects. In 1803, Edward Martin, a noted colliery viewer from the north of England who had established a practice at Morriston, reported on the coal resources of the Llangennech Estate[24] and followed this with an appraisal of the potential of the Bres/Wern coalfield in 1806.[25] Reports on proposed colliery developments at Gelligele and Penprys were prepared by William Bevan in 1816 and 1820 respectively[26] and Rhys (or Rees) Jones of Loughor reported on the Genwen Colliery in 1812,[27] on the coal seams to the north west of Felinfoel in 1822[28] and on the Morfa Baccas region in 1824.[29] Other reports may well have been commissioned but no reference to them has survived. Examination of the one full report which is available,[30] shows that it must have contributed significantly to a rationalisation of existing geological data, and there is little doubt that the advent to the area of men of the calibre of Edward Martin produced a rapid growth in the understanding of the structure of the Llanelli Coalfield.*

Although information was being accumulated at an increasing rate, and despite the fact that men possessing geological knowledge were at work in the area, the geology of the Coalfield was still a long way from being fully understood by the 1830s. This is evident from written statements and reports of 1835/36,[31] which show that, although the synclinal structure was appreciated and the outcrops of many of the coal seams quite accurately plotted, correct correlation of the seams throughout the area had not been achieved. The extent of knowledge of the Llanelli Coalfield at this time is represented in Figure 6.[32] This shows that the major

---

* Edward Martin published his *"Description of the Mineral Bason in the Counties of Monmouth, Glamorgan, Brecknock, Carmarthen and Pembroke"* in the Philosophical Transactions of the Royal Society of London in 1806. Here he summarised the structure of the South Wales Coalfield and showed that he understood the effects of faulting. His knowledge would have helped to clarify many of the previous misconceptions held in the Llanelli Coalfield. He was described in 1804 as *"the best collier in the Country"* (Nevill MS IV, letter from C. Nevill to R. Michell, 1 Dec 1804). On his death in 1818 he was described as being *"eminent for his extensive knowledge of collieries"* (*The Cambrian*, 15 May 1818).

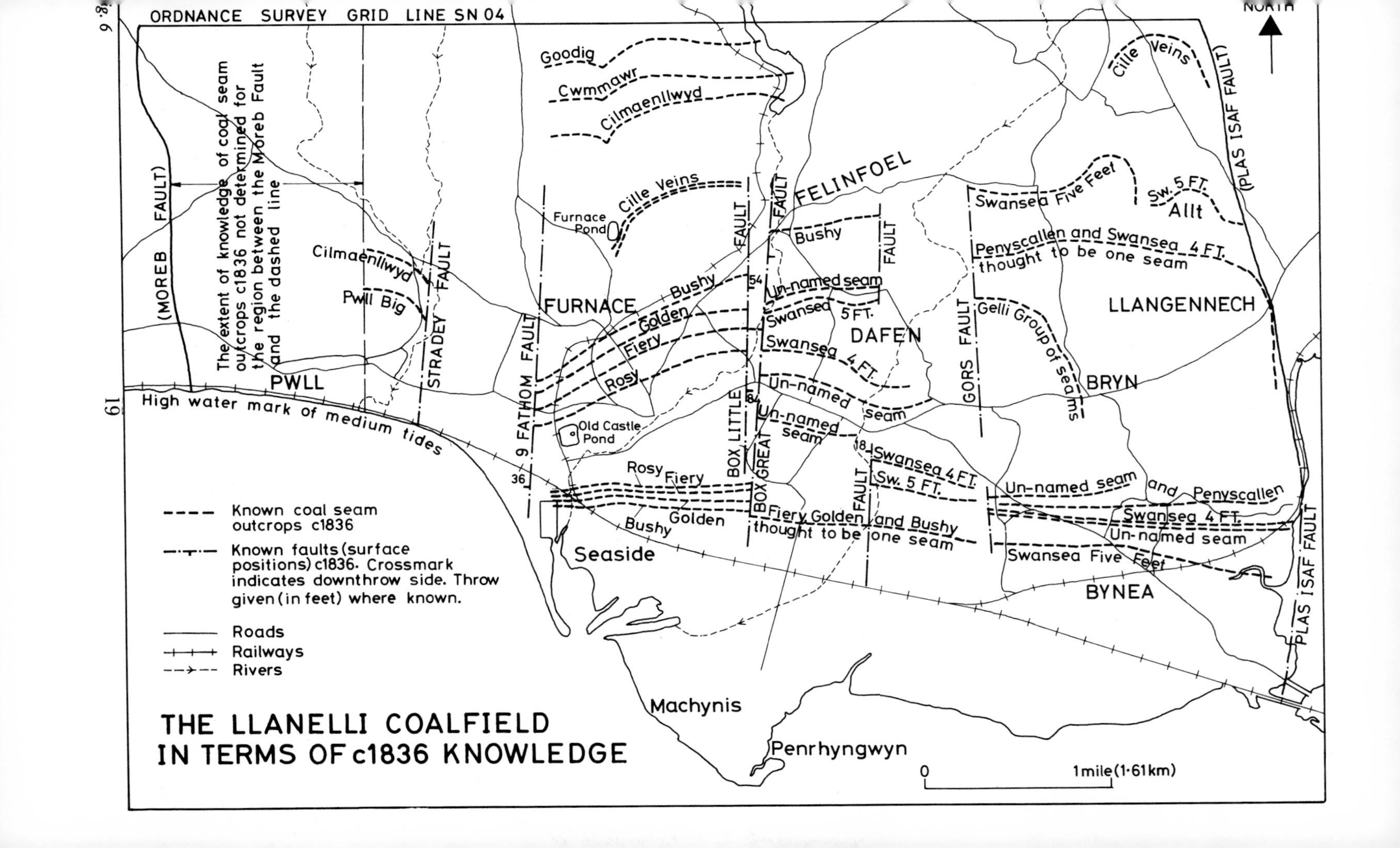
ORDNANCE SURVEY GRID LINE SN 04
NORTH
(MOREB FAULT)
The extent of knowledge of coal seam outcrops c1836 not determined for the region between the Moreb Fault and the dashed line
Cilmaenllwyd
Pwll Big
STRADEY FAULT
PWLL
High water mark of medium tides
9 FATHOM FAULT
36
Goodig
Cwmmawr
Cilmaenllwyd
Cille Veins
Furnace Pond
FURNACE
Bushy
Golden
Fiery
Rosy
Old Castle Pond
Rosy
Fiery
Bushy
Golden
Seaside
BOX LITTLE FAULT
54
84
BOX GREAT FAULT
FELINFOEL
Bushy
FAULT
Un-named seam
Swansea 5 FT.
DAFEN
Swansea 4 FT.
Un-named seam
Un-named seam
18
Swansea 4 FT.
Sw. 5 FT.
FAULT
Fiery Golden and Bushy thought to be one seam
GORS FAULT
Swansea Five Feet
Sw. 5 FT.
Allt
Penyscallen and Swansea 4 FT. thought to be one seam
Gelli Group of seams
LLANGENNECH
BRYN
Cille Veins
(PLAS ISAF FAULT)
Un-named seam and Penyscallen
Swansea 4 FT.
Un-named seam
Swansea Five Feet
BYNEA
PLAS ISAF FAULT
Known coal seam outcrops c1836
Known faults (surface positions) c1836. Crossmark indicates downthrow side. Throw given (in feet) where known.
Roads
Railways
Rivers
Machynis
Penrhyngwyn
0
1 mile (1·61 km)
THE LLANELLI COALFIELD
IN TERMS OF c1836 KNOWLEDGE

Fig. 6

north/south faults had been located over much of their lengths and that some of the ancillary faults were known.* The coal seam outcrops probably corresponded to the outcrop lengths located and worked by the mid 1830s. The structure of the Llanelli Coalfield was now known sufficiently to ensure that subsequent major colliery developments would be based on geological assessment, and would not be the speculative ventures which had been so common in Llanelli's bituminous and steam coal industry in previous years.

(III) COAL SEAM NAMES USED BEFORE 1836

Names were being allocated to Llanelli's coal seams by 1729. During the following 100 years it became common practice to allocate a name to any newly-discovered seam, and these names often related to the particular field, landshare or farm where the seam was located. However, the inability of the 18th and early 19th century colliery viewers and surveyors to correlate the seams throughout the area resulted in different seams being allocated the same name and, conversely, in the same seam being known by different names in various parts of the Coalfield. As all old documents and plans referred to the coal seams in terms of their contemporary regional names, it is essential to gain an understanding of this confused and complex situation, if a correct account of Llanelli's coal mining history is to be achieved.

All the coal seam names known to have been employed in the Llanelli Coalfield to 1836 are given in Table II, in relation to their period and region of use. Correlation of these names with known coal seams, in the sequence previously adopted in Table I, is also included and an explanation of this correlation is provided in Appendix A.

* Comparison of Figures 6 and 3 shows that present-day geological plans do not include ancillary north/south faults immediately west of the Box Fault (downthrowing some 30 to 50 feet to the east) and between Dafen Pill and Halfway (downthrowing some 18 feet to the west). These faults undoubtedly exist, with the former being shown on early 19th century plans (Llanelli Public Library plans 4 to 6, Jul 1806) and the latter being referred to in a mid 19th century document (LPL plan 23, Apr 1845).

TABLE II — COAL SEAM NAMES USED IN THE LLANELLI AREA UP TO 1836

Modern Seam Name(s)	Seam names used up to 1836*	Known years of use	Region
Gelli Group of Seams	Gelly-Gilly	1803	Bryn/Llangennech
	Gelly Little	1824	"
	Gellygele	c1825	"
	Gellicilau or Bryn	1836	"
	Gellywhiad or Small	1836	"
Un-named seam(s), Penyscallen, un-named seam(s)	Little	1805	Box/Capel
	Genwen Small	1812	Cwmfelin
	Little	1822	Box/Capel
	Small	1825	"
	Penprys (in error)	1825	Penprys
	Rosy (in error)	1835/36	Cwmfelin/Pencoed
Swansea Four Feet	Penyfigwm or Rosey or Great	1762	North crop between Box and Gors Faults
	Penyfigwm or Great	1776	"
	Penyfigwm	1808	Box Colliery
	Box	1813	"
	Great	1825	"
	Box Great or Nine Feet	1836	Box/Llandafen
	Fiery	1772	Cwmfelin/Pencoed
	Fiery	1802	"
	Old Penyfan or Daven Pill Fiery	1807	"
	Daven Pill Burry River Fiery	1808	"
	Old Fiery	1809	"
	True Fiery	1813	"
	Fiery	1813-1836	"
	Penprys	1803	Clyngwernen/Penprys
	Glanmurrog	1803	Glanmwrwg
	Glanmwrwg Great	1825	Clyngwernen/Penprys
	Penprys	1825	"
	St David's	1836	Gelly Farm
	Penprys or St David's	1836	St David's Pit
	Llangennech	1836	"
Un-named seam between Swansea Four and Five Feet Veins(?)	Golden (?)	1836	Cwmfelin/Pencoed

* Each name was followed by the word Vein unless otherwise specified e.g. Gelly-Gilly Vein.

Modern Seam Name(s)	Seam names used up to 1836	Known years of use	Region
Swansea Five Feet	Stradey (?)	1767	Penygaer
	Llanerch or Fiery	1805	,,
	Pinygare	1813	,,
	Penygare	1813	,,
	Warde's Fiery	1823	Box Colliery
	Warde's Fiery or Penygare	1827	,,
	Fiery	1835	,,
	Box Fiery	1836	,,
	Marsh (?)	1755-61	Ffosfach
	Great or Golden	1772	Cwmfelin/Pencoed
	Great	1802	,,
	Gwndwn Mawr Golden	1807	,,
	Old Gundunmaur or Golden	1808	,,
	Great or Old Golden	1812	,,
	Old Golden	1813	,,
	Great Fiery or Cwmfelin	1835	,,
	Fiery	1836	,,
	Cwmfelin	1836	,,
	Alt Cole (Allt Coal) (?)	1773-74	Allt
	Glyngwernen Upper and Lower	1803	Clyngwernen/Brynsheffre
	Glyngwernen	1825	,,
	Cornhwrdd	1836	Clyngwernen/Allt
Upper Swansea Six Feet or Rosy	Thomas David's or Ca-Main	1754-58	Between Box and Stradey Faults
	Rosey (?)	c1762	,,
	Stinking	1788	,,
	Rosy or Rosey or Rose	1798-1836	,,
	Penyfigwm or Great (in error)	1805-06	,,
	Little (in error)	1836	,,
	Carnarfon Upper	c1795-1836	Morfa Baccas
Lower Swansea Six Feet or Fiery	Fiery	1749	Between Box and Stradey Faults
	Fiery	1754	,,
	Llwynwilog	1754	Penyfan/Llwynwhilwg
	Fiery or Ca-Plump	1758	Between Box and Stradey Faults
	Fiery	1758-1836	,,
	Bowen's Real Old Fiery	1808	,,
	Old Fiery	1811	,,
	Lady Mansel's or Raby's Old Fiery	1819	,,
	Rosy (in error)	1836	,,
	Carnarfon Lower	c1795-1836	Morfa Baccas

Modern Seam Name(s)	Seam names used up to 1836	Known years of use	Region
Swansea Three Feet or Golden	Bwysva	1729	Bigyn/Penyfan
	Boisva	1754	"
	Parkycrydd or Killyveig	1758/59	North crop between Stradey and Box Faults
	Country coal and Dray Pitt or Engine	prior to 1776	"
	Old Golden	1794	"
	Golden	1794-1836	Between Stradey and Box Faults
	Golden or Fiery (in error)	1836	"
	Trosserch	c1825	North crop between Gors and Acorn Faults
Swansea Two Feet or Bushy	Biggin or Hill (?)	1754	Bigyn/Penyfan
	Cwm	1762-99	Furnace/Felinfoel
	Caereithin or Caerythen	1794-1836	"
	Bushy	1806-36	West of Box Fault
	Lower or Caithen	1819	"
	Bushy or Golden (in error)	1836	Bres Colliery
	Llanlliedi	1835/36	North crop between Box and Gors Faults
Un-named seams and Cille seams	Brickyard Seam	1822	Trebeddod
Pwll Little	Black Rock	1832	Pwll/Stradey
Hughes or Pwll Big	Pool	1765-68	Pwll
	Pwll	1836	"
Cilmaenllwyd	Stradey (?)	1762	Stradey Woods (?)
	Cwtta	1822	North of Trebeddod
	Stradey (?)	1836	Stradey Woods
	Cilymaenllwyd	1836	near Cilymaenllwyd
Cwmmawr	Hengoed	1822	Near Hengoed fawr
Un-named seam between Cwmmawr and Goodig Veins	Pantylliedy Fawr	1822	North West of Felinfoel
Goodig	No reference seen up to 1836		
Gwscwm	Trevenna Seam or Vein	1822	Trefanau Uchaf

CHAPTER 1 — REFERENCES

1 "The Geology and Geomorphology of the Country around Llanelli" by D. Q. Bowen, Llanelli Public Library Local History Research Group Series: No. 2 (in preparation).

2 Table I has been compiled from the information contained on 1:10560 scale Geological Survey plans (SN40SE, SN50SW, SN50SE, SS59NW, SS59NE); from Geological Survey Memoirs (The Geology of the South Wales Coalfield — Part VII, The Country around Ammanford, 1907; Part VIII, the Country around Swansea, 1907; Part IX, West Gower and the Country around Pembrey, 1907; Part X, the Country around Carmarthen, 1909; Special Memoir — Gwendraeth Valley and adjoining areas, 1968); Geological Survey vertical section 87 (1904); Natural Environment Research Council, Institute of Geological Sciences — British Regional Geology, South Wales (3rd ed.) by T. Neville George (1970).

3 To present a clear picture of the structure of the Llanelli Coalfield many of the representations on this plan and on subsequent Figures are not consistent with the standard designations adopted on Geological Survey plans.

4 The exploitation of Britain's early coalfields by an organised coal industry began some time in the 16th century. Llanelli's industry must have begun by the early 16th century, as it is known that coal mining was being actively pursued there between 1536 and 1539 ("The Itinerary in Wales of John Leland, in or about the years 1536-39" edited by L. Toulmin Smith, 1906, pp 59-60).

5 "The Rise of the British Coal Industry" by J. U. Nef, Vol II, Appendix N, pp 446-448.

6 Ibid, Appendix 0, pp 449-451.

7 Cawdor (Vaughan) 2/49 (4 Dec 1577). The "*Keven y Vinglawdd*" named in this lease is taken to be Penyfinglawdd which was once a large farm near Tyrfran.

8 "A survey of the Duchy of Lancaster Lordships in Wales 1609-1613", Board of Celtic Studies, University of Wales History and Law Series No. 12, p.262. The "*Brynbach*" described as being in Llanelly Borough is taken to be the Bryn area between the Town centre and Seaside.

9 Ibid, p.284.

10 Ibid, p.284 which refers to a "*Llwynhendû*". Also Carmarthenshire document 397 (14 Jun 1627) (CCL) which refers to a "*Lloyne hen ty*". Both locations are taken to be Llwynhendy.

11 Derwydd 216 (20 Sep 1618). "*Com Pencoed*" in the manor of "*Burwicke*" is taken to be the Pencoed area.

12 Ibid. "*Com y Velin*" in the manor of "*Burwicke*" is taken to be Cwmfelin.

13 Derwydd 204 (20 Feb 1622/23). A number of locations were given including "*Ty pen y koed*" and "*y kae dan y ddraynenbicka*" in the manor of "*Burwicke*". These locations are taken to be the Pencoed and Draenenbicca areas.

14 Derwydd 210 and 49 (24 Jan and 20 Oct 1634). These leases referred to "*Kraig Kaswthy*" which is taken to be Bigyn Hill.

15 Derwydd 687 (13 Sep 1662). The "*Keven y maes*" referred to is taken to be Cefnymaes.

16 Cwrt Mawr 979 (20 Sep 1663). The "*Bres Vawr and Bres Vach*" referred to are taken to be the Bres area from Llanelli Town centre to Bigyn Hill.

17 Stepney Estate 1100 (1 Aug 1729).

18 Cawdor (Vaughan) 13/372 (3 Sep 1733).

19 Cawdor (Vaughan) 21/623 and 21/624 (28 Jun 1754 and 22 Apr 1758 respectively).

20 Llanelli Public Library plan 3 (Apr 1772).

21 Mansel Lewis 1146, copy letter (1 Jan 1806).

22 Llanelli Public Library plan 3 (Apr 1772) was a report by John Thornton on the colliery lands between Cwmfelin and Pencoed.

23 Mansel Lewis London Collection 112, letter from Alexander Raby to a Mr Williams (18 Dec 1807).

24 Local Collection 37 (11 May 1824) quoting from "Reports of the Llangennech Collieries" by Edward Martin (21 Sep 1803).

25 Llanelli Public Library plans 4 to 7 (all dated July 1806).

26 Quoted in Local Collection 37 (11 May 1824).

27 Llanelli Public Library plan 8 (1 Jul 1812).

28 Mansel Lewis 2590, "A plan of part of the Carmarthenshire Railway etc..." by Rees Jones (Jan 1822).

29 Quoted in Local Collection 37 (11 May 1824).

30 Llanelli Public Library plans 4 to 7 (Jul 1806); Nevill MS IV, letter from C. Nevill to R. Michell (23 Jul 1806).

31 "*Extracts from the observations of B. Jones — gathered from the long experience of his father Rees Jones as a Mineral Surveyor etc... 1835*", contained in Thomas Mainwaring Commonplace Book (LPL); Nevill MS XVIII (Jan 1836) containing reports by Benjamin Jones on the coal districts under lease to the Llanelly Copperworks Company; GS 171/218 (1836 to 1842) entitled "*Notes and sections relating to the South Wales Coalfield by Sir William E. Logan (then Mr Logan) 1836 to 1842*" at the Institute of Geological Sciences, London. Rees (or Rhys) Jones of

Loughor had worked in the Llanelli Coalfield before 1800 and was the most experienced and knowledgeable of the local mineral surveyors. His son, Benjamin Jones, had worked as a civil engineer under I. K. Brunel before returning to Llanelli in the early 1830s to work initially as a colliery viewer and later as a solicitor (*Llanelly Guardian,* 19 Nov 1891). Comparison of the references given above shows that William Logan obtained much of his information from Benjamin Jones who, in turn, would have gained it from his father.

32 Figure 6 is based on William Logan's representation of the known seam outcrops and faults c1836 (contained in GS 171/218, op.cit.). Logan's representation was later incorporated on to the first 1 inch to 1 mile scale Ordnance Geological Survey plan of the area by Henry de la Beche published in the 1840s ("Wales and the Geological Map" by D. A. Bassett, Amgueddfa, 3 Winter 1969). I am indebted to Dr D. A. Bassett for bringing William Logan's notes to my attention.

CHAPTER 2

COAL MINING IN THE SIXTEENTH AND SEVENTEENTH CENTURIES

INTRODUCTION

Coal had been used industrially by smiths, lime and salt burners, brewers and dyers for at least 200 years before the 16th century, but it was generally considered inferior to wood as a fuel and domestic consumers hardly used it because dwellings were seldom provided with proper chimneys to draw off the sulphurous gases released by burning. During the 16th century, however, the demands of expanding industry — particularly iron-works which were great users of wood — an increasing population, and the growing tendency of landowners to turn wooded areas into more profitable arable and grazing land, led to serious deforestation. Governments became alarmed at the depletion of woodlands, which were then considered to be the nation's most important natural resource, and passed a number of Acts of Parliament designed to restrict the sale of wood. The Acts had the desired effect of making wood supplies scarcer and, therefore, dearer to buy, so that industrial and domestic consumers turned increasingly to the alternative fuel source, coal.[1]

The stage was thus set for a rapid expansion of the British coal industry. Although coal had been of secondary importance as a fuel up to this time, the Newcastle coal trade had been developing since the 13th century[2] and, locally, coal had been mined at Swansea since at least as early as the 14th century.[3] This known activity in other parts of Britain, combined with the knowledge that mining was well-established at Llanelli by the 1530s,[4] suggests that the Llanelli Coalfield was also being worked in an organised manner in the 15th century or earlier. No confirmatory documentary evidence of this possibility has been seen, however.

Be that as it may, when the general expansion of the British coal industry first occurred, Llanelli's coal resources proved comparatively easy to exploit because the limitations to working and trade, imposed by the primitive state of mining technology and the virtual non-existence of land transport systems, were compensated by the area's favourable geological and geographical attributes. These have been mentioned briefly in Chapter 1 but they were crucial to the expansion of the Llanelli industry in the 16th century and are, therefore, described again in greater detail. They were:-

(a) the coal was the type required. As coal began to replace wood the demand was mainly for bituminous coal and this was available at Llanelli.

(b) the coal was locatable because the synclinal structure of the Llanelli Coalfield brought the seams to the rock surface in areas where superficial deposits were absent or of minor thickness.

(c) the coal seams could be worked at a time when pumping systems were primitive and the problems of water accumulation prevented the exploitation of many coalfields. Llanelli's synclinal structure allowed shallow working to be effected in a number of seam outcrop areas, especially if they were areas of localised high ground where drainage adits could be employed.

(d) the coal occurred in close proximity to navigable rivers, pills and the sea-shore, allowing short haulage to the ships of the export trade,at a time when land transport systems had not been developed.

Before the end of the 16th century, an export trade in coal, although intermittent, was established between Llanelli and the West Country, the Channel Islands and France.

During the 17th century British and foreign manufacturers solved the technical problems of adapting their processes to the use of coal and it became Britain's main industrial and domestic fuel. By this time, Llanelli's coal industry was sufficiently well-established to capitalise on the increased demand for its product. There is evidence that improved mining techniques were being introduced to the Coalfield in the early 1600s, and the known activities of local men, such as Walter Vaughan and Owen Jenkine, point to an expansion in the area's coal industry. Many factors adversely affected the development of the British coal industry during this period, however, and Llanelli's trade would have been subject to these influences. The taxes levied on exported coal, regarded as a valuable source of income to the Crown during the reigns of James I and Charles I, retarded the growth of the coal export trade, and Llanelli's foreign trade, which was mainly with France, was particularly affected. The period of Civil War followed by the uncertain years of the Protectorate also restricted the expansion of the local coal industry, and recorded exports of coal from the area show that Llanelli's coal trade declined between 1620 and 1659. The Restoration, in 1660, marked a period of rapid expansion in the British coal industry and events in the Llanelli Coalfield mirrored the national trend. There was renewed activity in the granting of coal leases from 1660 onwards, and John Vaughan (Walter Vaughan's son), who had supported the Royalist cause during the Civil War and whose interests would not have flourished during the Protectorate, seems to have been the dominating figure in the resurgence of Llanelli's coal industry. Exports from the area increased steadily and reached their peak in 1687/88. The accession of William III in 1689 and the resulting Anglo-Dutch hostilities with France, Llanelli's main foreign market, radically affected the export trade, and Llanelli's coal industry entered a period of depression which was to last into the early 18th century.

(I) COAL MINING IN THE 16TH CENTURY

Although the claim has been made that coal was being mined at "*Caecotton-issaf*" (near Box Cemetery) in 1500,[5] the earliest written evidence, as we have seen, is that of John Leland who passed through the area between 1536 and 1539. His observations on the difference between Llanelli's bituminous coal and the anthracite of the Gwendraeth Valley have often been quoted but they bear quoting again:- "*At Llanethle, a village of Kidwelli lordship, a vi miles from Kidwelli, the habitans digge coles, elles scant in Kidwelli land. Ther be ii maner of thes coles. Ring coles for smith be blowid and waterid. Stone coles be sumtime waterid, but never blowen, for blowing extinguishit them. So that Vendwith Vaur coles be stone coles; Llanethle coles ring colis*".[6] They prove the early existence of an established industry, for which the claim has been made that its coal was being shipped from the Burry Estuary as early as 1551[7] and which, by 1566/67, had developed to such an extent that an export trade in coal was being carried on with the West Country, the Channel Islands and France.[8] In 1585 it was reported that the "*Port of Burry*" was the main Carmarthenshire creek for the export of coal[9] and further shipments, mainly to France, took place on an intermittent basis up to the end of the 16th century.[10]

Although the quantities of exported coal listed in the Welsh Port Books were small, they confirm that there was an established and organised Llanelli coal industry at that time. Little documentary evidence of 16th century activity has survived, however, the only reference seen being a Crown lease of 1580/81,[11] which related to land in Llanelli, and granted

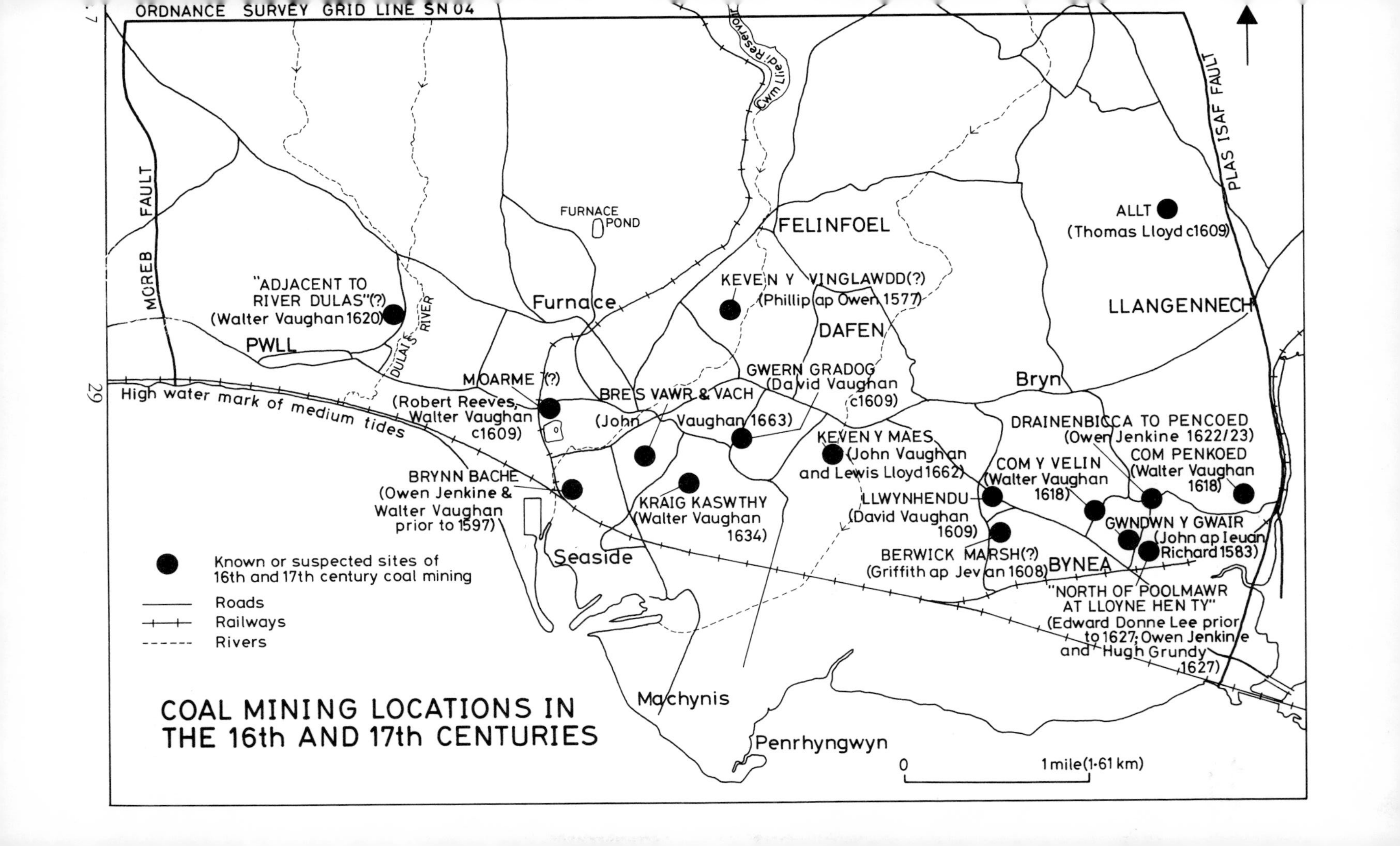
ORDNANCE SURVEY GRID LINE SN 04
PLAS ISAF FAULT
MOREB FAULT
Cwm Lliedi Reservoir
FURNACE POND
FELINFOEL
ALLT
(Thomas Lloyd c1609)
LLANGENNECH
"ADJACENT TO RIVER DULAS"(?)
(Walter Vaughan 1620)
Furnace
KEVEN Y VINGLAWDD(?)
(Phillip ap Owen 1577)
DAFEN
DULAIS RIVER
PWLL
MOARME (?)
(Robert Reeves, Walter Vaughan c1609)
GWERN GRADOG
(David Vaughan c1609)
Bryn
BRES VAWR & VACH
(John Vaughan 1663)
High water mark of medium tides
DRAINENBICCA TO PENCOED
(Owen Jenkine 1622/23)
KEVEN Y MAES
(John Vaughan and Lewis Lloyd 1662)
COM PENKOED
(Walter Vaughan 1618)
COM Y VELIN
(Walter Vaughan 1618)
BRYNN BACHE
(Owen Jenkine & Walter Vaughan prior to 1597)
LLWYNHENDU
(David Vaughan 1609)
KRAIG KASWTHY
(Walter Vaughan 1634)
GWNDWN Y GWAIR
(John ap Ieuan Richard 1583)
BERWICK MARSH(?)
(Griffith ap Jevan 1608)
BYNEA
Seaside
"NORTH OF POOLMAWR AT LLOYNE HEN TY"
(Edward Donne Lee prior to 1627; Owen Jenkine and Hugh Grundy 1627)
Machynis
Penrhyngwyn
Known or suspected sites of 16th and 17th century coal mining
Roads
Railways
Rivers
0
1 mile (1·61 km)
COAL MINING LOCATIONS IN THE 16th AND 17th CENTURIES

all the coal, lead and iron, under Duchy of Lancaster lands in the Lordship of Kidwelly,* to Sir George Cary, for 21 years at an annual rent of 10 shillings plus 20 shillings to the Queen. The low rental was typical of the Crown leases of the period, and its purpose was to encourage the mining of coal to alleviate the shortage of wood.[12] Two other leases may refer to the Llanelli area but the sites have not been positively identified. In 1577, Henry Vaughan of Trimsaran granted to Phillip ap Owen, of Bringroes, "*all manner of Vaines of Colles*" under Keven y Vinglawdd, described as being in Llanelli parish, for a period of 20 years in a bond of £2.[13] Keven y Vinglawdd could have been an early name for, or else was situated close to, Penyfinglawdd, which was once a large farm, between Llanerch and Felinfoel, where coal seams outcrop and superficial deposits are absent. In 1583, Griffith David of Llanelli granted to David John ap Ieuan Richard, of Pembrey, all the coal under Gwndwn y Gwair in Llanelli parish, in a bond of 100 marks.[14] This could have been Gwndwn, in Bynea, where coal seams outcrop in an area with no superficial deposits, close to known sites of early 17th century mining.

Surviving 16th century evidence, then, though sparse, does confirm the existence of the industry and that it had found export markets for its product.

(II) COAL MINING IN THE 17TH CENTURY

For the first twenty five years of the century Llanelli's coal exports, although still intermittent, increased in quantity.[15] The coal industry must have been expanding at this time, a conclusion supported by numerous extant, contemporary leases and references, which will be examined in detail to assess the state of the industry during this early phase.

The Crown lease, previously held by Sir George Cary, expired in 1601/02 and a dispute arose over its re-granting. Sir John Poynt of Acton claimed that the lease had been demised to Edmond Sawyer and Thomas Brymley "*for ever*", and that they had granted it to him. Phillip Vaughan, Robert Reeves, William Morgan, Owen Jenkine and David Lewis successfully disputed this claim,[16] with the result that the Crown leased all "*mines, quarries and pits of sea-coal and other coal*" as well as tin, lead, marl and metal, under Duchy of Lancaster lands in the Lordship of Kidwelly, to Phillip Vaughan of Trimsaran, in 1606, for a period of 21

* Many lands in and around Llanelli comprised part of the Duchy of Lancaster and belonged to the Crown. These lands were sold to the Vaughans of Golden Grove by Charles I in 1630.

years at a rent of 20 shillings per annum.[17] In view of the competition to gain this lease, Llanelli must have been considered a potentially valuable coalfield at this early time, and rivalries certainly existed between the people involved in mining the area's coal. Robert Reeves, who was mining on a common called Moarme (Morfa?) as a result of the 1606 Crown lease, alleged, c1609, that Walter Vaughan's workmen were hindering his miners by discharging the water from their mines into his[18] and, in 1613, John Griffith, who was working coal under his own lands at Llanelli, was charged with having entered the pits of his rival, William Vaughan, and having burned straw "*or some ill-flavoured stuff*" which nearly suffocated two workmen, his aim being to smoke out all Vaughan's colliers.[19] About the same time, Thomas Grent of Swansea complained that John David Griffith, Robert Reeves and others had refused to complete a lease of a coal mine at Llanelli, and had conspired against him.[20]

Not all relationships were as acrimonious as these and numerous other references show that a steady expansion of the industry was taking place on a more amicable basis. In 1608 Sir John Vaughan of Golden Grove and Dame Margaret, his wife, entered into an agreement with Griffith ap Jevan of Llangyfelach, by which Jevan was leased 12 acres (4.9 hectares) of land, previously occupied by John Griffith Howell and by John David Morris before him, "*with the right to the coal seams*".[21] Although the lands were not described, it is known that a yeoman named John David Morris lived and owned lands near Berwick Marsh (south of Llwynhendy/Bynea) in the early 17th century[22] and it is likely that the agreement related to this area. Between 1609 and 1613 David Vaughan worked coal at Llwynhendu (Llwynhendy) and at Gwern Gradog (the Penallt Road /Marble Hall area), while Thomas Lloyd of Llangennech mined at the Allt.[23] At this time, it was also recorded that the late Owen Jenkine (probably the Owen Jenkine involved in the dispute over the Crown lease of 1606) had worked a coal seam at Brynn bache, in the late 1590s, for the late Walter Vaughan.[24] The Walter Vaughan referred to was almost certainly Walter Vaughan of Golden Grove, who had died in 1597,[25] and Brynn bache is considered to have been the Bryn area, at the present junction of Albert Street, Queen Victoria Road and High Street, where the Rosy, Fiery and Golden Veins outcrop close to one another, under minor thicknesses of superficial deposits on localised high ground.[26] In 1611 the mines at Llanelli were being let at a royalty of 2s 2d and 1s 10d per wey (4 tons)* for coal from the "*upper vayne*" and "*lower vayne*" respectively, and, in 1613, it was reported that the pithead price of coal at Llanelli was

---

* Appendix F gives details of measures used in the Llanelli Coalfield.

10 shillings per wey. Also in 1613 an "*adventurer*" reputedly invested more than £1000 in a colliery at Llanelli, and it was reported that mining in Llanelli parish was being developed to "*the great reliefe of the country adjoyning*". About this time, too, skilled mining viewers were being brought into the Llanelli area to advise on the potential of collieries and to supervise the sinking of pits, and it is recorded that a yeoman, named John David Griffith, who had coal under his lands, consulted "*one that dealt in such business and had skill therein*".[27] In view of all this information, there can be little doubt that the Llanelli Coalfield was expanding during the early years of the 17th century. The competition for leases, the obvious rivalries between the participants, and the introduction of new mining techniques, provide a strikingly similar picture to the state of affairs that would exist 200 years later, when Llanelli's coal industry would experience its major early 19th century industrialisation.

It is probable that the expansion of the industry was helped by the advent to Llanelli, in 1616, of Walter Vaughan, the fourth son of the late Walter Vaughan of Golden Grove. He had married Ann Lewis of Llanelli before 1608 and they lived first at Talyclyn, near Hendy. Through his wife's inheritance of lands from her father, a release of lands from his brother, Sir John Vaughan of Golden Grove, and the purchase of many properties in the town and parish of Llanelli, Walter Vaughan became a major local landowner. When they moved to his wife's former home in Llanelli (probably an earlier building on the site of the present Llanelly House), he had already been involved in coal mining in the area for a number of years[28] and, from this time onwards, seems to have been the dominant figure in Llanelli's coal industry.[29] In 1618 he was mortgaged lands lying between Com y Velin and Com Penkoed in the manor of Burwicke, by Edward Williams, with the right to all "*coalworks and veins of coal*".[30] A description of the various landshares involved allows specific location to be made to the Cwmfelin, Ffosfach, Pencoed area,[31] where superficial deposits are absent or of minor thickness and where the Penyscallen, Swansea Four Feet and Swansea Five Feet Veins outcrop. Also in 1618, Griffith Thomas, yeoman, assigned a mortgage of the premises in Llanelli, where he and his wife lived, to Vaughan, who was to have all the "*coalmines in and under the premises*" at a rent of 22 pence for every "*weye*" of coal[32] and, in 1620, Thomas mortgaged lands bounded by the River Dulas (Dulais), and a piece of waste-land called "*coalbanke*", to Vaughan, with a lease of all the coal and minerals under the lands for a period of 21 years.[33] Although precise locations were not given in either case, the reference to the River Dulais suggests that the lands lay in the region of Stradey Woods.

In 1620 he applied, with others, for a patent for "*charking*" coal. This process, in which the impurities in coal were burned off to produce coke, to be used for ore smelting and metal manufacture in place of wood, was attributed to a Hugh Grundy, who had an iron-furnace at Tir Ponthenry. Nevertheless, the patent was granted to a company of seven people[34] which included Walter Vaughan, his brother Henry Vaughan of Derwydd and his brother-in-law John Protheroe of Nantyrhebog.[35] That the process was unsuccessful is not now important: what is important is that the episode is a pointer to the extent of Walter Vaughan's industrial activity, and some confirmation of the impression given by the leases that Walter Vaughan was one of Llanelli's early industrialists, and not just a local landowner.

Vaughan continued to expand his interests but, from about 1620, references survive relating to the coal mining activities of one Owen Jenkine (or Jenkin), who was a mercer in the town. The involvement in the industry of another Owen Jenkine, who died before 1613, has already been recorded. Because of the common practice of naming sons after a father, the Owen Jenkine who emerged around 1620 could have been the son of the deceased Owen Jenkine. If he were, he would probably have become involved, as a principal figure, after his father's death. Nevertheless, the first surviving reference to his activities was in 1620/21, when Ieuan ap Rees of Llanelli leased a fourth part of all the coal under his lands to Owen Jenkine, with permission to work the same.[36] No description of Rees's lands was given and it is assumed that they lay within the area studied. In 1622/23 Jenkine purchased lands and a mansion house called Ty pen y Koed, in the Drainenbicca/Pencoed region, from Hugh Howell, with all "*coalworks, mines and veins of coal belonging thereto*".[37]

Jenkine must have had adequate capital because, throughout the 1620s, he seems to have been as active as Walter Vaughan in the industry. Vaughan may have been concerned at the emergence of a rival, because he entered into negotiations with Jenkine for the purchase of his coal interests at Llanelli.[38] Vaughan was not entirely successful in his attempts to buy out Jenkine, because he was involved with both Jenkine and Hugh Grundy, in 1626, in a legal dispute over the conveyance of coal mines[39] although, in 1630, they did release a mine, veins of coal and a coal pit in Llanelli parish to him.[40] Jenkine continued to expand his coal mining interests and, in 1627, he and Hugh Grundy were leased coal and purchased coal interests in the Llwynhendy/Bynea region. The details of the purchase show that an Edward Donne Lee had constructed a drainage level to dewater coal mines in the Llwynhendy/Bynea region, prior to 1627, and this had been purchased from him by Hugh Grundy in 1627, for the purpose of working coal, leased to Owen Jenkine for 21 years by John David Morris and Griffith John.[41] The location of this activity is known with some certainty, because it was described as being to the north of Poolmawr at Lloyne hen ty. A plan of 1772 showed a Pwllmawr near to Cefn Berwick Farm,[42] and the village we now know as Bynea was previously part of Llwynhendy (the name Bynea being allocated only to Binie Farm and its lands bordering the River Loughor, north of the main Llanelli / Loughor road). Jenkine and Grundy's mines would, therefore, have been located just north of Bynea, and the 1772 plan certainly referred to many of the pits in this region as being old and disused. No further reference to Owen Jenkine has been seen, but the preceding evidence suggests that he played a significant part in the expansion of Llanelli's coal industry.

Walter Vaughan continued to enlarge his coal interests and, in 1629, he was granted the Crown lease of coal, tin and lead under Duchy of Lancaster lands in the Lordship of Kidwelly, for a period of 31 years, at an annual rent of 20 shillings and one tenth of all the tin and lead raised.[43] This lease had passed from Phillip Vaughan to a Symon Osbaldeston in 1616,[44] so, from 1580/81, when Sir George Cary held it, it had been held by four lessees. It is possible that one or more of these took the lease as a speculation, sub-leasing the coal,[45] but Walter Vaughan, deeply involved as he was in the industry, undoubtedly worked the coal himself. He obtained his last known lease of coal in the Llanelli area in October 1634, when he was granted land at Kraig Kaswthy (Bigyn Hill area) by Edward Lloyd, with the right to all water-courses and coal mines.[46] Walter Vaughan died on 7 April 1635.[47] Surviving records reveal him as the main developer of Llanelli's coal industry for 20 years, but he also built up, by inheritance and purchase, a large estate of lands, in and around Llanelli, which would play an important part in the development of the industry in the 18th and early 19th centuries.

Despite all the listed coal mining activities of Vaughan, Jenkine, Grundy, Lee, Reeves and others, which imply an expanding industry, recorded figures of coal exports from Llanelli showed a decline from 1620.[48] Charles the First's councillors looked to the coal trade to raise revenue for the Crown, levying a duty on coal exported to foreign destinations.[49] It is possible that, to avoid payment of duty, a proper return was not made of all coal exported at this time,[50] accounting for an apparent decline in the Llanelli industry until 1634/35, when Walter Vaughan died. This will be more fully discussed in Chapter 6.

In his will, Walter Vaughan made provision for his wife and children but his heir apparent was his eldest son, Francis, to whom the bulk of his estate was to pass on his widow's death.[51] He bequeathed his coal interests in Carmarthenshire and Glamorganshire to his wife (apart from his coal interests in Llannon parish which passed to his daughter Marie, wife of Henry Middleton). Francis Vaughan survived his father by only two years and Walter Vaughan's estate passed to his second son, John, in 1637,[52] the year in which the Civil War commenced. No evidence relating to Llanelli's coal industry has been seen for the 22 years of civil conflict and the subsequent Protectorate, but it is unlikely that it would have flourished. John Vaughan supported the Royalist cause and, although only suffering a fine for his allegiance,[53] the interests of the Llanelli Vaughans must have been adversely affected until the Restoration of Charles II, in 1660. By this time, John Vaughan was in full possession of his father's surviving coal interests—his mother having died c1653—and, within the first year of the accession of Charles II, he began to expand his interests by taking new leases of coal in the Llanelli area. In 1660 John Harry Lewis, yeoman, leased Vaughan *"all such Coale or Coale Mines as shall be in and uppon the Lands of the Said John Harry Lewis (lying northwards of the highway leading from the Town of Llanelly to Llwchwr and coasting eastwards towards the house of the said John Harry Lewis)"* for a period of 21 years at a royalty of 1s 6d per wey.[54] These lands cannot be located from this description but they lay within the area studied. In 1662 Vaughan entered into a partnership agreement with Lewis Lloyd of Llangennech, by which Lloyd leased to Vaughan the coal under Keven y Maes, for 3 years at a royalty of 2s 6d per wey, with Lloyd paying one half of the costs of sinking the pits and working the coal in return for one half of the profits.[55] Cefnymaes was situated just south of Halfway, in an area where the Swansea Five Feet Vein outcrops. In 1663, Charles Stepney of Llanelli, and his wife Elizabeth, leased Vaughan the coal under Bres Vawr and Bres Vach, at a royalty of 1s per wey,[56] and in 1665 Griffith Price of Llangyfelach leased him the coal in "*that only Vaine*", which Vaughan was then working at a royalty of 2s per wey.[57] No further reference to John Vaughan's coal mining activities has been seen and, on his death in 1669, his estate passed mainly to his wife Margaret and to his eldest son Walter, with the proviso that his wife's share of the estate would pass to Walter on her death.[58] It is probable that Walter Vaughan continued to work his father's coal interests but he died, on 12 October 1683, at the young age of 34 years.[59] Walter Vaughan had not married and his will specified that his inheritance should pass to his mother for the duration of her life and, on her death, to his four sisters or their heirs.[60] From 1683 onwards the Vaughans' Llanelli estate was, therefore, controlled by a woman and it is possible that she did not exploit its coal interests as energetically as her late husband or son.

Only one reference appears to have survived to confirm that others, besides the Vaughans, were active in the Llanelli Coalfield between 1660 and 1699—Henry Mansell of Stradey stating, in his will of 1677, "*First I will that my debts and several charges shall be paid and discharged out of the rents and profits of my estate and Coulworke*",[61] implying that coal mining was being pursued on Stradey Estate lands. Nevertheless, Llanelli's coal industry underwent significant expansion after 1660, because it was reported, in 1675, that Llanelli "*derives a considerable trade in coal etc.*"[62] and coal exports steadily increased, reaching a peak of some 8,000 tons in 1687/88. Llanelli's foreign trade was predominantly with France, however, and the hostilities with that country, which followed the accession of William III in 1688, radically affected Llanelli's coal exports, which fell, by the end of the century, to just over one quarter of the peak years of 1687/88. Both the coastal and foreign exports diminished during this period, and it would appear that the loss of the French trade led to a general decline in Llanelli's coal industry at this time. (See Chapter 6).

Llanelli's coal industry, then, affected by national and international factors, expanded intermittently during the 17th century, reaching peak production in the 1680s but suffering recession during the remainder of the century, although output in 1699 was still higher than in the years before 1680. During the century, the industry was consolidated and its export markets, some of which were to last another 200 years, established. Workings, too, were no longer restricted to seam outcrop areas. With the arrival of skilled mining viewers to the area, improved mining methods resulted which, together with the employment of pumping devices, made it possible to win coal below the natural water level, and pits could be sunk some way down the dip of the seam, increasing the amount of coal which could be won from a colliery.

No evidence has survived to suggest that purpose-built shipping places and transport systems were provided, and small ships must have sailed up the local rivers and pills to ground at safe locations, where coal could be loaded by hand from carts or pack animals. Apart from the fact that Hugh Grundy was involved in mining coal in the Bynea area and in smelting iron at Ponthenry (two activities which may have been unrelated), no evidence has been seen to suggest that metalliferous or other industries were locally consuming Llanelli's coal. It would, therefore, seem that the stimulus for the expansion of Llanelli's coal industry in the 17th century was solely that of export trade demand.

CHAPTER 2 — REFERENCES

1 The statements made in the preceding paragraph have been mainly derived from "Annals of Coal Mining and the Coal Trade" by R. L. Galloway (1898) and "The Rise of the British Coal Industry" by J. U. Nef (1932).
2 J. U. Nef, op.cit.
3 R. L. Galloway, op.cit., quoting from "Charters granted to Swansea" by G. G. Francis (1867).
4 "The Itinerary in Wales of John Leland in or about the year 1536-1539", pp. 59-60, edited by L. Toulmin Smith (1906).
5 "The History of Llanelly" by J. L. Bowen (1886) quoting an unidentified source. (LPL).
6 "The Itinerary in Wales of John Leland etc...", op.cit.
7 J. U. Nef, op.cit., Vol I, p. 53.
8 "Welsh Port Books 1550-1603" by E. A. Lewis (1927).
9 E. A. Lewis, op.cit., p.xxxiii.
10 Ibid.
11 "Records of Court of Augmentations relating to Wales and Monmouthshire" by E. A. Lewis and J. Conway Davies, p. 258.
12 J. U. Nef, op.cit., Vol I, p. 144.
13 Cawdor (Vaughan) 2/49 (4 Dec 1577).
14 Cawdor (Vaughan) 65/6635 (11 Dec 1583).
15 E. A. Lewis, op.cit.; J. U. Nef, op.cit.; "The Welsh Coal Trade during the Stuart Period 1603-1709" by B. M. Evans, M.A. Thesis, Un. of Wales Aberystwyth (1928).
16 "Exchequer Proceedings concerning Wales. In Tempora James I". Board of Celtic Studies, History and Law Series No. 15. Compiled by T. I. Jeffrey Jones, p. 125.
17 Cawdor (Vaughan) 21/602, copy lease dated 25 Jun 1606.
18 "Cadets of Golden Grove" by Francis Jones, Trans. of the Hon. Soc. of Cymmorodorion, 1971, Part II. Moarme has been interpreted as being Morfa and, if this is the case, Morfa Bach common near to the present Sandy area is the most likely situation.
19 J. U. Nef, op.cit., Vol I, p. 339, quoting Star Chamber Proceedings, James I, 288/5.
20 "Star Chamber proceedings relating to Wales" by I. O. Evans, quoting Star Chamber Proceedings, James I, 155/5.
21 Penlle'rgaer B. 17.2 (1 Sep 1608).
22 Carmarthenshire document 387 (27 Aug 1622) (CCL).
23 "A Survey of the Duchy of Lancaster Lordships in Wales 1609-13", Board of Celtic Studies, Un. of Wales History and Law Series No. 12, transcribed and edited by Wm. Rees, p. 284.
24 Ibid.
25 Francis Jones, op.cit.
26 Reference was made to the fact that Brynn bache lay within Llanelli Borough. Although no charter had been granted Llanelli was apparently termed a Borough "*by prescription*" — See "The Borough of Llanelly" by Hopkin Morgan, Carmarthenshire Local History Magazine, Vol. II, 1962.
27 J. U. Nef, op.cit., Vol I, pp. 54, 324 & 417, quoting Star Chamber Proceedings, James I 155/5 and 288/5.
28 Francis Jones, op.cit.
29 It is always difficult to decide whether the apparent domination of an activity by one of the major local landowners is due merely to the fact that his estate's papers have survived to the present day, whereas those of the other industrialists, who may have just passed through the area, have not; this certainly being the case at Llanelli in the 18th and 19th centuries. Walter Vaughan does seem to have been particularly active in Llanelli's coal mining industry, however.
30 Derwydd 216 (20 Sep 1618).
31 Llanelli Public Library plan 3 (April 1772) and its associated schedule.
32 Derwydd 689 (29 Sep 1618). The location of the premises was not given and it is assumed that they lay within the area studied.
33 Derwydd 199 (23 Jun 1620).
34 "History of Coal Mining in Great Britain" by R. L. Galloway (1882), David and Charles edition (1969), p. 41.
35 Francis Jones, op.cit.
36 Derwydd 680 (17 Jan 1620/21).
37 Derwydd 204 (20 Feb 1622/23).
38 Derwydd 723 (14 Mar 1622/23).
39 Derwydd 252 (4 Apr 1626).
40 Francis Jones, op.cit.
41 Derwydd 233 (27 Apr 1627) and Carmarthenshire document 397 (14 Jun 1627) (CCL).
42 Llanelli Public Library plan 3 (Apr 1772).

43 Cawdor (Vaughan) 21/604, copy lease dated 3 Jul 1629.
44 Cawdor (Vaughan) 21/603, copy lease dated 28 Jul 1616.
45 J. U. Nef, op.cit., Vol. I, p. 323, remarked on similar happenings in other parts of Britain. This problem virtually disappeared at Llanelli after 1630 because Charles I sold the lands comprising the Lordship of Kidwelly in order to raise money and they passed into the possession of the Vaughans of Golden Grove. (See Francis Jones, op. cit.).
46 Derwydd 49 (20 Oct 1634).
47 Francis Jones, op. cit., gives the date of Vaughan's death as 7 Apr 1635. Edwinsford 3524 (1 Jul 1647) gave a date of 7 Apr 1634. Despite the fact that this document was written only 12 years or so after the event, the date of 1635 would seem to be correct in view of the fact that Vaughan was leased coal at Kraig Kaswthy in October 1634.
48 B. M. Evans, op. cit.
49 J. U. Nef, op.cit., Vol II, pp 222-223.
50 Discussion on this point is given in Appendix F.
51 Museum Collection 417 (14 Mar 1634).
52 Francis Jones, op.cit.
53 Ibid.
54 Cilymaenllwyd 77 (15 Nov 1660).
55 Derwydd 687 (13 Sep 1662).
56 Cwrt Mawr 979 (20 Sep 1663). Elizabeth Stepney was John Vaughan's widowed sister-in-law who had married Charles Stepney after the death of her first husband Francis Vaughan.
57 Penlle'rgaer B. 15.15 (14 May 1665).
58 Francis Jones, op.cit.
59 "Llanelly Parish Church" by A. Mee (1888), p.lv.
60 Edwinsford 2762 (23 Aug 1682).
61 MS4.64 — "Carmarthen Registry, Wills Transcripts" (CCL).
62 "Britannia. Volume the First, or an Illustration of the Kingdom of England and the Dominion of Wales" by John Ogilby (1675).

CHAPTER 3

COAL MINING IN THE EIGHTEENTH CENTURY

INTRODUCTION

By 1700 coal was Britain's major resource and it was inevitable that the expansion of the industry would continue, irrespective of the interruptions caused by internal politics and international conflict. Increasing industrial and domestic use of coal and the rapid growth of population, which further stimulated demand, led to increased investment and activity throughout the coalfields, as investors realised that high profits could be made from coal mining. The introduction, in 1712, of Thomas Newcomen's steam engine for pumping water out of coal mines, thus enabling workings to be taken to greater depths than ever before, together with the development of transport systems, purpose-built shipping places and smelting-works erected on the coalfield,[1] produced significant changes, and a vast increase in output, in many coalfields where they were introduced during the first half of the century. But this modernisation was not introduced, at that time, to the Llanelli Coalfield— which had already been worked for some 200 years and was well placed for the export trade — because the local developers who ran the industry had insufficient capital, knowledge or experience to initiate the changes which were occurring elsewhere, including nearby Swansea and Neath.

Although Llanelli's coal industry emerged from the depression of the late 17th century, it underwent little real expansion before 1750. Records are scarce for the period 1700-1749 but those extant show that the Stepney Family, by intermarriage with the Vaughans of Llanelli, became the most active landowners, in the industrial context, in Llanelli. Among others, John Stepney, Hector Rees, John Allen and Thomas Bowen, all local men with limited capital, were active at this time but the available evidence suggests that their involvement in the industry was on a similar scale to that which had gone before.

This pattern changed abruptly from 1749 onwards, as industrialists from outside the area came to Llanelli, erected steam engines, built canals, railroads and shipping places and quickly advanced the technology of the industry. Henry Squire, David Evans and John Beynon from Swansea, Chauncey Townsend and his son-in-law, John Smith, from London (who already possessed colliery interests at Swansea and Llansamlet) together with Sir Thomas Stepney, the 7th Bart., who had interests in shipping and who was attempting to advertise Llanelli's advantages as an industrial and trading centre, were in the forefront of this first phase of real industrialisation. They operated mainly in the Bynea to Pencoed area, where the first steam engine, canal, railroad and metalworks on the Llanelli coalfield were all erected. Also involved in the local coal industry during this important period were the previously-active Hector Rees, John Allen and Thomas Bowen, together with John and Edward Rees, Charles Gwyn, Daniel Shewen, Robert Morgan and other smaller operators. Most of the area's coal was leased to Chauncey Townsend, however, and he either chose not to, or else found himself unable to, work this coal on the scale envisaged by the local landowners when he was granted his leases between 1752 and 1762. His inheritors also failed to work his extensive coal interests in the Llanelli area after his death, in 1770, and, as a result, the first phase of Llanelli's real industrialisation faltered and the degree of expansion which could have been expected for this period was never realised. Llanelli's coal industry fell once again into recession, although the Mansells of Stradey, Thomas Jones of Llwyn Ifan, John Thomas of Panthowell, William Hopkin, John Jenkin and other mainly local people continued to exploit the resources on a limited scale and a new metal industry, a small iron foundry, might have been established at the Wern, by William Yalden, in 1784.*

This state of affairs persisted until the mid 1790s, when a second phase of real industrialisation commenced. It was caused mainly by a second influx of entrepreneurs foreign to the area, from 1794 onwards, although some local people were involved. John Givers and Thomas Ingman were the first of these new industrialists and they erected an iron furnace and foundry at Cwmddyche (the present Furnace area) in 1793/94; David Hughes and Joseph Jones sank the Carnarfon Colliery, erected a steam engine and constructed a canal at Morfa Baccas in 1794; William Roderick, Thomas Bowen and Margaret Griffiths commenced mining in the Bres/Wern area in 1794/95 and constructed a canal to their shipping place at Seaside; and Alexander Raby arrived in 1796/97, to take over Givers and Ingman's interests and to exploit the Mansells' coal resources on an unprecedented scale as far as Llanelli was concerned. They were

---

* Llanelly Parish Church by A. Mee (1888) p. xxvi. No documentary confirmation of Mee's claim has been seen.

soon followed, in the early 19th century, by John Symmons of Paddington, Charles and Richard Janion Nevill of Birmingham, George Warde of Berkshire, the Pembertons from the north of England, the Earl of Warwick, John Vancouver and others, who played their part in Llanelli's rapid industrial expansion.

Surviving information about these activities is far greater than that available for the 16th and 17th centuries, and a detailed chronological treatment would obscure the main trends of the development of the industry. Llanelli's coal industry in the 18th century will, therefore, be considered in terms of the activities of, firstly, the main land-owning families and, secondly, the individuals and partnerships who actually carried out the mining, with the inter-relationships between them incorporated into the various accounts. The main land-owning families — the Llanelli Vaughans and their successors the Llanelli Stepneys, the Mansells of Trimsaran and Stradey and the Vaughans of Golden Grove — will be considered only in as much detail as is necessary to allow their involvement in, and influence upon, the local coal industry to be established. For this purpose, Appendices C, D and E provide sufficient details of the ownership of lands and the devolution of the estates to allow an understanding of the complexities of the granting of coal leases to be gained. The individuals and partnerships will be considered in the order in which they first made their appearance in Llanelli's coal industry and not in the order of their importance, so that some form of chronology may be maintained.

(I) THE MAIN LAND-OWNING FAMILIES

(1) *The Vaughans and Stepneys of Llanelli*

The Llanelli Vaughans, as described in Chapter 2, had acquired substantial lands within and outside the Llanelli Coalfield. The estate was held by Margarett Vaughan, widow of John Vaughan, at the beginning of the 18th century, when the coal interests were in a depressed state because of the loss of the export trade with France and the fact that Margarett Vaughan, who was almost 80 years old, was perhaps not as actively involved with her mining interests as her husband John and son Walter had been. She died on 26 January 1703/04[2] and her late son Walter's will, which stipulated that the estate should pass equally to his sisters or their heirs on his mother's death, became effective.

When he made his will in 1682, Walter Vaughan had four surviving sisters — Jemimah, Anne, Mary and Margarett. Jemimah died on 7 March

1687/88[3] but three of her children — Margarett, Dorothy and Rachell — were alive in 1704.[4] The Llanelli Vaughans' estate, therefore, passed to six female descendants on Margarett Vaughan's death in 1703/04. They agreed to apportion the estate between them and, on 2 August 1705, the "*Partition Deed of the Llanelly Estate*" was signed by all parties concerned. By its terms, the Vaughans' estate was divided into four parts but the woods, timber, coal mines and damage to lands caused by coal mining continued as the common property and expense of all the beneficiaries.[5] The signatories to the deed were the six female descendants and the husbands of those who were married. One-quarter shares in the estate were, therefore, allocated to Margarett Vaughan's three surviving daughters: to Anne and her husband Griffith Lloyd; to Mary (widow of Sir Rice Williams); and to Margarett and her husband Sir Thomas Stepney. The final one-quarter share was allocated to the late Jemimah Vaughan's three surviving daughters: to Margarett and her husband Thomas Phillips; to Dorothy (widow of John Parker); and to Rachell and her husband Pendry Vaughan. With respect to the commonality of ownership of coal, it was agreed that "*Costs, charges and expense of digging and sinking pitts*" and "*all other costs and charges necessary or incident to the working of the said coale mines before mentioned shall from time to time be borne and paid equally share and share alike*". It was also agreed that a compensation of 4 shillings per acre would be paid to each beneficiary, for land "*wasted and spoiled*" by carrying on the coal mines. Jemimah Vaughan's three daughters drew up a further partition deed in 1709 and divided their one-quarter share into three one-twelfth shares in the estate.[6] The large estate, which had been accumulated by the Llanelli Vaughans in the 17th century, therefore, underwent considerable devolution in the first decade of the 18th century, with the lands passing into the ownership of six different people, as shown in Appendix C. Despite the attempts to ensure that the Estate's coal interests would remain undisturbed, by holding them as common property, the devolution of the Estate led to complications and disputes, over the granting of coal leases and in the payment of royalties, throughout the 18th and 19th centuries.

Sir Thomas Stepney, 5th Bart. of Prendergast, Pembrokeshire, had married Margarett Vaughan (daughter of Margarett Vaughan who died in 1703/04) in 1691.[7] They lived at Llanelli before the turn of the century,[8] moving, in the early 18th century, into the present Llanelly House, which he built on the site of a previous house.[9] There is no evidence that he was himself engaged in mining but, living in the area, it is reasonable to assume that he and his wife were nominally in control of the coal interests of the Vaughans in Llanelli. A payment of royalties to

Plate 2 — Llanelly House, c1860. Built by Sir Thomas Stepney, 5th Bart., c1714, probably on the site of an earlier mansion. The left hand side of the house corresponds to the present shop frontages of Llanelly House in Vaughan Street.

Sir Thomas in 1709, for coal worked under his lands by a John Rees, indicates that he acted as lessor and agent for the other beneficiaries.[10]

Sir Thomas and Margarett Stepney had only one child, John,[11] who was born in 1693.[12] His aunt, Anne Lloyd, willed or entailed her one-quarter share in the Llanelli estate to him in October 1705,[13] soon after the partition deed had been signed. Anne Lloyd died in 1706/07[14] and her husband, Griffith Lloyd, died before 1724,[15] at which time John Stepney would certainly have inherited his aunt's one-quarter share of the estate. There is unauthenticated evidence that he was involved with others in granting a 20 year lease of coal at Ffosfach to Rowland Davies of St Clears in 1723[16] and, by 1729, John Stepney was personally engaged in mining coal under lands in the Bigyn Hill region, which he had inherited from Anne Lloyd. In August 1729 he came to an agreement with Thomas Price of Penllergaer regarding the working of coal under "*Caswddy Hill and Bwysva*", where both parties owned intermixed landshares, and where continuity of working was difficult unless an understanding was reached between them. Stepney and Price granted the coal under certain lands to each other, for a period of 21 years at a royalty of 3 shillings per wey, and agreed not to cause any damage to the drainage level used to dewater the workings, which had recently been constructed by Thomas Price in partnership with Hector Rees.[17] Reference in the agreement to the fact that the coal lay under Caswddy Hill (Bigyn Hill), Bwysva (not located), Dorgraig (a landshare immediately south of the present Bigyn School)[18] and Coed Hirion (the old name for the present Tŷ Isaf),[19] locate John Stepney's coal mining activities to the Bigyn Hill/Penyfan area, with drainage of the workings (most likely in the Golden Vein) achieved by a drainage adit driven into the south side of the hill (see Figure 8). The period over which John Stepney worked is not known, and no further reference to his coal mining activities has been seen. Lady Margarett Stepney died in 1733[20] and Sir Thomas Stepney, 5th Bart., died in 1744/45,[21] at which time John Stepney became Sir John Stepney, 6th Bart., possessed of one-half of the Vaughans' Llanelli estate by inheritance of his mother's one-quarter share. The sixth Baronet outlived his father only a short time and died in 1748, aged 55 or 56.[22] (Plate 3)

At the time of his death, Sir John Stepney, 6th Bart., was a widower and he was survived by three children — Thomas, Mary and Justina Ann. Thomas succeeded his father in 1748, took possession of one-half of the Vaughans' Llanelli estate and became known as Sir Thomas Stepney, 7th Bart.[23] Llanelli was about to undergo the first phase of its real industrialisation at this time, and Sir Thomas, although only 23 years of age, became deeply involved in a number of activities which would help to promote

Plate 3 — Site of mining operations on the outcrop of the Golden Vein, at Dorgraig by Thomas Price, Hector Rees and John Stepney c 1729. Dorgraig was immediately to the right of Bigyn School, shown on the hill crest.

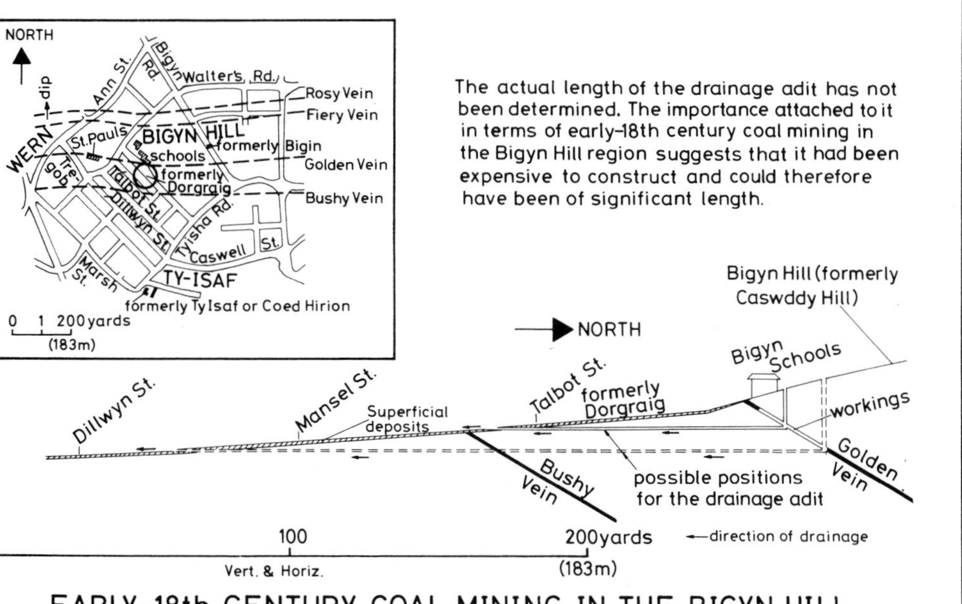

EARLY-18th CENTURY COAL MINING IN THE BIGYN HILL AREA BY JOHN STEPNEY, THOMAS PRICE AND HECTOR REES

the trade of the area. He was active in the granting of coal leases, helped to establish the first metalworks on the Llanelli Coalfield and purchased full or part interests in a number of export trade vessels. The extent of his involvement in Llanelli's mid 18th century industrialisation can be assessed by summarising his known activities. In 1748 he probably leased coal in the region of Llwyncyfarthwch to Robert Morgan of Carmarthen[24] and, in 1749, he leased the coal under his lands in the Bynea/Pencoed region to Henry Squire, David Evans and John Beynon of Swansea for a period of 31 years.[25] About the same time, together with other inheritors of the Llanelli Vaughans' estate, he granted coal in the Pencoed region to Thomas Bowen.[26]

Stepney's correspondence from 1748 onwards, leaves no doubt that his main interest lay in promoting the shipping trade of the area and he was personally involved as a shipowner. In November 1748 an Andrew Jones, at the Cornfactor, London, wrote to him "*Have been inform'd by my good friend Mr Morgan Rice that you had lately built a fine New Vessel abt 200 tons and was likewise engaged in several others and that your Honour wanted the opinion of some of our Traders... As there are no other owners but your Self concern'd, I doubt freight cannot be procured at your place except you become adventurer of both ship and cargo*".[27] In 1749 Stepney wrote to a Mr Sheppard, Master of the Commissioners at Plymouth, and requested "*please to advise me further with regard to your Coal Contracts which if you cleared would be gladly concerned in — we have coals of several kinds here and I believe as cheap as anywhere*". His approach must have succeeded because, three weeks later, he wrote to Sheppard saying "*By the first opportunity shall send you a specimen of fiery vein coals and flatter myself that ye quality thereof will be esteemed, which will be of service to us in our joint contract in the Spring*".[28] In another letter, which was probably a standard advertising letter sent to merchants and agents, he requested information on the present price of "*Smith's coals at your Market*" and gave the price of coal at Llanelli as "*coals such as Newcastle 4s 3d per ton*".[29] In 1750 a Christopher Noble, at London, wrote to Sir Thomas suggesting that coal exports to Oporto would yield a good profit[30] and he was soon exporting coal to Oporto, Lisbon, Cornwall, Cork and Brest.[31]

In 1752 Sir Thomas and the other inheritors of the Llanelli Vaughans' estate, and the Mansells of Trimsaran and Stradey, granted all unleased coal under their lands in the Llanelli area (apart from the Mansells' lands west of the Lliedi River) to Chauncey Townsend for 99 years, and also promised him further coal when leases expired.[32] Sir Thomas, and the other landowners, expected a major exploitation of their coal resources by

Townsend and were effectively granting him a monopoly of the Llanelli coal trade. At the same time, Stepney was trying to complete the industrialisation process by introducing metal smelting to the area. (The modern trend was to site such works near suitable coal but, unlike nearby Swansea and Neath, Llanelli had not developed in this respect). In 1752, Stepney's agent wrote to a Mr Debonair "*Lead may be smelted cheaper than in Swansea or Neath as coal is to be had on the spot. Sir Tho:s will send him the charges in Erecting and a Calculation of the Profits. He will be 1000l* (£1,000) *himself concerned*".[33] A year later Sir Thomas wrote on the same matter to a Peter Bryan, at Cognac, "*As for lead I give you the price and charges here underneath and I have some intentions of erecting a Smelting House for Lead soon, which if completed I would readily make an annual contract with you.*"[34] As a result, a lead smelting works was built at Pencoed in 1754 or 1755 but it is not known if Sir Thomas was eventually involved in it. It was erected on his land and used his coal, however, because Stepney wrote to a Mr Barrow Smith, in 1755, requesting money owed by the "*Pencoyd Company*" for land rent and supplied coal.[35] The works was shown on plans of 1761 and 1772,[36] which referred to it as "*Smelt House*" and "*Lead House*" respectively and, as far as can be ascertained, it was the first metal works established on the Llanelli Coalfield. (Substantial remains still exist in 1979). Stepney's industrial interests did not flourish, however, due, in no small measure, to Townsend's failure to begin the anticipated large-scale exploitation of local coal, but the Pencoed leadworks also ran into difficulties, shortly after its establishment,[37] and the safety of the Burry Estuary for large export trade vessels was brought into question when the master of a ship refused to stay at an accepted safe anchorage, off Penclawdd, and withdrew back to deeper water, off Whitford, thereby prolonging loading times and increasing costs.[38] Stepney's foreign trade did not realise its early potential and merchants at Lisbon and Bordeaux informed him, as early as 1752, that no profits would result if he sent cargoes to them.[39] (Plate 4)

Sir Thomas Stepney continued to build his estate in terms of landownership, and, in 1758, purchased a further one-quarter share in the Llanelli Vaughans' estate from Thomas Williams.[40] (This share had passed to Williams on the death of his brother Sir Nicholas Williams one of the original signatories to the 1705 Partition Deed).[41] He had also gained additional lands, mainly outside the area studied, by the inheritance of his wife Elizabetha Eleanora Lloyd, daughter of Thomas Lloyd of Derwydd and Danyrallt,[42] and was undoubtedly a major landowner in Carmarthenshire, on a scale exceeding that of the head of the Llanelli Vaughans prior to the 1705 Partition Deed. He continued to lease coal not

Plate 4 — Remains of the Pencoed Lead House. Erected 1754/55 and the first metal works in the Llanelli area. Much of the original structure was removed during the construction of the Llanelli/Llandilo Railway.

already granted to Townsend, leasing coal under Brynsheffre[43] and, possibly, the Trostre region[44] to Robert Morgan of Carmarthen before June 1759; in 1760 he leased coal under lands at Pencoed Isaf to Abel Angove, John Wedge and John Thomas;[45] in 1765 he leased the coal under Hendy and Penllech (the present Pwll area) to Edward Rees of Cilymaenllwyd[46] and granted certain coals under Hendy to John Thomas of Panthowell;[47] and, in conjunction with the inheritors of Jemimah Vaughan's one-quarter share in the Llanelli Vaughans' estate, he leased coal under Trostre, Llwyncyfarthwch and Cae Cefen y Maes to Thomas Bowen in 1768, probably leasing Penrhyngwyn Point to him for coal shipping purposes at the same time.[48]

Although Sir Thomas Stepney, 7th Bart., had been a most important figure in the first phase of Llanelli's real industrialisation, being involved in shipping and perhaps the lead smelting works at Pencoed, no evidence has been seen to suggest that he was personally involved in the coal industry, other than by granting leases of his coal, until the latter years of his life. About 1768, he took a lease of Richard Vaughan's coal at the Allt, Llangennech;[49] in December 1771 he approached William Clayton, of the Alltycadno Estate, for permission to make a level at Brynsheffre and construct a waggonway over Clayton's lands to the Loughor River; and, in

August 1772, requested permission to construct a quay, opposite Box tenement by the side of the River, for shipping the coal.[50] Sir Thomas Stepney probably did not mine coal because he paid Richard Vaughan only the sleeping rent of £21 for the "*Allt Collery*" in 1772,[51] and his plans to construct the waggonway and quay were not realised on his death, at 48, on 4 October 1772.[52] He had been a leading figure in promoting the industrial development of the Llanelli area in its critical years in the early 1750s, but his efforts were probably nullified by the reluctance or inability of Chauncey Townsend to work the extensive coal areas leased to him.

Sir Thomas Stepney's eldest son, John, aged 29, succeeded to the baronetcy. He had been the Member of Parliament for Monmouth since 1767[53] and his interests and entire way of life had undoubtedly been outside Llanelli for many years before he succeeded. Within a year of becoming the 8th Baronet, he sold the Stepneys' Prendergast Estate in Pembrokeshire to meet certain financial terms of his great-grandfather's will (the 5th Baronet had specified that his grand-daughters, Mary and Justina Ann, were each to receive £2,000 and a further £2,000 as marriage settlements and this clause had not been fully administered)[54] and showed no interest in his father's industrial activities because, in 1773/74, he negotiated with Thomas Jones of Llwyn Ifan, who wished to purchase the colliery at the Allt.[55] Stepney spent most or all of his time away from Llanelli and, in 1775, went on a diplomatic mission to Dresden as Envoy Extraordinary and Minister Plenipotentiary to the Court of the Elector of Saxony, remaining there until 1782, when he moved to Berlin. He returned to Britain in 1784, retired from Parliament due to ill-health in 1788,[56] but did not come to live at Llanelli. He was involved, together with the remaining inheritors of the Llanelli Vaughans' estate, in granting coal leases to John Smith in 1785 and 1786[57] and may have kept the colliery at the Allt in operation,[58] although it is likely that these negotiations and activities were conducted by his agents, because all subsequent evidence points to a lack of interest in Llanelli on the part of the 8th Baronet. In 1787 he advertised part of his estate at Llanelli for sale,[59] and wrote to his agent, William Hopkin, expressing the hope that £23,000 would be realised if all the lands were sold.[60] This letter to Hopkin confirmed Stepney's absence from Llanelli, because he stated "*nor do I know at all when I shall be in Wales*" and his friendship with the Earl of Cholmondeley, to whom his estate would pass on his death, was also confirmed by his request, to Hopkin, to send woodcocks to the Earl at his Piccadilly address. It is not known if any of the advertised lands were sold at this time, but a new sale, which included all of Stepney's lands at Llanelli, was advertised in 1791. The sale notice was entitled "*A particular of a valuable freehold estate situate in the Borough and parish of Llanelly in the County of Carmarthen, of the present yearly rent and duties of £842.5.5, but very improveable, the property of Sir John Stepney, Baronet, proposed to be sold entire by Private Contract*",[61] and there is no doubt that Stepney was prepared to sever his connections with the area. The Estate was not sold at this time, because coal leases were subsequently granted, by Sir John, of most of the lands itemised in the 1791 particular of sale,[62] but lands at Llangennech were purchased by John Symmons in the late 1790s/early 1800s.[63] Unlike his father, who had been a leading figure in the industrialisation of the Llanelli Coalfield, Sir John Stepney spent his life outside Llanelli and the interests of his large estate, and perhaps those of the Llanelli area, must have suffered because of his absence.

(2) *The Mansells of Trimsaran and Stradey and the Shewen Family*

Edward Mansell, the only son of Henry Mansell and Frances Stepney, succeeded to the Stradey Estate, on his father's death, c1673. He married Dorothy Lloyd, widow of Theophilus Lloyd and daughter of Philip Vaughan of Trimsaran, in 1684, shortly after she had inherited her father's Trimsaran Estate. In 1696/97 Edward Mansell was created a baronet as "*of Trimsaran*", and became known as Sir Edward Mansell, 1st Bart.[64] At the beginning of the 18th century Mansell was the owner of extensive lands, in and around Llanelli, through his own and his wife's inheritance. In particular, many of the lands west of the River Lliedi were in his possession, although his ownership extended throughout the Llanelli Coalfield.

It has already been seen that Sir Edward's father possessed "*coulworkes*", although it is not known if he had been actively engaged or was just a lessor. (See Chapter 2). Similarly, no evidence has survived to prove that the first Baronet had coal mines of his own, although he always reserved the right to sink pits and carry away the coal whenever he leased farmlands to other people. In 1708 he leased Trostre, Parkymynydd and other lands to Griffith David but retained "*full liberty to sink any pitt or pitts*"[65] and, in 1711, he, and his wife Dame Dorothy, reserved the same right, and also "*full liberty of ingress, egress and regress and landing place*" for the coal "*ye nearest way to sea over any part of ye demised premises*", when leasing Brinbach, Randir-y-gilbach and Wayne bach (considered to be the Bryn area near Seaside) to David.[66] In 1713 Mansell again reserved the right to sink pits when leasing Tir y Fran, Kilyvig, Caer Gorse, Twmpath y Wayn, Cae y Grose and Caemain to Rowland Davies, although Davies was allowed to extract sufficient coal for lime-burning purposes,[67] and, in 1715, Trebeddod was leased to Evan William of Llanelli, with Mansell retaining the liberty to mine the coal.[68] The inclusion of the coal reservation clause in all these leases, implies that Sir Edward Mansell, 1st Bart., was involved in mining, or intended to mine, his own coal but no proof of this appears to have survived. The first Baronet died in March 1719[69] and his wife, Dame Dorothy, survived him only a short time, dying on 13 September 1721.[70]

The first Baronet was succeeded by his eldest son, also named Edward, who became known as Sir Edward Mansell, 2nd Bart., of Stradey and Trimsaran. The second Sir Edward married three times. His first marriage to Letitia Catlo ended in divorce in 1730, his second wife Anne Philips died in 1740 and, in the same year, he married Mary Bayley of Hereford.[71] There were no children from these marriages. No evidence has survived to suggest that the 2nd Baronet was personally involved in the local coal industry, and the first recorded instance of his involvement as a coal lessor dates to 1750, when he granted the coal under his lands in the hamlet of Berwick to Henry Squire, David Evans and John Beynon.[72] This was followed by the major leases of the coal under his lands to the east of the Lliedi River to Chauncey Townsend, in 1752.[73] Sir Edward Mansell, 2nd Bart., died on 9 May 1754 and, because he had no children, the baronetcy passed to his 23 year old nephew, Edward Vaughan Mansell, son of his late brother Rawleigh. The second Baronet's will stipulated that the Trimsaran Estate was to remain in his wife's possession, passing, on her death, to his sisters' children.[74] Dispute arose between the widow (Lady Mary Mansell) and the third Baronet regarding the apportionment of certain lands, which were claimed to belong to both the "*Trimsaran or Vaughan*" Estate and the "*Stradey or Mansell*" Estate, and a settlement was not reached between them until an order of the Court of Chancery was obtained, in 1766.[75]

Sir Edward Vaughan Mansell, 3rd Bart., therefore, inherited only the Stradey Estate but this had a rent roll of some £1,000 per annum.[76] The 3rd Baronet had an extravagant life style, however, and it is likely that he began accumulating heavy debts soon after 1754, despite the substantial income provided by his inherited estate. He became associated with a Swansea attorney named Daniel Shewen, about this time, Shewen marrying Mansell's sister, Bridget, in December, only 6 months after his wife's death in June 1757.[77] The family ties were further strengthened when Daniel's brother Joseph married Mansell's sister, Anne, on 27 February 1759.[78] Daniel Shewen took control of Mansell's financial affairs, probably at the time of his marriage to Bridget, and it seems that he took advantage of Mansell, who was described as being "*a person of weak intel-*

lect and incapable of transacting business".[79] In 1758 and 1759 Shewen obtained from Mansell a lease and release of all the coal under Stradey Estate lands at Penygaer, Bryngwynmawr and Bryngwynbach and of the Golden Vein under the entire Stradey Estate, for a period of 60 years at a royalty of 1s 3d per wey of 10 tons.[80] This was an extremely low royalty for the time and it is likely that Daniel Shewen misled Mansell, although their family relationship could have been a contributory factor. Shewen worked coal in the Llanelli area as a result of his leases, because he complained, in 1761/62, that at least £800 worth of his coal was awaiting shipping but bad winds had kept vessels away,[81] and it was recorded that a pit at Bryngwynmawr, re-opened in 1813/14, had originally been sunk by Daniel Shewen.[82] It is likely that Shewen's mining operations were small scale ventures, however, because a legal opinion, given in 1805, stated *"but Mr Shewen, not having the means to carry on the work properly as it could not be worked advantageously without a fire engine, which would have cost from 12 to 1500l, he desisted from working at all"*.[83]

Shewen quickly realised that he lacked the capital to take full advantage of the extensive coal areas leased to him at preferential royalty levels, and, in 1762, he surrendered his 1758/59 leases so that all the coal under the Stradey Estate could be granted to Chauncey Townsend. The terms of the resulting lease to Townsend leave little doubt that Shewen was firmly in control of Mansell's affairs. On the surrender of Shewen's leases, Sir E. V. Mansell leased the coal under the entire Stradey Estate lands in the Llanelli Coalfield to Townsend, on 2 August 1762, for 99 years at a royalty of 3 shillings per wey and also promised him the coal at Ffosfach, already granted to Squire, Evans and Beynon, on the expiry or surrender of their lease.[84] On the same day that the lease was signed, Townsend agreed to pay Shewen 1s 3d per wey for all coal raised by him, which had previously been under lease to Shewen.[85] So, in addition to taking virtual control of Mansell's financial interests, Shewen had also managed to establish himself as a middle-man between the Stradey Estate and the industrialists who wished to exploit its coal resources. At this particular time, he would have had the greatest expectations of receiving a substantial personal income from Chauncey Townsend's coal mining activities.

Sir E. V. Mansell's wife, Anne, died in January 1763,[86] and, within a few weeks of her death, Mansell, domiciled in London, applied for a licence to marry Elizabeth Stevens, described as being of Southwark.[87] It seems that Mansell had become involved with a man named Morris, who planned to get his hands on Mansell's inheritance through this marriage, because Daniel Shewen took immediate steps to prevent it taking place. Shewen wrote to a Charles Owen in London, requesting him to take out a writ of

Habeus Corpus, to remove Mansell to Fleet Prison for non-payment of debts, later explaining that he did so to prevent Mansell marrying a *"Woman of no Fortune"* and to keep him away from *"this vile fellow Morris"*.[88] As a result, Sir E. V. Mansell was committed to Fleet Prison between 15 and 27 February 1763, ostensibly for non-payment of a debt of £800,[89] but actually because Daniel Shewen wished to retain full control of Mansell's affairs. There is evidence that Chauncey Townsend may have been involved in this incident because, prior to Mansell's imprisonment, it was arranged that Townsend's son-in-law (John Smith) was to raise the money to secure Mansell's release from prison.[90] Mansell's total debts amounted to some £12,000 at this time, however,[91] and, once in prison, he found it difficult to get out. It is not known whether this was because his many creditors demanded payment when they realised he was in prison, or whether it was caused by lack of action on Shewen's part, but it resulted in Mansell having to spend some 9 months in the Fleet. He finally secured his release in November 1763, when he conveyed the Stradey Estate to appointed trustees (Thomas Pryce, Alexander Small and Gabriel Powell), who were empowered to employ the rents or sell the lands to pay off his debts and to provide him with an annuity of £200.[92]

Daniel Shewen's actions must have adversely affected his reputation, because Joseph Shewen, probably with Chauncey Townsend's backing,[93] took over the management or receivership of the Stradey Estate and Mansell went to live with him at Clifton, near Oulney, Bucks., on leaving Fleet prison.[94] Mansell married Mary Shewen, Joseph Shewen's daughter by a first marriage, probably in December 1764,[95] and the intertwining of the Mansell and Shewen families by marriage was further strengthened. A sale of some of the Estate's lands was proposed in 1767[96] and, in 1769, the trustees conveyed the Stradey Estate to Thomas Watts, of the Sun Fire Office, London and William Hamilton, of Lincoln's Inn, as security for a loan of £13,000, which was to be used to pay off Mansell's debts.[97] Daniel Shewen still retained an interest in the affairs of the Stradey Estate because his 1762 agreement with Chauncey Townsend remained valid, and he was corresponding with his brother, Joseph, over coal mining in the Llanelli area and other allied matters, as late as 1769.[98] Daniel Shewen died in June 1770,[99] when sole control of Mansell's affairs passed to Joseph Shewen, who was, curiously, brother-in-law and father-in-law to Mansell. Sir E. V. Mansell borrowed a further £1,000 from Watts and Hamilton in 1771[100] and took up residence at Stradey, together with his wife and Joseph Shewen, in 1772.[101]

In the meantime, Chauncey Townsend had hardly worked the Stradey Estate coal and, after his death in 1770, his inheritors showed a similar reluctance to exploit the extensive coal areas leased to them. Daniel Shewen had been greatly angered by this inactivity,[102] because his prospects of personal revenue had never been realised, and Sir E. V. Mansell and Joseph Shewen were faced with the same state of affairs. Joseph Shewen realised, as his brother had before him, that the Stradey Estate possessed great industrial potential and that the only way of developing it, in the prevailing circumstances, was to become personally involved. He managed to persuade John Smith, and the other inheritors of Chauncey Townsend's interests, to surrender the coal under Stradey Estate lands to the west of the Lliedi River, and an agreement to this effect was signed in October 1776.[103] Mansell borrowed a further £2,500 from Watts and Hamilton in April 1777,[104] probably for the purpose of financing his mining activities, and Joseph Shewen and Sir E. V. Mansell commenced coal mining operations on the Stradey Estate. They apparently raised and sold some 1,500 weys (7,500 tons) of coal annually, but their capital did not allow the purchase of the steam engines, which would have significantly increased their output.[105]

This limited scale of working did not produce enough revenue to solve the financial problems of the Stradey Estate and Watts and Hamilton exhibited a Bill of Complaint against Mansell in the High Court of Chancery, in March 1785, because considerable arrears of interest had accrued on the £16,500 borrowed. Mansell's answer to the Bill was heard on 7 July 1786 and it was arranged that one of the Masters of the Court would evaluate the exact sum owed. Watts and Hamilton agreed to return Mansell's mortgaged estate to his control on payment of the debt, but warned that, in default of such payment, they would foreclose upon Mansell and his heirs and appoint their own receiver of rents. Mansell drew up his will in October 1786, whilst the evaluation was under way, and stipulated that, on his death, his estates, and encumbrances upon them, were to be administered by Daniel Jones of Glanbrane, Glamorganshire and by Joseph Shewen, with any revenue paid to his wife, Dame Mary Mansell. He further stipulated that, after his wife's death, any revenue was to be paid to their children, Edward Joseph Shewen Mansell and Mary Martha Ann Margaret Mansell, and to their heirs after them. A clause in the will limited the granting of coal leases to a period of 3 lives or 21 years, whichever was the shorter. The evaluation of the Stradey Estate was completed by 18 February 1788, but both Sir E. V. Mansell and Joseph Shewen died before any proceedings concerning Mansell's debts could be heard, and the suit became abated.[106] Joseph Shewen died on 19 December 1788[107] and Sir E. V. Mansell died eight days later on 27 December 1788,[108] when his son became Sir Edward Joseph Shewen Mansell, 4th

Bart. Dame Mary Mansell was left with the 3rd Baronet's virtually bankrupt estate, described as being *"not nearly sufficient for the payment of his Debts due on Bonds and Simple Contract"*.[109]

Dame Mary Mansell proved her husband's will, took posession of the Stradey Estate and continued working the collieries commenced by her father and husband. Watts and Hamilton filed a Bill of Revivor against Dame Mary, her children and Daniel Jones, and the Court of Chancery decreed, in June 1790, that their former decision, made in 1786, should be adhered to. At this time, Daniel Jones, probably fearful of becoming involved in the Mansells' debts, declined to act as an administrator of the Estate and withdrew from all further participation in the Mansells' affairs. Watts and Hamilton were owed £21,950.19.5 ½d by December 1791, but agreed to hold their final demand and, in June 1792, when the debt had increased to £22,322.16.1 ½d, Dame Mary and her children stated, through their legal counsel, that they were unable to pay this sum. On 13 June 1795 the Court of Chancery, with the agreement of Watts and Hamilton, ordered that the Stradey Estate be valued to pay off the debts by a sale of some of its lands.[110] William Black of Epping was appointed valuer and his report, dated November 1795, concluded that the value of the lands exceeded the debts and that the yearly rents amounted to £1,159.19.0. He also reported that the Estate's coal resources could be worked profitably if steam engines were employed, but, at the time of his survey, the collieries were standing idle and had become filled with water.[111] As a result of his report, an Act of Parliament was obtained in 1796. This allowed the sale of some of the outlying or detached parts of the mortgaged Estate and, to attract industrialists to develop its resources, allowed coal leases to be granted for periods of up to 91 years, instead of the 21 years specified in Sir E. V. Mansell's will. Hugh Smith and John Pearce were appointed as trustees, to apply the money arising from the sale in reduction of the debts.[112]

No quick sale of outlying lands resulted[113] but the clause in the Act, allowing coal leases to be granted for up to 91 years, soon yielded benefits. John Givers and Thomas Ingman had established an iron furnace on Stradey Estate lands at Cwmddyche (the present Furnace area) in 1793/4, and had arranged to receive their coal from the Mansells.[114] They were financially assisted by Alexander Raby, who, on the passing of the Act, took over Givers and Ingman's ironworks,[115] obtained 91 year leases of all the Mansells' coal to the west of the Lliedi River,[116] and embarked on the establishment of an industrial empire on Stradey Estate lands. Raby planned to expand the ironworks, to work the coal with steam engines, and to build a network of tramroads which would connect all his activities with one another and with the dock he intended to construct at Seaside.

As the century closed, the Mansells of Stradey must have had every hope of solving their long-standing financial problems. The sale of their outlying lands would pay off their debts, and royalties from Raby might even make them rich. Sir Edward Joseph Shewen Mansell, 4th Bart., was not to benefit, as he died, unmarried and intestate, on 6 April 1798 and the baronetcy became extinct.[117] The Estate devolved upon his sister, Mary Martha Ann Margaret Mansell, who married her first cousin, Edward William Richard Shewen (son of Joseph Shewen and Anne Mansell), thus continuing inter-marriage between the Mansell and Shewen families.[118] The revenue of the Estate, however, was still paid to Dame Mary Mansell.

No plan, showing the locations of pits and levels from which the Mansells and Shewen worked the Estate's coal on a small scale, has survived. However, Bowen says that Dame Mary Mansell was working coal in 1791 at Penyfinglawdd, Sandy, Morfabach, Caeperson, Caerhalen, Waunelly and Cilfig, with a water wheel employed for pumping near Caemain.[119] She probably inherited these collieries from Sir E. V. Mansell and Joseph Shewen, some of the pits possibly being sunk by Daniel Shewen between 1758 and 1762. Early plans showed un-named pits at some of these locations and the following entries, shown on the plan in Appendix I, can be tentatively ascribed to the activities of Daniel Shewen, Sir E. V. Mansell, Joseph Shewen and Dame Mary Mansell between 1758 and c1794:- pits at Penyfinglawdd Farm (25)* and Weinelly (47) shown on a plan of c1822;[120] pits at Caerhalen (32), Caeperson (48) and Cilfig (33) shown on a plan of 1814;[121] water-wheel pit at Caemain (46) shown on a plan of 1787.[122] It is also suspected that the level at Caereithin (23), later named the Old Slip colliery by Alexander Raby, was worked by the Mansells and the Shewens. The pit sunk at Bryngwynmawr by Daniel Shewen has not been positively located, but a pit or level on the outcrop of the Swansea Five Feet Vein at Bryngwynmawr Farm (28) was shown on a plan dated 1800[123] (Fig. 9).

(3) *The Vaughans of Golden Grove*

The Vaughans of Golden Grove owned lands within the area studied and were involved in the granting of coal leases and wayleaves throughout the 18th century. Their ownership of lands at Llanelli fell into three

---

* The number in brackets is that given to the pit on the Ordnance Survey location plan in Appendix I.

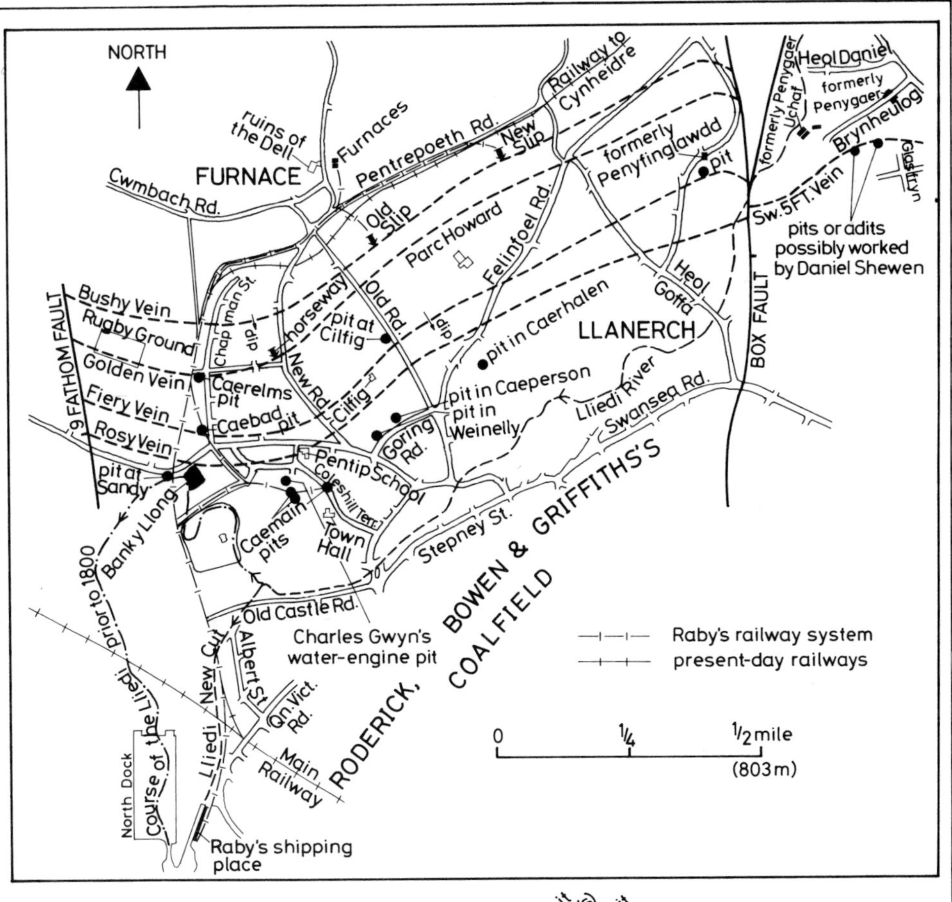

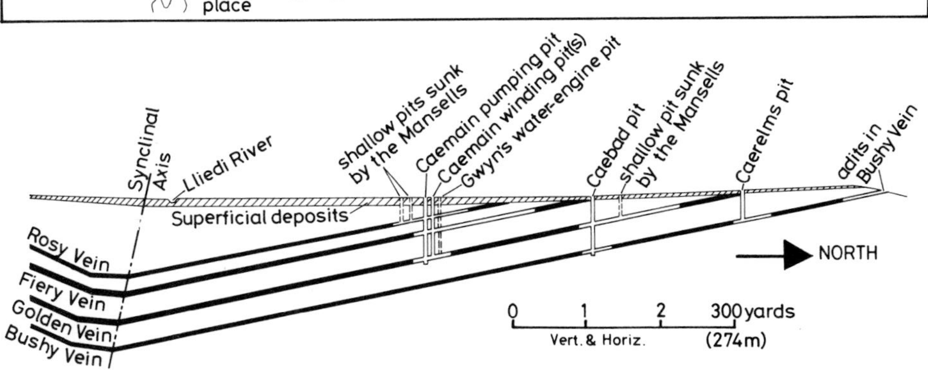

COAL MINING ACTIVITIES OF DANIEL SHEWEN, CHARLES GWYN, SIR E.V. MANSELL, JOSEPH SHEWEN, DAME MARY MANSELL AND ALEXANDER RABY PRIOR TO 1800

distinct categories as far as coal mining activities were concerned. These were:-

1. lands which belonged to the Estate before the 18th century;
2. lands which were inherited by the Estate holder late in the 18th century — two-twelfths of the original estate of the Llanelli Vaughans;
3. lands over which the Estate holder had manorial jurisdiction, as Lord of the Manor, and which included the seashores between high and low water marks and the common lands in and around Llanelli.

Unlike the Llanelli Vaughans, the Stepneys and the Mansells, the Vaughans of Golden Grove were not involved in industrial activity in the Llanelli Coalfield, but a brief account of their main line of inheritance is required to understand their role as coal lessors.

In 1700 the Golden Grove Estate was owned by Sir John Vaughan, 3rd Earl of Carbery.* He died on 16 January 1712/13 and, his sons having died in infancy, his only surviving child, Lady Anne Vaughan, inherited at the age of 23: the Earldom also became extinct. Being unmarried and the owner of the largest estate in South-West Wales, Anne attracted many suitors, from whom she chose the 28 year old Charles Pawlett, Marquis of Winchester and son and heir of the Duke of Bolton.

They were married on 1 August 1713 but it was immediately apparent that Pawlett, a spendthrift, had married Anne for her money, to settle his debts and finance his expensive life style. His attempts to obtain whatever money he could from Anne and the Estate, led to strained relationships between them and, soon, to a deed of separation, drawn up on 28 July 1714 only 12 months after marriage. Although a reconciliation took place a few months later, Pawlett continued to spend heavily. His succession to the Dukedom in January 1721/22, when Anne became the Duchess of Bolton, made no difference to his life style, so that Anne had to raise, by Act of Parliament in 1724, a massive mortgage of £20,000 on the Estate to pay his debts. The Duke left Anne in 1729 to live with an actress in London and they apparently never saw each other again but, having a continuing interest in the Estate under their marriage settlement, the Duke still tried to obtain whatever money he could from its income. Anne died on 20 September 1751 and her husband on 26 August 1754.[124]

---

* Sir John Vaughan's grandfather had been created the first Earl of Carbery by Charles I in 1628. Carbery was situated in County Cork, Ireland.

The Duke's plundering of its finances and resources undoubtedly led to a deterioration in the affairs of the Golden Grove Estate, but the mineral resources were not sufficiently exploited to compensate for this, probably because Anne seldom lived on the Estate and left matters in her agents' hands. This was certainly so with the Estate lands in the Llanelli Coalfield. Only two leases appear to have been granted by Pawlett and his wife in the first half of the century: to Mary Lloyd, in 1718, of coal under the Allt in Llangennech Parish for 21 years;[125] and to John Allen of Llanelli, in 1733, for the duration of the life of the Duke of Bolton, of coal under the common called Gwern Caswthy (Wern) and under the "*waste overflowed by the sea-tides lying westwards of Gwern Caswthy*".[126]

Anne Vaughan died childless and willed the Golden Grove Estate to her distant cousin John Vaughan of Shenfield, Essex, who had been her constant ally during her disputes with her husband. Vaughan was 58 years old when he succeeded to the Golden Grove Estate, and already owned lands throughout Carmarthenshire, through inheritance and marriage. Although permanently domiciled in Essex, he set about restoring the Golden Grove Estate and took a particular interest in his inherited manorial rights, which were a potential source of high revenue from royalty payments. The Llanelli Coalfield was just beginning its first phase of industrialisation, and Vaughan displayed a keen interest in what was going on and regularly corresponded with his agent regarding the mining of coal under his lands at Llanelli. He probably leased two veins of coal at Llanelli to Robert Morgan of Carmarthen in 1751,[128] and, in 1754, as soon as John Allen's lease of coal at the Wern expired, he granted the same coal to Thomas Bowen of Llanelli.[129] In 1755/56 he attempted to lease the Marsh Vein (probably the Swansea Five Feet Vein at Ffosfach) to Squire, Evans and Beynon and negotiated with them regarding wayleave payment for their waggon way to Spitty Bank.[130] Also in 1755/56, he negotiated with John and Hector Rees over the construction of a quay in the Pwll area[131] and, in 1758, leased the coal under the common of Murva Vach (Morfa Bach) to Thomas Bowen.[132] By 1759 Samuel David was mining his coal at the Allt,[133] and, in 1761, acting through his agents, Vaughan attempted to lease the Marsh Vein to Chauncey Townsend.[134]

John Vaughan had, therefore, embarked upon the development of the Golden Grove Estate holdings in the Llanelli Coalfield. It is difficult to decide whether this was due to Vaughan's efforts, or whether it was an inevitable consequence of the industrialisation of the Llanelli area from 1749 onwards, but there can be little doubt that John Vaughan's business-like interest in his inherited estate was a vast improvement compared with the previous plundering of its resources by the Duke of Bolton. John

Vaughan died, aged 71, on 27 January 1765 and the Golden Grove Estate passed to his only son Richard Vaughan.[135]

Richard Vaughan was 39 years of age when he succeeded to the Golden Grove Estate, having already obtained a one-twelfth share in the original Llanelli Vaughans' estate through his wife, Margaret Elizabeth Philips, who had inherited it from Margaret Vaughan, daughter of Jemimah Vaughan.[136] Vaughan's succession co-incided with Chauncey Townsend's decision to invest substantial capital into the coal areas leased to him at Llanelli and, in September 1765, a draft agreement was drawn up between them allowing Townsend to construct a canal, from his colliery at Genwen over the common called Dolevawr in the hamlet of Berwick.[137] This agreement was finalised in July 1766[138] and Townsend commenced his mile long Yspitty Canal. In 1768, Richard Vaughan leased the coal under the "*waste lands or sea*" at Hendy and Penllech (the present Pwll area) to Edward Rees[139] and, in the same year, he probably leased the coal at the Allt, Llangennech to Sir Thomas Stepney, 7th Bart.[140] The first phase of Llanelli's industrialisation came to a halt after 1770, however, and Richard Vaughan derived little financial benefit from Estate lands in the Llanelli Coalfield thereafter, as he died in 1781, when the Golden Grove Estate passed to his 24 year old son, John Vaughan.[141]

John had also inherited a further one-twelfth share in the original Llanelli Vaughans' Estate from his great-uncle Richard Phillips (allocated to Rachel Vaughan, daughter of Jemimah Vaughan in 1709), so he held two-twelfths of the original estate, William Langdon holding another one-twelfth and Sir John Stepney, 8th Bart., the remaining three-quarters.[142] The finances of the Golden Grove Estate had never recovered from the crippling mortgage of £20,000 raised to pay the Duke of Bolton's debts, however, and between 1783 and 1785 some of its lands were sold to raise money.[143] John Vaughan received only small rentals from the Estate's industrial interests at Llanelli[144] but, in 1785 and 1786, he joined Sir John Stepney, 8th Bart., and William Langdon in granting coal leases to John Smith.[145] Little working resulted from these leases and, by 1790, Vaughan received only one coal rental, the colliery at the Allt, from all his Llanelli lands.[146] The financial problems of the Golden Grove Estate continued and Vaughan tried to obtain a mortgage of up to £6,000 on part of the Estate, between 1791 and 1795.[147] By this time, the second phase of Llanelli's industrialisation had commenced, and Vaughan became involved with the other landowners in the Llanelli area in granting coal leases. In September 1794 he leased the coal under the common of Morfa Baccas to David Hughes and Joseph Jones, granting them permission to construct a canal,[148] and, in October, the coal under the common of the

Wern to William Roderick.[149] Coal under Morfa Bach Common was leased to Alexander Raby in 1799.[150] At the end of the 18th century the Golden Grove Estate was still financially insecure, and John Vaughan tried to sell more land to meet the demands of creditors.[151] In 1804 the Estate passed into the hands of John Campbell (Lord Cawdor). (This is discussed in Chapter 4).

(II) THE INDIVIDUALS AND PARTNERSHIPS

The Llanelli Vaughans, the Stepneys and the Mansells (together with the Shewens), then, mined their own coal only on a limited scale throughout the 18th century, their main involvement in the development of Llanelli's coal industry being as land-owners and coal-lessors. Those who were leased the coal took the financial risks, and Llanelli's 18th century industrial growth depended greatly on the success or failure of their enterprises. Individuals and partnerships were involved in the mining, the scale of their operations ranging from very small one-pit concerns to large multi-pit complexes. Evidence has survived to trace the history of the larger enterprises from 1749 onwards, but very little information has been found concerning the smaller and earlier ventures. Nevertheless, each individual or partnership, known or suspected to have been involved in the coal mining industry at Llanelli in the 18th century, will be dealt with.

(1) *John Rees*

Thomas Mainwaring records that a John Rees paid £11.11.6 "*Coal Money*" to Sir Thomas Stepney's agent, Mrs Bridgett Jones, on 31 July 1709.[152] John Rees's colliery may not have been in the Llanelli Coalfield and it is not definitely known who John Rees was. Bridgett Jones's account also contained reference to Hector Rees and it is suspected that John Rees of Cilymaenllwyd (Hector Rees's father) was the person involved in mining Stepney's coal.[153] Hector Rees was subsequently involved in coal mining at Pwll and at the Bigyn, and, if his father had been mining before him, it is likely that his operations would also have been in the Llanelli area.

(2) *Mary Lloyd*

The Marquis of Winchester and his wife Anne leased all the coal at the Allt, Llangennech to Mary Lloyd on 12 December 1718, for 21 years.[154] The Swansea Five Feet Vein outcrops on high ground at the Allt and had been worked a century before by a Thomas Lloyd (See Chapter 2). Nothing is known about Mary Lloyd and the period during which she worked is undetermined.

(3) *Rowland Davies*

Thomas Mainwaring records that John Stepney and others leased the coal at Ffosfach to Rowland Davies of St Clears, on 23 July 1723, for 20 years at a royalty of 3 shillings per wey.[155] It is not known if Rowland Davies worked the coal, but information contained on a plan of 1772[156] confirmed that old workings had been encountered in the Swansea Five Feet Vein in the Ffosfach region, during mining operations between c1750/1760.

(4) *Hector Rees and Thomas Price*

Thomas Price of Penllergaer and John Stepney of Llanelli (later 6th Baronet) both owned intermixed lands in the Bigyn/Penyfan region. Continuity of mining was difficult in these circumstances and Price and Stepney entered into an agreement, in 1729, designed to help both parties in their separate coal mining operations. They leased the coal under certain landshares to each other and agreed not to cause any damage to the existing drainage adit, which was used to dewater workings in the Golden Vein. The drainage adit was described as "*the gutter or level lately made by the said Thomas Price in partnership with Hector Rees*".[157] This shows that Hector Rees (of Cilymaenllwyd?) and Thomas Price of Penllergaer had formed a partnership, just prior to 1729, to mine coal in the Golden Vein and, perhaps, in other seams in the Bigyn Hill region (Fig. 8). No further reference to their joint venture has been seen and it was possibly short lived. It was probably taken over by John Allen about 1733. (See Fig. 8 and Plate 3)

(5) *John Allen*

In 1733 the Duke of Bolton (Charles Pawlett), as Lord of the Manor and Commote, leased all the coal under the "*common called Gwern Caswthy near the town of Llanelly adjoining Caswthy Hill*" and under the "*waste overflowed by the sea-tides lying westward of Gwern Caswthy and adjoining the land of Sir Edward Mansell*" to John Allen of Llanelli. The lease was for the Duke's lifetime, at a royalty of 3s 6d per wey.[158] Allen had, therefore, been leased the coal under the Wern, and under the old seashore west of the Wern, (the low ground immediately west of Albert Street where the new cut to the River Lliedi runs) (Fig. 10). Although no confirmatory evidence has been seen, it is assumed that Allen commenced mining shortly after obtaining the lease. About the same time, he was leased coal under lands owned by Sir Thomas Stepney, 5th Bart., in the Wern/Bigyn region, because records have survived of royalty payments, of 3 shillings per wey, paid to Stepney by Allen in 1740 and 1741 for "*coal wrought from under Crofswthy*".[159] Crofswthy (Caeswddy) was the same location as the previously itemised operations of John Stepney, Hector Rees and Thomas Price and, because Allen was mining Stepney's coal, it is reasonable to assume that he had taken over John Stepney's collieries and, perhaps, those of Rees and Price. If this was the case, then John Allen would have possessed most of the coal along the southern outcrop of the Rosy, Fiery, Golden and Bushy Veins to the west of the Box Fault.

Little information has survived of Allen's activities but he seems to have been involved in Llanelli's coal industry for many years. He was mining at Caeswddy in 1740/41 and Cawdor Estate accounts show that he had two or three water engines employed at his collieries in 1749 [160] His lease from the Duke of Bolton expired on Pawlett's death in 1754, however, and the same coal was leased to Thomas Bowen of Llanelli, who would have taken over Allen's collieries on the common lands and seashore.[161] Allen did not cease operations at this time, though, and was engaged in coal mining in the Llanelli area, probably on Stepney lands, in 1757.[162] His involvement probably ended in 1761, when it was reported that "*Mr Allen's works were drownd at Llanelly*".[163] No subsequent reference to John Allen has been seen.

He was involved in Llanelli's mining industry for some 28 years, and it is possible that his contribution to Llanelli's 18th century industrialisation is seriously under-estimated, due to little record of his activities having survived. He formed the only real link between the local developers of the early 18th century and the industrialists who came to Llanelli after 1749, to whom his experience and inherited knowledge of the Llanelli Coalfield would have been of immense value in the important years of growth in the 1750s. Bowen records that he opened many pits at the Wern and at Caeau Gleision* (Greenfields)[164] and plans of 1846[165] show that he sank a pit (54) on the old seashore (immediately west of Albert Street) and drove a heading westwards from it under the sea. The heading was described on the plans as "*direction of old heading*

---

* The area identified with "*Greenfields*" at the present time lies close to the axis of the Llanelli syncline and it is unlikely that Allen would have been able to sink pits of sufficient depth to reach the coal seams. The Caeau Gleision referred to may have lain further west.

made by Mr Allen in his colliery" and other old pits and water-wheels shown in the immediate vicinity (51, 52, 55, 59), could also represent activity on Allen's part. It is known that a number of pits worked by Roderick, Bowen and Griffiths after 1794 were re-openings of Allen's old collieries,[166] and it is likely that a number of the un-identified entries along or just down the southern outcrop of the Rosy, Fiery, Golden and Bushy Veins west of the Box Fault (Fig. 10 & the plan in Appendix I), can be attributed to John Allen. (Plate 5)

(6) *Thomas David*

As we saw, in Chapter 1, the Rosy Vein to the west of the Box Fault was termed Thomas David's Vein, in the mid 1750s. Leases of 1754 and 1758[167] certainly used this name, but no information regarding Thomas David himself has been seen (David was a very common surname at Llanelli in the 18th century and we know that more than one Thomas David lived in the area).[168] He must have been personally involved in coal mining in Llanelli for one of the exploited seams to be named after him

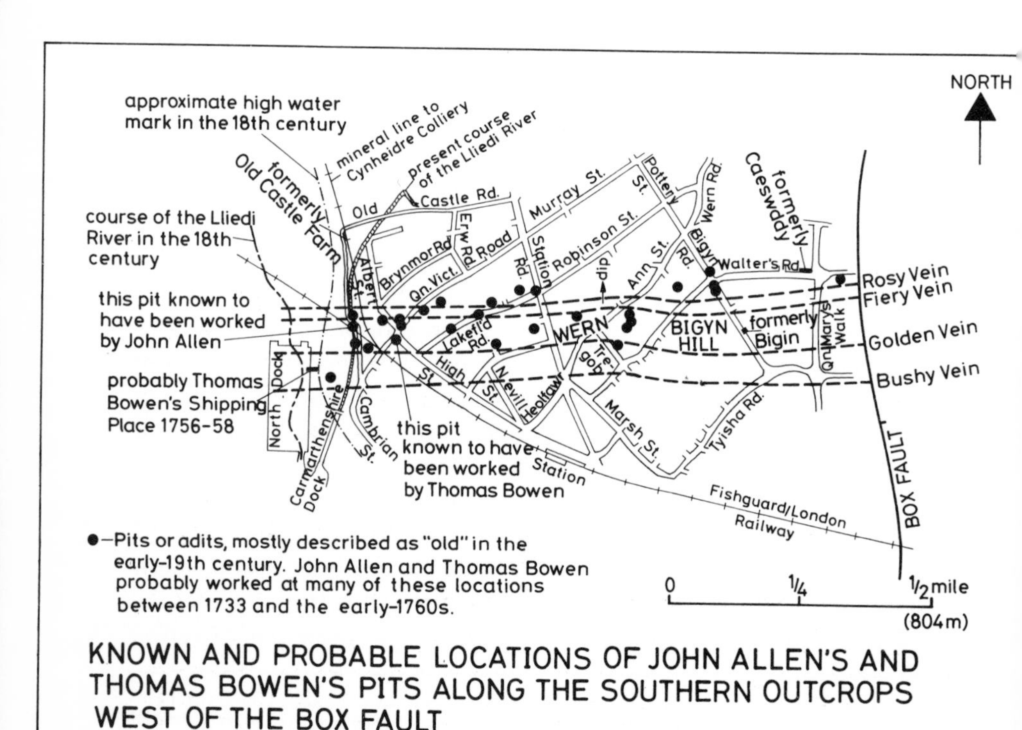

Fig. 10

Plate 5 — Site of mining operations by John Allen and Thomas Bowen, 1733 to c1758. The old sea-shore west of Albert Street. The Lliedi River did not run through this area until 1845, when the Lliedi New Cut was made. The higher ground on the left was known as Bryn.

and he would have worked before 1754. Where he mined is not known, but it would have been west of the Box Fault.

(7) *Mary Rees*

In 1747 Sir John Stepney, 6th Bart., leased all the coal under Clyn-gwernen Ucha (the present Clyn-gwernen) to Mrs Mary Rees, "*the Old Nurse*", for the duration of her life and those of her sons, John and Thomas, at a royalty of 2 shillings per wey.[169] Very little is known about Mary Rees but the low royalty payment leads one to suppose that the coal was for lime-burning and domestic purposes at Clyngwernen and not for commercial sale.

(8) *Thomas (?) Popkin*

An entry on a plan of 1772 says that old workings, encountered in the Swansea Four Feet Vein at Cwmfelin, had been effected by "*Old Popkins by virtue of a Horse Engine*".[170] We are not certain of Old Popkins' identity nor of the period during which he mined. Justina Ann Stepney, daughter of Sir John Stepney, 6th Bart., married a Thomas Popkin in 1747,[171] however, and the Stepneys owned lands in the Cwmfelin/Pencoed region. Thomas Popkin died in 1770,[172] but he may have mined coal at Cwmfelin after marrying Justina Ann and could well be the Old Popkins referred to.

(9) *Robert Morgan*

John Innes records that Robert Morgan of Carmarthen was leased coal under Llwyncyfarthwch by Sir Thomas Stepney, 7th Bart., c1748.[173] About 1750 he was granted the coal in two seams, at unspecified locations in the Llanelli area, by John Vaughan of Golden Grove.[174] Morgan was a major industrial figure in West Wales, owning an iron furnace at Carmarthen and iron forges at Cwmdwyfran and Whitland.[175] His interest in Llanelli may have arisen through his desire to supply his iron-works with coal from his own collieries but it is possible that he was considering establishing other metalworks on the Llanelli Coalfield, to avoid coal haulage costs. It seems that Morgan did not initially mine this coal, because rentals of the Golden Grove Estate, between 1751 and 1754, recorded nil payments from Morgan, although an annual sleeping rent of £16.8.4 had been specified.[176]

By 1755 Morgan must have become convinced that it was economically sound to site his metalworks near suitable coal, and negotiated for the purchase of the newly-erected, but financially troubled, lead works at Pencoed, or to build an iron works on Sir Thomas Stepney's lands at Llangennech. Stepney wrote on this matter to one of the partners in the lead works, in February 1755, and stated "*sell it* (i.e. the Pencoed lead works) *to Mr Morgan of Carmarthen at the real worth. He is now preparing and determined to build an Iron work on my Estate near Langennech if he does not purchase your work, which he seems much disposed for*" [177] It appears that Morgan did not purchase the Pencoed lead works, because no further reference to it has been seen, but he secured a coal supply for his intended works in the Llangennech area. He was leased the coal under Stepney's lands at Brynsheffre, before 1759, for 21 years at a royalty of 3 shillings per wey[178] and, in October 1759, was granted the coal under Cae yr Arglwddes at Llangennech, by Maria Eliza Catherina Lloyd of Alltycadno, for 21 years at a royalty of 4 shillings per wey.[179] There is also unauthenticated evidence that Morgan obtained a lease of coal in the Trostre region from Sir Thomas Stepney in 1759.[180]

Morgan probably mined only minor quantities of coal as a result of his four or five leases in the Llanelli area, and no evidence has survived to confirm that he erected an iron works on Stepney's lands at Llangennech. The agent to the Alltycadno Estate wrote in 1761 "*Mr Robert*

Morgan thought he had a prospect of Good work when he had a lease for working in Parkyrarglwyddeys, but after laying out a good deal of money about trialls, he has not only given that over as worth nothing, but also Sir Thomas Stepneys adjoyning to it".[181] This statement is confirmed by rentals of the Golden Grove Estate which show that, whereas only the annual sleeping rent was paid for the two veins leased between 1758 and 1760, small quantities of coal were worked in 1760 and 1761. The rental for the year ending Michaelmas 1762 contained the statement "*Late Mr Robert Morgans — A Colliery £11.0.0*".[182]

We are not certain why Morgan abandoned his plans to establish metalworks and to mine coal in Llanelli. His lack of success in developing the colliery at Llangennech could have been a major reason, but it is also possible that he concentrated his efforts into his existing concerns. He erected a tin mill at Carmarthen in 1759 and, when the first tinned plate was produced there in 1761, severed his connections with the Kidwelly Tinplate Works.[183] It is probably more than coincidental that this was the year in which his interest in the Llanelli Coalfield seems to have ceased.

(10) *Henry Squire, David Evans and John Beynon*

The first of the new industrialists to bring capital and technical innovation to Llanelli in the mid 18th century were Henry Squire, shipwright, David Evans, apothecary, and John Beynon, all of the town of Swansea. Their partnership was leased coal under lands in the Cwmfelin/Pencoed region by Sir Thomas Stepney, 7th Bart., in December 1749[184] and, in February 1750, they were leased coal in the same area by Sir Edward Mansell, 2nd Bart., and Dame Mary.[185] Both leases were identical in terms of contract, the coal being granted for 31 years at a royalty of three shillings per wey of "*24 loading carts to the weigh, the same with established measure at Swansea and Neath*". Squire, Evans and Beynon had, therefore, brought modern techniques from the industrialised Swansea and Neath area and they quickly set to work upon the exploitation of both the Swansea Four Feet and Swansea Five Feet Veins in the area we now call Bynea. The major winning in the 7 feet thick Swansea Four Feet Vein was effected from a water-level pit (202)[186] (See Appendix B) and that in the 9 feet thick Swansea Five Feet Vein from a pit equipped with a steam engine (199). Other pits were also sunk or reopened to help or supplement the two main coal winning pits. A waggon road, constructed of iron-plated wooden rails built into stone sleepers, ran from the steam engine pit to a shipping place on Spitty Bank, 1¼ miles

(2kms) away. Thomas Mainwaring says that the steam engine was erected about 1750[187] and was the first steam engine used in the Llanelli Coalfield. The waggon road was certainly built by 1751, because it was called "*Mr Squire's Waggon Road*" on a plan made in that year.[188] It was the first purpose-built railroad or railway in the Llanelli area and, with it, Squire, Evans and Beynon had introduced modern industrial techniques to the Bynea area and launched the Llanelli Coalfield on the first phase of its real industrialisation.

Although no records of the partnership's affairs have survived, they apparently worked without interruption throughout the 1750s. In 1755/56 they negotiated with John Vaughan of Golden Grove over the "*Marsh Vein*" (Swansea Five Feet Vein) near Ffosfach, but no lease resulted and dispute arose with Vaughan over the wayleave payment required for the trespass of their waggon road over his lands.[189] Their differences were settled and Vaughan's rentals from 1757 onwards show that Squire, Evans and Beynon agreed to an annual payment of two and a half guineas (£2.12.6) for their waggon road.[190] The partnership made an important acquisition, in May 1760, when Sir E. V. Mansell leased Binnie (Bynea Farm and its lands) to Henry Squire for 21 years at an annual rent of £38.[191] Possession of these lands, bordering the Loughor River, allowed the establishment of shipping places and, perhaps more importantly, provided on-the-spot facilities for rearing and grazing the horses and oxen required for working the pits and hauling the coal.[192]

Squire, Evans and Beynon were, therefore, successful in their exploitation of the coal resources of the Bynea area, but Daniel Shewen's control of Sir E. V. Mansell's financial affairs led to problems which were not of the partnership's making. When Shewen had persuaded Sir E. V. Mansell to lease all the Stradey Estate coal to Chauncey Townsend, in 1762, he had also included a clause which promised the coal granted to Squire, Evans and Beynon, in the Ffosfach region, to Townsend on the expiry or surrender of their lease, although it still had 19 years to run.[193] As already seen, under the 1762 lease, Shewen was to receive 1 shilling and 3 pence for every wey of that coal, previously granted to Shewen, which Townsend raised. Although Squire, Evans and Beynon's Stradey Estate coal at Bynea had never been granted to Shewen, he must have considered that the acquisition of a fully-proven, working colliery would encourage Townsend to start his long-awaited development of the Llanelli Coalfield, to Shewen's financial benefit. He took steps to regain Squire, Evans and Beynon's lease by default, demanding full payment of royalty money from them, when he assessed their finances to be low. Shewen wrote to Mansell, in October 1763, requesting him to "*forward your answer to Squire and Evans that it might be put in time to oblige them to deposit the Land money in Court which I imagine they'll find a Difficulty, and no small one, to make it up*".[194] (The mention of only Squire and Evans suggests that Beynon had withdrawn from the partnership or had died by this time). Squire and Evans raised money to pay royalties of £304.15.0, for coal they had worked under Stradey lands in the Bynea region between 1760 and 1763,[195] and Shewen's attempts to dispossess them failed. When Joseph Shewen took over the management of Sir E. V. Mansell's affairs from his brother Daniel, in late 1763/early 1764, the threat to the partnership diminished.

Squire and Evans probably worked their collieries throughout the 1760s, but Henry Squire died or withdrew from the partnership before 1772, a plan dated to that year referring to their mining activities in terms of "*Doctor Evans*" alone.[196] (Doctor Evans and David Evans were one and the same person; a contemporary letter referred to a "*Dr Evans*" and then to the same man as "*Mr David Evans, Apothecary, Swansea*").[197] Evans must have mined throughout the 1770s because John Smith, Chauncey Townsend's son-in-law, who wished to gain the lease of the Bynea Farm, complained to Sir E. V. Mansell, in 1778, that Evans was extracting coal not on lease to him, under Stradey Estate lands in the Bynea area.[198] Evans, by now, must have been in financial difficulty because, by 1779, he owed five years rent to the Golden Grove Estate, for the trespass of his waggon road.[199] In 1781/82 the waggon road was described as "*not used*",[200] so it is likely that Evans relinquished all interest in mining in the Bynea area, on the expiry of the partnership's original coal leases and the lease of Bynea Farm in 1780/81.

Squire, Evans and Beynon had, between them, worked from 1749 to 1781, the full duration of their leases, and had introduced the steam engine and the waggon road or railroad to the Llanelli Coalfield. Their activities in the Bynea area are fairly well known. They worked the Swansea Four Feet Vein from a water level pit (202) with drainage being achieved through other pits (193 to 197, 200). The Swansea Five Feet Vein was worked from the Ox Pit (209),[201] from the 32 fathom and 21 fathom pits (198, 210) and from their steam engine pit (199) (Fig. 11). This pit was later called the "*Little Engine Pit*",[202] the word "*Little*" being added after Chauncey Townsend had erected his larger steam engine at Genwen c1767/68. This name has persisted for over 200 years, the area around the site of Llanelli's first steam engine still being referred to as "*Engine Fach*" by people at Llwynhendy and Bynea, although the name does not appear on modern Ordnance Survey plans. (Plate 6)

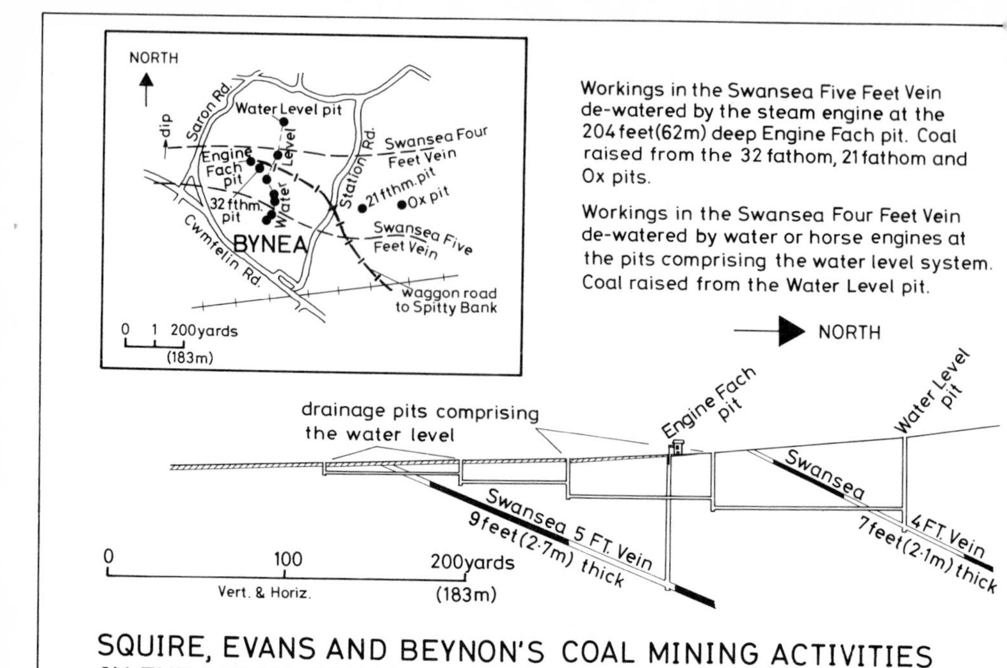

SQUIRE, EVANS AND BEYNON'S COAL MINING ACTIVITIES IN THE BYNEA REGION BETWEEN 1749 AND 1781

Fig. 11

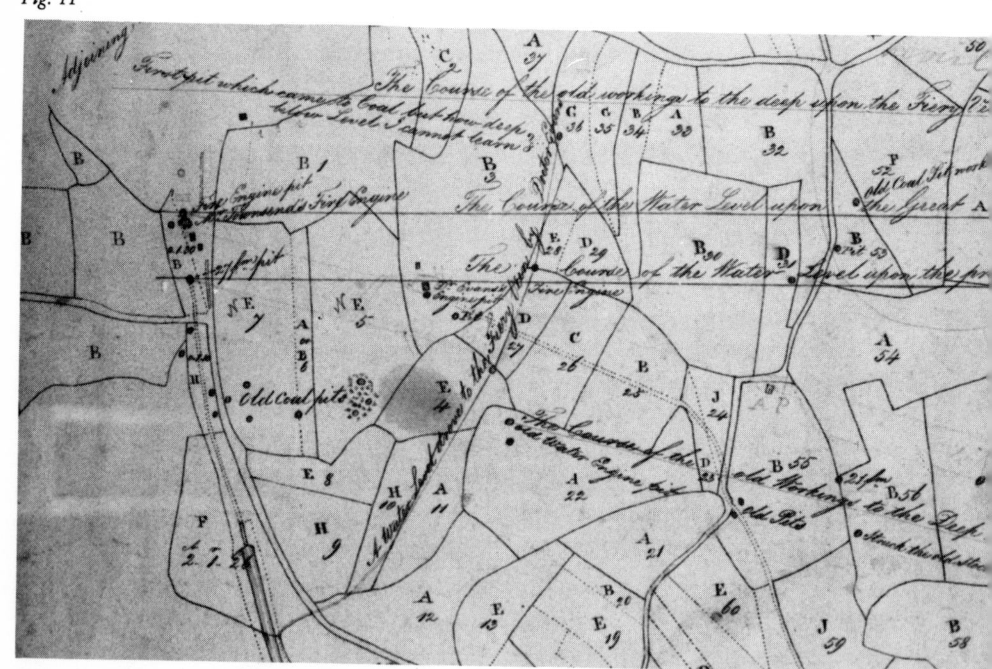

Plate 6 — Part of John Thornton's 1772 plan showing Squire, Evans and Beynon's and Chauncey Townsend's mining sites at Bynea. "*Dr. Evans's Fire Engine*" was at Engine Fach and "*Mr. Townsend's Fire Engine*" was exactly on the site of the existing old engine house at Genwen.

(11) *Thomas Bowen*

Between 1748 and 1752 Sir Thomas Stepney, 7th Bart., and other inheritors of the Vaughans' original Llanelli Estate, leased the coal under Pencoed and adjacent lands to Thomas Bowen of Llanelli.[203] Bowen commenced operations soon after obtaining the lease. He was shipping coal by April 1752, carrying it to large ships, anchored in deep water off Penclawdd, in keels.[204] Bowen worked the Swansea Five Feet Vein in this region by means of a water-engine pit at Ffosfach (218), but abandoned other sinkings to the same vein (224 to 227) because of flooding. He also formed a shipping place or quay near the main channel of the River Loughor at Ffosfach.[205]

Bowen expanded his activity in 1754, when he was leased the coal in the "*Biggin or Hill*," "*Boisva*" and "*Llwynwilog*" Veins, in the Wern/Bigyn region, and that under the seashore west of the Wern, by John Vaughan of Golden Grove, for 21 years at a royalty of 3 shillings and 6 pence per wey.[206] This coal had previously been granted to John Allen, but his lease had expired on the Duke of Bolton's death, in 1754, so Bowen took over Allen's collieries on lands over which Vaughan held manorial jurisdiction. This lease covered two separate regions along the southern outcrop of the Llanelli Syncline. It is considered (See Chapter 1) that Bowen had been granted the Golden and Bushy Veins in the Bigyn/Wern area, together with the Fiery Vein at Penyfan/Llwynwhilwg, and also the Golden and Bushy Veins under the seashore west of the Wern (the low ground immediately west of Albert Street). Bowen probably commenced working in both these regions but encountered difficulties, because Vaughan enquired, in March 1756, "*what objections and hindrance to Bowen's work at Llanelly?*".[207] Estate rentals, too, show that Bowen worked less than 100 tons of coal from under Vaughan's lands in the year ending Michaelmas 1757.[208] Nevertheless, Vaughan must have been convinced that Bowen would be successful, because he leased him the Rosy and Fiery Veins under the common of Morfa Bach, and under the seashore west of Old Castle Farm lands, in April 1758, for 21 years.[209] This area lay immediately north of the seashore west of the Wern and Bowen, therefore, had possession of all the coal under the common lands and seashore to the west of Llanelli. Bowen's activities in this region are known to some extent, because plans, prepared in 1846,[210] made reference to both Bowen's and John Allen's previous interests (Fig. 10). It seems that Bowen opened or inherited one or more of the pits on the old seashore (51, 53 to 55), and he certainly worked a pit at the junction of High Street and Queen Victoria Road (59), because it was described on the 1846 plans as "*Old Coal Pit when* (where?) *they struck water in Mr Bowen's coal-works*". A shipping place close to the pits and on the old course of the River Lliedi, shown on the plans, must be the quay that Thomas Bowen later said he had built in 1756[211] (See Chapter 6). Bowen's activities in the Wern/Penyfan/Llwynwhilwg region are not known in the same detail, although Innes and Hopkin Morgan record that Bowen was working at Penyfan in 1750, moving his coal by canal to a shipping place at Penrhyngwyn Point, near Machynis.[212] 1750 is too early, but the statement is otherwise consistent with the facts. Bowen had been leased coal at Llwynwhilwg (near Penyfan) and was shipping his coal near to Penrhyngwyn in 1757, because a plan of that year showed a coalworks belonging to him at Dafen Pill[213] (Fig. 12). Bowen may have started to build his canal in the 1750s but this cannot be confirmed (See Chapter 5).

Bowen had invested capital, and had made genuine efforts to mine coal under Golden Grove lands, but he was unsuccessful. Estate rentals showed that he paid no royalty after the small amount paid in 1756/57, implying that no coal was worked after that time.[214] Some verification of his failure was provided by Bowen himself, in 1802, when he stated that he had made

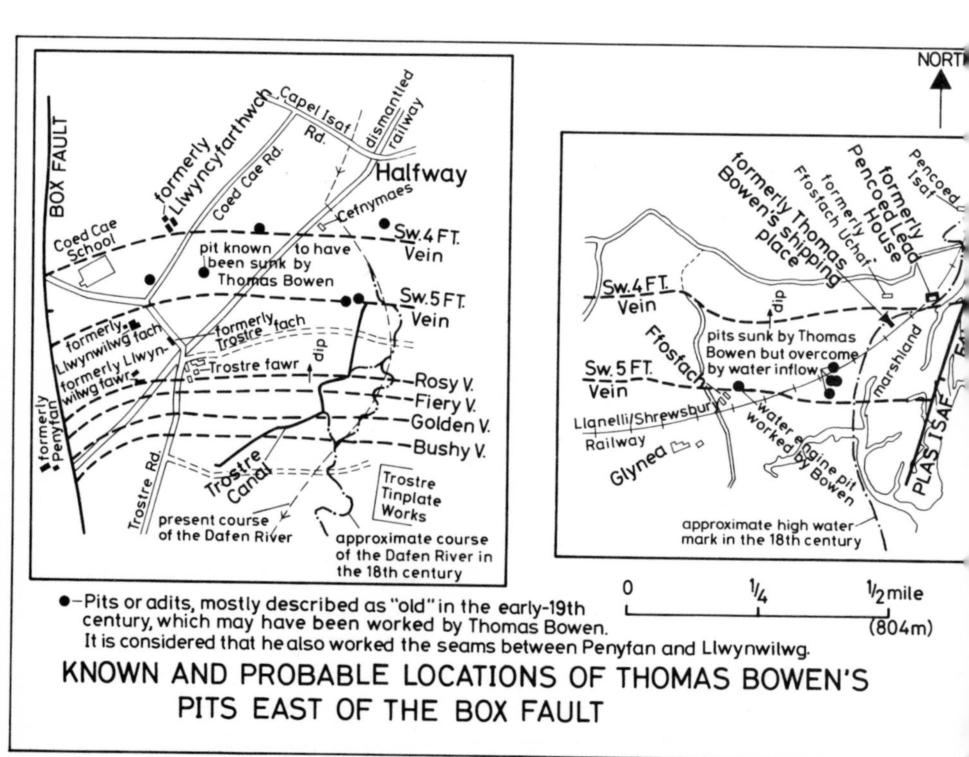

Fig. 12

use of his quay (on the Lliedi River) for only two years after 1756;[215] by Daniel Shewen who referred to "*Bowen's work failing*" in June 1763;[216] and by an unauthenticated report that Bowen's colliery at Penyfan was discontinued due to an accident to the water-engine employed to drain the workings.[217]

Despite his adverse experiences at Llwynwhilwg/Penyfan, Bowen must have been impressed with the coal-shipping potential of the Dafen Pill region, because he leased coal under Trostre, Llwyncyfarthwch, Cae Cefen Y Maes and probably Llwynwhilwg from Sir Thomas Stepney, 7th Bart., and others, in November 1768, for 17 years at a royalty of 3 shillings per wey.[218] He probably abandoned his operations at Pencoed about this time[219] to concentrate his efforts into the establishment of a large mining and coal export venture based around Dafen Pill. He must have set to work immediately because, by April 1769, he had abandoned his first pit or pits because of flooding, and had sunk a new coal winning pit. It was stated at the time "*Mr Bowen I hear has got a fresh pit into coal on the same estate as before but cannot make any head thereof without making a Water Course over part of Sir Edward's Estate which is Drawsdre or Llwynwhilwg Vach*".[220] Bowen soon built the water-course (probably for impelling water-engines at his pits) and a canal, because he paid rent to the Golden Grove Estate for the trespass of a canal and a water-course from 1771 onwards.[221] Bowen also paid rent to Sir Thomas Stepney, 7th Bart., and then to Sir John Stepney, 8th Bart., for Penrhyngwyn Point from 1772 onwards.[222] There is little doubt that Bowen had constructed the Trostre Canal to barge the coal from his works at Trostre, Llwynwhilwg and Cefnymaes into Dafen Pill, which ran past Penrhyngwyn Point near the main course of the Burry River. Bowen probably used Penrhyngwyn Point as his coalbank but this is not confirmed (See Chapters 5 and 6).

The locations of some of Bowen's pits in this region are known. A pit on the outcrop of the Swansea Four Feet Vein at Cefnymaes (92) was described as "*Mr Bowen sunk this pit, coal small being on the crop*"[223] and two pits on the outcrop of the Swansea Five Feet Vein (121, 122), situated exactly at the northern extremity of the Trostre Canal,[224] must have been Bowen's sinkings. It is also considered that one or more of the un-named pits and levels, shown on post 1830 plans, in the Penyfan/Llwynhilwg/Trostre region near the southern extremity of the Trostre Canal, were worked by him. It is likely that Bowen stopped working coal from the 1768 lease in 1781 or 1782, because the canal and watercourse, which were in use in May 1781,[225] were listed as "*not used*" in the Golden Grove Estate rentals in, and after, the year ending Michaelmas 1782.[226] Bowen was then about 67 years old[227] and he may have decided to retire.

Bowen, then, mined coal throughout the Llanelli area between c1750 and 1782 and his knowledge of the Llanelli Coalfield must have been unequalled when he retired. William Roderick later used this knowledge, in 1794, when he persuaded the 79 year old Bowen to enter into partnership with him to restart mining operations in the Bres/Wern area.

(12) *Chauncey Townsend; John Smith; Charles and Henry Smith*

Chauncey Townsend of St Peter le Poor, London, an English industrialist, must have learned of the potential of the Llanelli Coalfield when developing his coal mining interests in Neath and Llansamlet and entered into negotiations with the local landowners over leases of their coal, before 1752. These negotiations must have been lengthy, because William Jones of Loughor was engaged to survey and map the area around Berwick Chapel (Llwynhendy/Bynea) for the preparation of the leases[228] and, in 1751, produced a plan, which would have taken a considerable time to complete, because it covered the lands from Halfway to Loughor Bridge in detail.[229]

The negotiations resulted in Townsend being granted leases of extensive areas, in and around Llanelli, by the major landowners, in October 1752. Sir Edward Mansell, 2nd Bart. and Dame Mary leased coal under *"Heolehene, Lloynhendy, Maeserdaven, Baradus, Parkymonith, Penevan, Tyrymorva and Dufryn"*[230] and Sir Edward also leased coal under *"Keven, Berwick, Techon, Baccass, Teer y Morva, Maeserdaven and Dufrin"*,[231] on 24 October 1752. The same day, Sir Thomas Stepney, 7th Bart., and the other inheritors of the Vaughans' Llanelli Estate, leased all the coal under their lands in the Borough hamlet and in the hamlets of Berwick and Westfa. They also agreed to re-lease the coal under their lands at Pencoed to Townsend when Thomas Bowen's lease expired.[232] All three leases were for 99 years, at a royalty of 3 shillings per wey, but no reference was made to a minimum royalty payment or to sleeping rent. The leases all specified that steam engines would be employed to work the coal.

These leases gave Townsend a near monopoly of the Llanelli Coalfield, only coal under Stradey Estate lands west of the River Lliedi, that already leased to others and that belonging to small freeholders being outside his control. The major landowners must have been convinced that Townsend would invest substantially in the Coalfield, or they would not have combined to lease him most of their coal and even promise him coal already leased to others. Townsend commenced operations soon after obtaining his leases because, within a few weeks, he obtained permission from a yeoman named Owen Howell to construct waterways and waggon ways over Maesarddafen lands,[233] being no doubt attracted, as was Thomas Bowen, by the natural advantages of the Dafen Pill region. Charles Hadfield considers that Townsend built a canal at Maesarddafen, following his agreement with Howell, but no confirmatory evidence of this has been seen.[234] However, it is known that he sank a shallow pit to the Swansea Four Feet Vein (168), between Llwynhendy and Maesarddafen, the coal from which could well have been transported over Maesarddafen lands to Dafen Pill,[235] and he was certainly shipping coal from Dafen Pill in 1757, because a plan of that year designated the channel, connecting the Pill to the main course of the River Loughor, *"Esq. Townsend's Barge Way"*.[236] Further evidence that Townsend worked at Llanelli in the 1750s is a record of a burial in 1754, which stated *"Wm Bowen, labourer, killed in Mr Townsend's works"*,[237] but the locations of his early pits are not known with certainty and it seems that his operations were on a small scale.

Although Townsend had not embarked on the large-scale exploitation of the Coalfield expected by the landowners, he enlarged his extensive coal holdings still further, in 1762, when Sir E. V. Mansell, 3rd Bart., granted him all unleased Stradey Estate coal, including that under lands west of the River Lliedi, and promised coal already leased to Squire, Evans and Beynon.[238] The lease was granted not because Mansell was satisfied with the scale of Townsend's mining activity between 1752 and 1762 but, as already seen, because of Daniel Shewen's plans for personal financial gain. The development of the Coalfield now depended almost entirely on Townsend but he did not expand his own operations, although some of his newly-acquired coal, west of the River Lliedi, was sub-leased to Charles Gwyn in 1762 or 1763. With the Coalfield almost unworked, Llanelli's industrialisation, which had shown such promise in 1749/50, ground to a halt. Shewen, denied his anticipated royalties, complained about Townsend's inactivity and Gwyn's ineptness, in 1763[239] and in 1765, when he wrote to Sir E. V. Mansell *"I wish you'd write to Mr Townsend a few lines to let him know that by all accounts from the Country, the manner in which he carries on your Collieries is astonishing to the whole Neighbourhood, particularly at a time when there's such a demand for coal and that if he does not think them worth carrying on with some spirit you desire he would give up his lease"*.[240]

There are a number of possible explanations for Townsend's failure to develop the Coalfield between 1752 and 1765. He had invested heavily in coal and copperworks in Glamorganshire,[241] and in leadmines in Cardiganshire and Carmarthenshire,[242] and the ventures may have taken all his available capital. He seems to have been in financial trouble in 1760, because a letter said "*I should think money is very scarce with Townsend, for there is not a penny in wages in Cwm Ystwyth*"[243] (one of his leadmines in Cardiganshire). The absence of a sleeping rent or a minimum royalty clause in his leases meant that there was no financial penalty if he failed to work and, if he were short of money, he could remain idle with impunity. Daniel Shewen, on the other hand, believed that Townsend held back because he anticipated a sale of Stradey Estate lands to settle Sir E. V. Mansell's mounting debts and, as he wished to purchase, did not develop his undertakings, to avoid paying the high prices that colliery lands of proven commercial value would command. Shewen wrote in 1763 "*Our coal works must answer well when the money affair is settled but am clear they are not worked properly purely in expectation of a Sale of the Estate when they expect them for a song*".[244] A third possible explanation was put forward, in a legal opinion of 1805, which said "(Townsend) *was very desirous of ingrossing all the coalmines he could get in Carmarthenshire which lay near Glamorganshire to prevent a competition in the markets where he disposed of his coal*",[245] the conclusion being that he never intended to develop the Llanelli Coalfield, but secured the leases to prevent others from developing it and setting up a rival trade. Whatever the true reason, Townsend raised little coal before 1765, thereby directly inhibiting Llanelli's industrial growth, which was further inhibited by Thomas Bowen's and Squire, Evans and Beynon's natural reluctance to lay out capital on collieries they would lose when their leases expired.

By late 1765 Townsend was either in a position to finance, or else decided to begin, his long-awaited industrialisation of the Coalfield. He concentrated on the Bynea area and planned to work the nine feet thick Swansea Five Feet Vein with a large steam engine, transport the coal down a mile long canal and ship it from a pill, near Yspitty, which ran into the Loughor River. He drew up a draft agreement with Richard Vaughan of Golden Grove, in September 1765, regarding the construction of the canal and this was finalised, in July 1766, when Townsend was leased land at "*Dolevawr*" and agreed to start work immediately.[246] For his intended colliery, Townsend obtained a lease of coal under Genwen from Thomas Richards of Swansea and Thomas Richards of Jamaica, in April 1766,[247] and commenced sinking pits and erecting a steam engine. The scale of his investment was referred to by his grandson, some 44 years later, when he stated "*I also know that, previous to the year 1770, my grandfather had sunk pits, erected fire engines, and had made a Canal for leading Coals down to the water-side near Llanelly in Carmarthenshire. I am unable, from any document which I have ever acceded to, to ascertain precisely what that expenditure was, but I am very certain from the depth of the pit, from the view of the engines, from the Canal, that it must have been a considerable expenditure which was made there".*[248] Townsend planned to dewater his workings by a large steam engine and sank his pumping pit (182)[249] a depth of 240 feet (73m) to the Swansea Five Feet Vein (then termed the Old Golden or Great Vein). His main coal raising pits were the 27 fathom pit (183) and a pit (181) west of the pumping pit, although other nearby pits were also used (Fig. 13). A plan of 1772[250] termed the steam engine both the "*Llwynhendy Fire Engine*" and "*Mr Townsend's Fire Engine*" and showed that a waggon road ran from the two main coal raising pits to the top of a mile long (about 1.6km) canal (See Chapter 5 — the Yspitty Canal), which ended at a pill called Townsend's Pill. It would seem that Townsend carried his coal in waggons to the canal, where it was loaded into barges which took it to a coalbank at the side of Townsend's Pill. The coal was then loaded directly into small ships, which

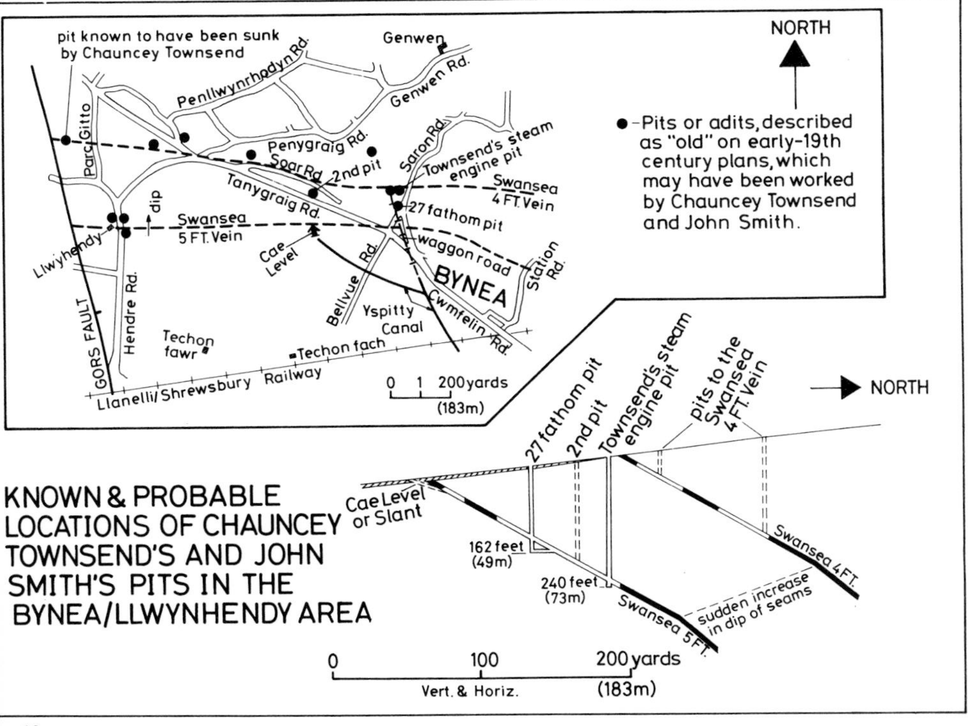

Fig. 13

Plate 7 — Site of Chauncey Townsend's Fire Engine at Genwen. The existing engine house was built by General Warde in 1806/07 in place of Townsend's original engine house. Warde used Townsend's original pumping pit.

had been able to sail up the pill, or into keels which carried the coal out to larger vessels at safe anchorage in the Estuary. Townsend did not commence working at Bynea until late 1768/early 1769, because he wrote to Sir E. V. Mansell, in September 1768, saying "*My fire Engine is in great forwardness, the Boyler I hope ready to put up and hope and expect to work that Coall very soon. I should have been at work now but for a rascall of a Mason who built the houze so ill it fell down*".[251] Even after the colliery became operative, the work did not proceed smoothly, because Daniel Shewen reported, in April 1769, that some person had sabotaged the engine by breaking the flywheel and throwing stones into the pumps,[252] and additional unauthenticated evidence exists suggesting that continuity of working at Bynea was difficult, because of the intermixed Trimsaran Estate and Stradey Estate landshares.[253] These lands had remained the subject of a dispute between Dame Mary Mansell of Trimsaran (the 2nd Baronet's widow) and Sir E. V. Mansell, 3rd Bart., despite an order of the Court of Chancery, in 1766,[254] and Townsend was caught in the middle of this conflict. (Plates 6 & 7)

From the evidence, it appears that Townsend worked only small quantities of coal by 1770, despite his investment of significant capital after 1766. Townsend drew up his will in March 1770[255] and died in the same year.[256] His inactivity between 1752 and 1766 had undoubtedly hindered the expansion of Llanelli's coal industry at a critical time, and the previously accepted view, that he was responsible for Llanelli's early industrialisation, should be reassessed.

Townsend left his possessions to his children. His son, James, inherited a two-fifth's share and his three daughters, Elizabeth, Charlotte and Sarah, a one-fifth share each.[257] Elizabeth's husband, John Smith, who had already been involved in Townsend's business affairs, took over the principal management of the collieries and may have continued working at Bynea, because a scheme for sinking the pumping pit (182) to 360 feet (110m), to secure more coal, was proposed in 1772.[258] It is iikely that Townsend's inheritors were uncertain about the financial viability of their father's coal interests at Llanelli, because they considered selling them. David Evans (Dr Evans) wrote to Joseph Shewen on this matter, in April 1775, and said "*I am credibly informed Mr Smith does not intend paying the Country a visit until the latter end of the Summer and that he intends disposing of his Coaleries, particularly that in my neighbourhood and that he has made some overtures to Sir Edward and yourself upon that account*".[259] Shewen and Mansell were faced with the likelihood that Smith would be as inactive at Llanelli as Townsend had been, and they negotiated with him for the surrender of the coal under Stradey lands, west of the River Lliedi, so that they could exploit it themselves.[260] Mansell's overtures co-incided with a period of industrial depression and Smith was happy to comply with his request. Smith must have thought that some of Llanelli's coal could be profitably developed, however, because he obtained control of the two-fifths share in Chauncey Townsend's industrial interests, held by James Townsend,[261] and also informed Mansell, in early 1776, that he intended working the Swansea Four Feet Vein to the west of Bynea.[262] The surrender was effected in October 1776 when John Smith, John Oliver Willyams (Charlotte Townsend's husband) and Sarah Townsend returned the Engine Vein, Cwm Vein, Country Coal or Dray Pitt Vein (i.e. the Golden and Bushy Veins), and all other coal seams westwards of the River Lliedi, to Mansell. It was stated that Mansell had already repossessed the Country Coal or Dray Pitt Vein and was raising coal from it, and would regain the Engine Vein as soon as possession could be legally obtained from Charles Gwyn. The agreement stated that Smith intended working the Swansea Four Feet Vein, but he also had an escape clause which stated that he would work the coal "*as soon as there shall be a Demand and sale for the Coal and Culm to be raised therefrom, sufficient to pay the Expenses thereof and a reasonable Proffit which there is not at Present, nor (as it is apprehended) will there be for several years to come*".[263]

Smith continued to express his intention of working coal at Llanelli. In 1778 he told Mansell that he wished to form a shipping place between Spitty Bank and Bynea Farm[264] and, in 1780, a year before David Evans's lease of Bynea Farm expired, he announced his intention of mining coal, constructing canals and waggonways and forming shipping places, if that lease was re-granted to him.[265] Mansell must have been convinced that Smith would fulfil his promises and he granted Bynea Farm to him in October 1781, five months after Evans's lease expired. They agreed that Smith could work coal to the east of his steam engine, as soon as he had constructed a canal between Ffosfach and Bynea Farms, to communicate with a shipping place to be made on Mansell's lands near Spitty Bank.[266] The other major landowners were equally convinced that John Smith would invest heavily in the Coalfield and agreed, in 1785, to accept a surrender of the lease, originally granted to Chauncey Townsend in 1752, and to grant a re-lease, which would include coal under lands held by Thomas Bowen in his 1768 lease, which was about to expire, as Smith evidently wished to add the Dafen Pill area to his existing holdings. The surrender and re-lease were finalised on 27 June 1785, when Sir John Stepney, 8th Bart., John Vaughan and William Langdon leased the coal under the lands comprised in the previous lease of 1752, together with the coal under lands in the region of Dafen Bridge (Halfway), to Smith for 76 years. He agreed to sink two pits near Dafen Bridge "*One for a Coal pit the other for an Engine Pit*", within two years, and said that he intended constructing a canal, for the passage of coal barges from the top of Dafen Pill to a shipping place he would form at Penrhyngwyn.[267] Six months later, in January 1786, Smith obtained a legal assignment of the three-fifths interest in Chauncey Townsend's coal leases, held by James Townsend and by Thomas Biddulph and Sarah (Townsend) his wife, giving him a four-fifths holding, through his wife's (Elizabeth Townsend's) one-fifth share.[268] This was followed, in November 1786, by a lease of lands in the hamlet of Berwick (Wain Techon, Morva yr Ynis and Dole Vawr y Bynie) from Sir John Stepney, John Vaughan, Sir E. V. Mansell and others for the purpose of extending his existing Yspitty Canal westwards to Dafen Bridge.[269] (Plate 8)

Despite his expressions of intent and his acquisition of leases, John Smith exploited Llanelli's coal resources to a lesser extent than even Chauncey Townsend had. The Engine Pit at Genwen was not deep-

Plate 8 — Dafen Bridge — the present Halfway. Called Dafen Bridge because the main road crossed the Dafen River here by a bridge; the present name was allocated because the site was half-way between the toll-gates at Capel and Pemberton. John Smith's unfinished pit (later General Warde's Daven Colliery) was just behind the houses on the right.

ened,[270] the pits at Dafen Bridge were commenced but not completed—one (119) later being described as "*pit begun by John Smith and finished by General Warde*"[271]—and the canals between Ffosfach and Bynea Farms, from Dafen Pill to Penrhyngwyn, and from the Yspitty Canal to Dafen Bridge were not constructed (See Chapter 5). It was said, in the early 19th century, that none of the coal under Stradey Estate lands to the east of the River Lliedi had been worked since 1766,[272] although a direct query, made in 1828, regarding the quantity of coal raised by Smith from under the Stradey Estate since 1781, yielded the reply "*he worked a small quantity of Coal, on the Chappel Farms* (Capel Isaf and Capel Uchaf?) *probably after this date — and which was the only coal he did work although he commenced sinking several pits*".[273] It is known that Smith extended the Yspitty Canal a short distance westwards, to ship the coal from workings in the Swansea Five Feet Vein (177, 178), between 1776 and 1785, (See Chapter 5) but this seems to have been his only significant activity, in the Coalfield, between Townsend's death in 1770 and his own death in 1797.[274] As with Townsend, it is difficult to give a convincing explanation for John Smith's inactivity in the Llanelli Coalfield, especially when he took over the management of a major working colliery at Genwen, into which significant capital had been invested. A legal opinion of 1805, already quoted,[275] said that Smith, like Townsend before him, had kept the Llanelli Coalfield in as low a state of production as possible to prevent any rivalry to his Glamorganshire coal interests. There is support for this opinion. Smith carefully paid the rent for the Yspitty Canal and Penrhyngwyn Point up to his death, although he hardly made use of them.[276] Another possible explanation could be that disputes between Townsend's beneficiaries prevented Smith from managing and working the inherited collieries effectively. He certainly took the step of legally obtaining three-fifths of Townsend's original coal leases from James and Sarah Townsend, through recourse to the law in 1786, and the remaining one-fifth share (Charlotte's) was to be the subject of legal arbitration after his death.[277] A third possible explanation is that the coal export trade did not flourish between 1770 and 1790, and investment into Llanelli's coal industry, which was based predominantly on export trade demand, was inhibited. The 1776 agreement between Smith and Mansell[278] stated quite specifically that there was little expectation, for several years, of an economic climate which would yield profits from coal mining investment, and continuing hostilities with France depressed shipping trade throughout most of the time that Smith controlled the Llanelli Coalfield (See Chapter 6).

On John Smith's death, in 1797, his four-fifth's interest in the Llanelli coal leases passed to his two sons, Charles and Henry Smith.[279] They did not work at Llanelli, but took steps to acquire full possession of their grandfather's leases in the area by contesting the right of John Oliver Willyams (the late Charlotte Townsend's husband) to retain a one-fifth share in Townsend's interests. After arbitration, Willyams was required to assign this share in "*the lease of the Llanelly Colliery*" and in coal mines at Llangyvelach and St John, Glamorganshire to them, in December 1799.[280] As the century ended, Charles and Henry Smith were, therefore, as firmly in control of the Coalfield as their grandfather had been, some 40 or 50 years earlier, and they seemed equally disinclined to work the coal. Whatever their motives or circumstances, there is little doubt that Chauncey Townsend and his heirs, by their inactivity, had held back the major industrialisation of the Llanelli area by almost half a century.

(13) *Hector Rees; John Rees; Edward Rees; John Rees (?) of Cilymaenllwyd*

Hector Rees and his son John Rees, of Cilymaenllwyd, approached John Vaughan of Golden Grove, in 1755, for permission "*to build a wharff near the Pool at the Black Rock in Pembrey Parish*"[281] and, a year later, the same scheme was described as "*turning the River Dulais and Building a Quay at the Black Rock*".[282] A plan of 1757 itemised "*Coaleries*" within the Burry Estuary and described one of them as being "*at the Pool, belonging to Hector Rees, Esq.*"[283] Hector and John Rees were, therefore, involved in coal mining in the Pwll area as early as 1755 and, as we know that Hector was mining in the Bigyn Hill region before 1729, he was possibly active in Pwll for years before 1755. John Rees died between 1756 and 1760[284] and Hector Rees soon afterwards in October 1760, aged 77.[285] Hector Rees left the lands comprising the Cilymaenllwyd Estate partly to his youngest son Edward and partly to his grandson John (son of John Rees),[286] who was only eleven years old at the time of Hector's death.[287]

Edward Rees took charge of the mining interests of the Cilymaenllwyd Estate and expanded his operations in 1765, when he was leased the Pool Vein (Pwll Big Vein), under Sir Thomas Stepney's lands at Hendy and Penllech (the Pwll area), for 61 years.[288] Rees was also granted permission to raise coal, worked from under other people's lands, through the pits on Stepney's lands, paying a wayleave of 8 pence per wey.[289] Rees enlarged his coalfield at Pwll, in January 1768, when he obtained another 61 year lease of the coal under John Vaughan's lands at Hendy and Penllech and was granted permission to mine "*the Pool Vein or any other vein or veins which may take its course under the Waste Lands or Sea, Eastward or Westward of the Black Rock*"[290] by Vaughan, Lord of the Manor.

With the 1765 and 1768 leases, an unseen lease, granted before 1765, of coal under lands of a Mrs Williams[291] and the Cilymaenllwyd Estate's ownership of lands in the area, Edward Rees possessed most of the coal in and around Pwll. Thomas Mainwaring says that a steam engine was erected to drain the workings in the Pool Vein, shortly after the 1768 lease was granted.[292] If this is true, Edward Rees introduced the steam engine to the Llanelli Coalfield outside the Bynea area, and his engine, installed about the same time as Chauncey Townsend's, would have been the second or third engine to operate within the area. It is also considered that, as the quay or wharf at Black Rock (later to be called Pwll Quay — See Chapter 6) had been formed by this time, Edward Rees was involved in extensive coal mining operations at Pwll. However, little evidence has survived to confirm the period over which he worked there or to definitely locate his pits. Thomas Mainwaring says that the colliery with the steam engine operated until about 1800,[293] but Stepney Estate and Golden Grove rentals between 1772 and 1782 contain no record of royalties paid by Rees, under his 1765 and 1768 leases, although his working of the "*Foy Colliery*" (probably Ffoy, near Pontyates), with others, in 1782 was mentioned.[294] We do not know when Edward Rees died but his nephew, John Rees (grandson of Hector), probably took control of the mining interests of the Cilymaenllwyd Estate before the end of the 18th century. The steam engine colliery was probably the same colliery as the 19th century Pwll Colliery, as Mainwaring said that this was a re-opening of Edward Rees's colliery.[295] If this were so, then one or more of the pits at Pwll (1 to 3) could have been original sinkings by Edward, or even by his father Hector (Fig. 14). Other unidentified entries in the Pwll/Cilymaen-llwyd area could represent activity on their part. (Plate 9)

(14) *Samuel David*

We know nothing of Samuel David save that he paid rent to the Golden Grove Estate for "*the colliery on the Allt*", between 1758 and 1763.[296] The lease has not been seen, but it would have been granted in 1757/58, because John Vaughan had complained, in October 1756, that nobody wanted to lease the "*Allt Colliery*" from him.[297] We know that Sir

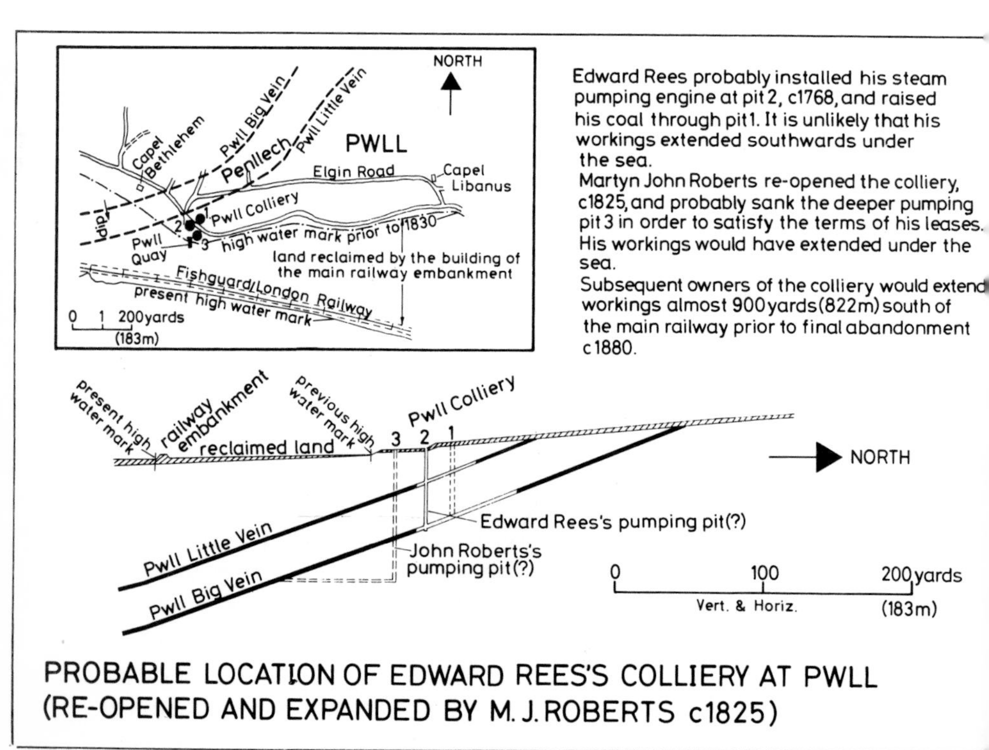

Fig. 14

Plate 9 — Site of the Pwll Colliery. Worked by Edward Rees c1765 and by M. J. Roberts c1825. The existing engine house may have been built by Roberts but could have been erected by Mason and Elkington when they took the colliery over in 1865.

Thomas Stepney, 7th Bart., was leased the coal at the Allt by Richard Vaughan, c1768, and David must have ceased mining by that time. Nothing is known about David's operations, but one or more of the unidentified entries to the Swansea Five Feet Vein on the Allt could have been worked by him. A Samuel David, described as a shopkeeper, held the tenancy of a landshare near the Allt (Cae yr Arglwddes) in 1759[298] and was probably the Samuel David involved in coal mining.

(15) *Abel Angove, John Wedge and John Thomas*

Thomas Mainwaring states that Sir Thomas Stepney, 7th Bart., leased the coal under Pencoed issa (Pencoed Isaf) to Abel Angove, John Wedge and John Thomas in 1760.[299] This lease is not recorded in Stepney Estate rentals[300] and it is not known if Angove, Wedge and Thomas worked any coal.

(16) *Charles Gwyn*

It is not known when Charles Gwyn first became involved in Llanelli's coal industry, but he was one of the witnesses to the signing of Chauncey

Townsend's 1752 coal leases.[301] His relationship to the principal parties has not been determined but he was probably Townsend's local representative or agent at Llanelli. Innes states that Gwyn opened a pit at Llwynhendy about 1750,[302] possibly a reference to his supervision of Townsend's known early activities in the Bynea region, but evidence has survived to confirm that Gwyn was Townsend's agent after 1762. The 1805 legal opinion stated that Chauncey Townsend, after obtaining the lease of coal under the Stradey Estate, in August 1762, "*employed an Agent of the name of Chas. Gwyn to work that part of the Vein called the Parkycrydd or Killyvig or the Engine Vein (now called the Golden Vein) which lay contiguous to the coal of Sir Thomas Stepney and others*".[303]

The manner in which Gwyn supervised Townsend's interests greatly concerned Daniel Shewen, particularly because he was paid a royalty on coal raised, and he wrote in April 1763 "*I have by this day's post wrote a long letter to Mr Townsend with regard to Gwynn's behaviour in Sir Edward's Collieries*"[304] and again, in June 1763, "*he* (Townsend) *and I have corresponded for some time on account of the Collieries and he seems very thankfull for informing him of the negligence of the people concerned*".[305] Despite Shewen's complaints, Gwyn retained control of the colliery but worked only some 2,500 tons of coal per year up to the early 1770s.[306] When the surrender of Stradey Estate coal west of the River Lliedi was being negotiated between Sir E. V. Mansell and John Smith, in 1775/76, they tried to re-possess the Golden Vein by bringing an action of ejectment against Charles Gwyn. This failed, suggesting that Gwyn was an under-tenant of Townsend's rather than his agent, and he was finally given compensation for the surrender of what was termed "*the Water Engine work*".[307]

Charles Gwyn had, therefore, been active in the Llanelli Coalfield, as Chauncey Townsend's agent or under-tenant, possibly from 1752, but certainly between 1762 and 1776. The location of Gwyn's colliery during this latter period is known with some certainty. It was situated west of the River Lliedi and worked the Golden Vein by means of a water engine. We also know that Sir E. V. Mansell, and then Dame Mary Mansell, worked this colliery after Gwyn, between 1776 and c1793.[308] A plan of 1787[309] showed a working water engine pit (46) on Stradey Estate lands west of the River Lliedi, where the Golden Vein could be reached, and this was almost certainly the site of Charles Gwyn's colliery. (Fig. 9). (Plate 10)

(17) *John Thomas and Griffith Bowen*

John Thomas of Panthowell was granted the use of certain coal pits at Hendy (Pwll) by Sir Thomas Stepney, 7th Bart., for 31 years from 1

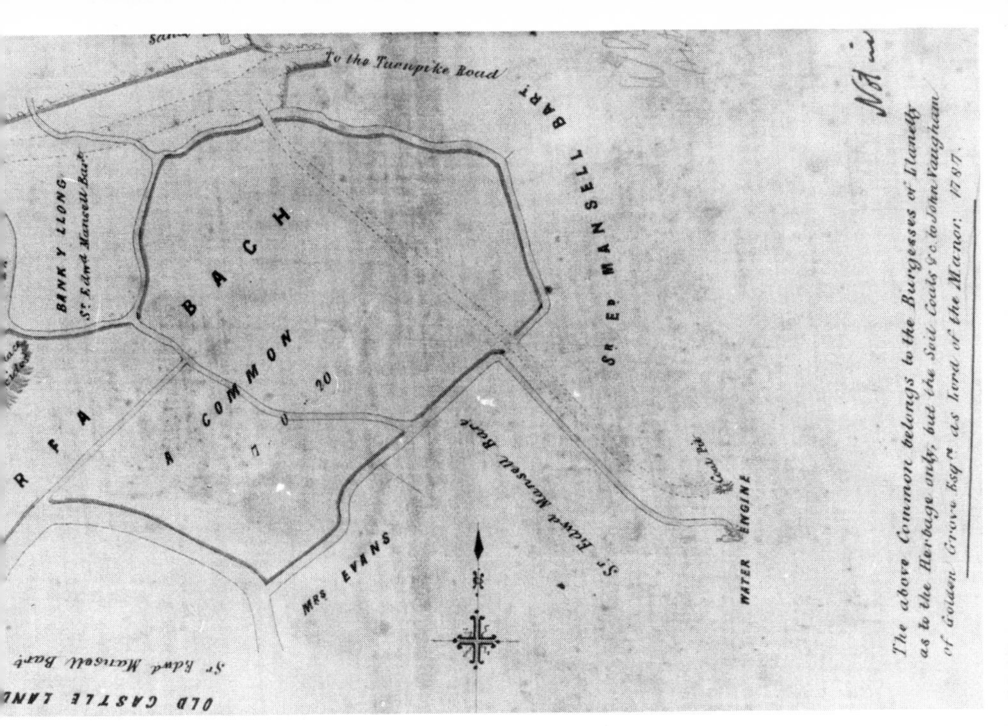

Plate 10 — Plan of Morfa Bach, 1787. The water engine pit was worked by Charles Gwyn and the Mansells. (Bottom) Aerial view of the eastern part of the site today. The playing field behind Coleshill School was the site of Charles Gwyn's water engine, and later of Raby's Caemain Colliery. Pentip School was built on the Caemain Colliery tip. The capping to a collapsed shaft of the Caemain Colliery shows as the white patch in the playing field.

January 1765, at an annual rent of £3.12.0.[310] This allowed Thomas to raise coal, worked under his own freehold lands, through the pits on Stepney's lands, and he was also given permission to transport this coal over Stepney's property.[311] The annual rent was paid in 1773 and 1776 by *"Mr Griff Bowen represent of John Thomas"*,[312] implying that Thomas had either leased his coal to Bowen or he was his agent. The period over which Thomas or Bowen worked the coal under Panthowell (Pant-Hywel) lands before 1765 and after 1776 is not known, nor is it known if John Thomas was the same person as the Thomas involved with Abel Angove and John Wedge, mentioned previously.

(18) *Ned Morgan*

Daniel Shewen wrote to his brother, Joseph, in April 1769, saying *"some wicked person or persons"* had made two attempts to destroy Chauncey Townsend's steam engine, and had damaged it to such an extent that work had been at a stop for over three weeks. Shewen said *"The fault is laid to little Ned Morgan who has a large quantity of coal for shipping worked under part of Penyvan and he found he should have no Carriers whilst the Engin work was carry'd on... Gwyn had him burnt in effigee on Easter Monday"*.[313] We know nothing more of Morgan's activities.

(19) *Thomas Jones and John David (?)*

In August 1773, Thomas Jones of Llwyn Ifan approached William Roderick, agent to the Alltycadno Estate, and asked that he be allowed to complete the colliery at Llangennech, which Sir Thomas Stepney, 7th Bart., had started just before his death, in October 1772.[314] In October 1773, Roderick informed William Clayton, the owner of the Alltycadno Estate, that Jones was also involved in negotiations with Sir John Stepney, 8th Bart., for Golden Grove Estate coal at the Allt, which had been leased to Sir Thomas Stepney c1768 and, in January 1774, Roderick reported that Jones and Stepney were about to finalise their agreement.[315] By November 1774 a lease of Clayton's coal at Brynsheffre was awaiting signature, and Jones requested permission to implement Sir Thomas Stepney's plans, with the construction of a waggon road from Brynsheffre down to an intended shipping place adjacent to a deep part of the River Loughor opposite the tenement of Box.[316] Clayton signed the lease granting Thomas Jones coal under Brynsheffre for 21 years at a royalty of 2 shillings per wey.[317] Although there is no evidence, Stepney probably assigned or sub-let the Golden Grove Estate coal at the Allt to Jones about this time.

Jones quickly set to work, paying royalties for coal mined at Brynsheffre in 1775/76.[318] Rent was also paid to the Stepney Estate, in the same year, for "*coal at the Allt*",[319] by a John David, who was possibly Thomas Jones's partner or agent. Jones continued mining and, in 1777, started to build a waggon road from his pits to Llangennech Marsh.[320] It is also considered that he constructed a short canal, from the waggon road to Llangennech Pill, at this time (See Chapters 5 and 6). Rentals of the Alltycadno Estate show that Jones worked only some 2,500 tons of coal as a result of the Brynsheffre lease between 1775 and 1782. He made no royalty payment after 1782 and probably stopped mining by 1783.[321]

From the scale of his operations, Thomas Jones must have spent a large sum of money, to no avail, in his attempts to develop coal resources at Llangennech. The actual locations of his pits are not known but one or more of the unidentified entries in the Brynsheffre/Allt region could be his.

(20) *Hyldebrand Oakes*

Hyldebrand Oakes of Hampton Court Park was granted a 7 year lease of Llanelly House, in May 1776, by Sir John Stepney, 8th Bart.,[322] and, in November 1776, he was leased Skybor Issa by Sir E. V. Mansell, 3rd Bart., for 63 years with the right to work coal under the demised property.[323] (A plan of 1822[324] showed Skybor Issa to be the present Ysgybor Isaf Farm near Felinfoel). The mention of coal in the latter lease may have related to minor extraction for lime-burning purposes only, but, as Oakes was apparently wealthy, it is possible that he intended settling in the area and investing in the local coal industry. No further reference has been seen to Oakes and it is not known if he actually worked coal.

(21) *William Hopkin and John Jenkin*

It is not known when William Hopkin of the parish of Llangunllo, Radnor[325] came to Llanelli, but he was probably agent to Sir John Stepney, 8th Bart., when he surveyed and mapped the lands to the west of the area covered by William Jones's plan, for the 1785/86 leases to John Smith.[326] Nothing is known of Jenkin's activities before 1788.

As already seen, the Golden Grove Estate coal at the Allt, Llangennech was leased to Sir John Stepney, 8th Bart., and sub-let to Thomas Jones of Llwyn Ifan. Jones ceased operations c1783 and the sleeping rent fell into arrear. The arrears were paid in Michaelmas 1788, an entry in the Estate rental stating that *"the Holding is rented from that time to Messrs. Hopkin and Jenkins"*.[327] The arrangement was formalised in January 1789, when Hopkin and Jenkin were leased the *"Allt Colliery"* by John Vaughan for 25 years and given permission to transport coal over Llangennech Marsh to the River Loughor.[328] Thomas Jones also assigned his lease of the coal at Brynsheffre to them,[329] so Hopkin and Jenkin were undoubtedly working Jones's old collieries. In February 1793, William Clayton granted them a lease of the coal under *"Talyclyn Ycha, Talyclyn issa and Brinshaffrey"* for 31 years but, as they had already commenced mining at Talyclyn,[330] it was effective from 29 September 1792.[331] Hopkin and Jenkin improved on Thomas Jones's transport and shipping system by constructing new waggon roads from Brynsheffre and, perhaps, from the Allt, and by extending the canal at Llangennech southwards, to a point on the River Loughor opposite the tenement of Box, where they may have formed a shipping place (See Chapters 5 and 6).

Although mining for himself, Hopkin continued as an agent to the Stepney Estate,[332] and also became involved in surveying the route of a proposed canal, from Spitty Bank to Llandilo or Llandovery in 1793.[333] He probably continued to work as agent and surveyor because his own mining ventures failed. He and Jenkin surrendered the lease of the Allt Colliery back to John Vaughan in 1794/95[334] and they probably also ceased operations at Brynsheffre and Talyclyn then, as we know that their lease from William Clayton was surrendered prior to 1805, many years before it expired.[335]

Hopkin and Jenkin, therefore, worked coal at the Allt and Brynsheffre (also at Talyclyn) between 1788 and c1794, and must have invested significant capital into the construction of waggon roads, a canal extension and, perhaps, a shipping place but, like Thomas Jones before them, they failed to develop the coal resources at Llangennech. Their pits are not definitely located, but one or more of the entries at Brynsheffre and the Allt could have been worked by them. No further reference has been seen to Jenkin but Hopkin remained active in the Llanelli area, and became one of the Llanelly Enclosure Commissioners and also the agent to the Box Colliery, in the early 1800s.[336]

(22) *John Hopkin*

John Daniel, gentleman, leased the coal under Penllwyn yr Odin and Gardde (just north of the Llwynhendy to Bynea region) to John Hopkin, yeoman, in 1793 for 60 years.[337] The Penyscallen and Swansea Four Feet

Veins could have been worked here, but no evidence has survived to confirm any activity by him.

(23) *Rev. David Hughes and Joseph Jones*

In September 1794 John Vaughan, as Lord of the Manor, leased coal under the common of Morfa Baccas to the Rev. David Hughes of Ffosfach and Joseph Jones of Brincarnarfon, for 40 years at a royalty of 6 shillings per wey. They agreed to provide a steam engine at a pit sunk 120 feet (37m) deep and were granted permission to build a canal, from the collieries over Morfa Baccas common to the River Loughor.[338] Both John Vaughan and Sir John Stepney, 8th Bart., reserved the right to use the canal, implying that it was to pass over Stepney's lands or that Stepney had granted coal in the area to Hughes and Jones. Hughes was the freeholder of Tyrtene Farm[339] (then near the seashore close to Tir Bacas Farm) and Jones was probably the freeholder of Bryn Carnarfon Farm. Both farms bordered Morfa Baccas common and Hughes and Jones held the coal under an extensive area south of the village of Bynea.

They sank the Baccas Colliery (later to be called the Carnarvon Colliery) (228, 229 and possibly 230 and 231) to the Rosy Vein[340] and constructed a canal over Morfa Baccas to Pill y Ceven which led into the main channel of the River Loughor (See Chapters 5 and 6). The colliery was sited on and to the south of the Carnarfon Fault, however, where the geological structure of the Coalfield, which did not follow the simple synclinal structure existing to the north of the fault, was not then understood. This lack of knowledge of the coal seam occurrences made working difficult, if not impossible, at that time and, coupled with the death of the managing partner (probably Joseph Jones), led to the closure of the colliery after a short time. It was subsequently reported that "*the Managing partner of the concern suddenly died and the London proprietors did not think proper to continue the workings of the colliery which they did not understand*".[341] Mainwaring also records that "*the works were soon discontinued owing to disagreement among the proprietors.*"[342] There is no record of royalty payment by Hughes and Jones, the only recorded payment being rent for the trespass of their canal in 1797.[343]

Hughes and Jones failed to develop the coal to the south of the Carnarfon Fault and their venture must have been costly. Their failure in the Baccas or Carnarvon colliery was not peculiar to them. It was unsuccessfully worked at various times throughout the 19th century and was finally re-opened and abandoned as late as 1927. The ivy-covered stack to the colliery still exists as a prominent landmark, south of the Llanelli to Swansea road across Bynea flats. Hughes and Jones's canal can also be traced for much of its length over the salt marshes, although sedimentation has concealed its true identity (See Chapter 5). (Plate 11)

(24) *William Roderick, Thomas Bowen and Margaret Griffiths*

William Roderick of Brynhavod in the parish of Llangethen (near Llandilo) had been involved, as agent to the Alltycadno Estate, in Llanelli's coal industry as early as 1770.[344] He thus knew the potential of the partly-developed Coalfield and became directly involved in mining, when the second phase of the industry's development began in the mid 1790s. In October 1794 Roderick obtained a lease of all the coal under the Demesne lands south and south east of the River Lliedi (most of the Llanelli Town area south of the river), and under Llanerch, Talsarne and Corsawddy (Caeswddy), from Sir John Stepney for 60 years at a royalty of 6 shillings per wey, the minimum yearly royalty payment, after the first two years, to be for 1500 weys (15,000 tons) of coal, even if a lesser quantity were raised. He was also granted the right to erect copper or iron works and to construct watercourses and canals, and agreed to erect an engine of sufficient power to raise coal from the Fiery Vein, within 28 calendar months of the granting of the lease.[345] On the same day, John Vaughan leased Roderick all the coal under the Wern for 60 years, at a royalty of 6 shillings per wey, and stipulated that an engine of sufficient power to raise coal from "*the utmost depth*" of the Fiery Vein (i.e. from the bottom of the syncline) was to be erected within 18 months and that the minimum yearly royalty payment, after two years, should be for 1,000 weys (10,000 tons) of coal.[346]

Roderick had, therefore, been leased the coal in the Rosy, Fiery, Golden and Bushy Veins under a large tract of land under Llanelli Town, Bres, Wern, Bigyn Hill, Talsarnau and Llanerch. He had undoubtedly been helped in the negotiations over the leases by Thomas Bowen, who had mined along the southern outcrops of the seams some 40 years earlier, and who understood the structure of the Coalfield better than any other person. Roderick's confidence in Bowen's knowledge and ability may have led him to accept the strict time and minimum royalty payment clauses in the leases. Roderick took the 79 year old Thomas Bowen and Margaret Griffiths (Bowen's widowed daughter)[347] into partnership, retained a one-half interest for himself and allocated a quarter share each to Bowen and Griffiths.[348] Shortly after the granting of the leases, they

Plate 11 — Site of the Baccas or Carnarvon Colliery. A colliery unsuccessfully worked by Hughes and Jones in the late 18th century and by others in the 19th century. Re-opened and abandoned as late as 1927. The ivy covered stack locates the site.

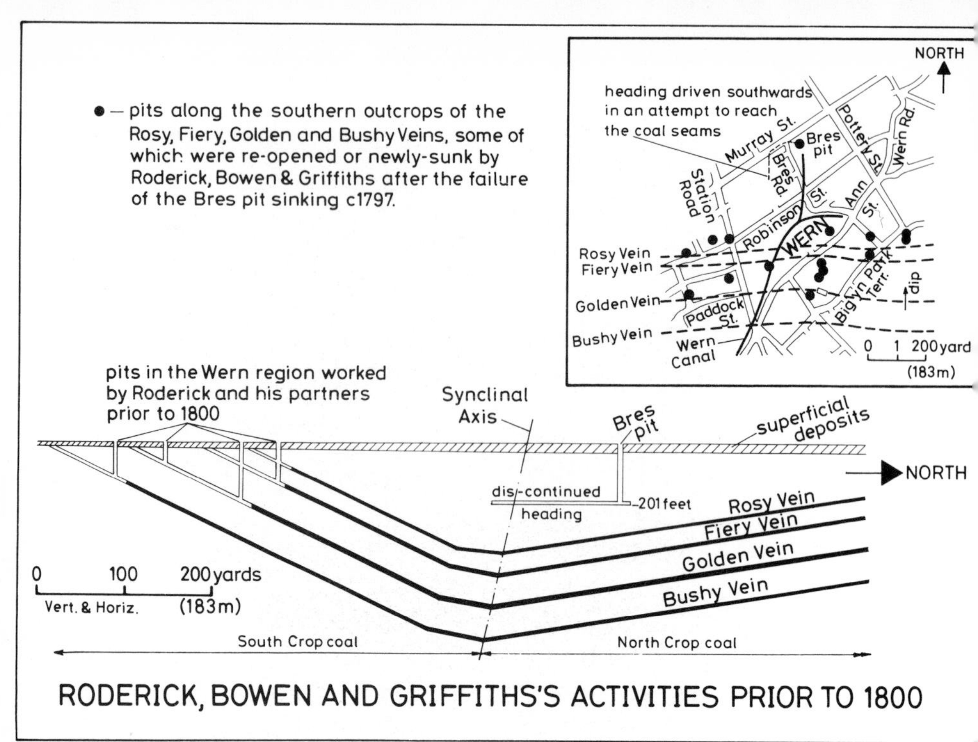

Fig. 15

began sinking a pit on Bresfawr (49 — later to be called the Bres pit) where they assessed the synclinal axis to be, to meet the clause regarding mining to the "*utmost depth*" of the Fiery Vein, and commenced the construction of a canal from the pit, across the Wern, to a shipping place or dock on the seashore at Llanelli Flats. The first coal seam (the Rosy Vein) lay almost 300 feet (91m) below the ground surface at the Bres and this was too deep for the power of the water engine used to sink the pit. They tried to overcome flooding for two years but sank the pit to a depth of only 201 feet (61m).[349] A heading was driven southwards from the pit bottom, to try to meet up with the dipping seams but was discontinued before coal was reached (Fig. 15). They must have realised then that they would not succeed at the Bres unless a steam engine was employed and, having insufficient capital to purchase one, they abandoned the sinking and moved to the Wern, where the coal seams were not so deep.[350] Most of the coal at shallow depth in the Wern/Caeswddy region had been worked out, however,[351] and they sank new pits to reach the unworked parts of the seams. It also appears that they re-opened a number of old pits, previously worked by John Allen and perhaps by Thomas Bowen himself.[352] They raised coal with this change of plan, but their previous difficulties

Plate 12 — Alexander Raby.

had brought dire financial trouble. Considerable capital had been outlaid on the abortive sinking of the Bres pit, on the sinking and re-opening of other pits, and on the formation of a three-quarters of a mile long canal and a dock, but only 725 weys (7,250 tons) of coal were raised between January 1795 and September 1800,[353] which, when sold, would have realised a sum just equal to the minimum royalty payments stipulated in the leases. They would never fully recover from this disastrous start to their enterprise.

(25) *Alexander Raby*

Alexander Raby was an English capitalist involved in metal and coal industries in various parts of England and Wales before he came to Llanelli. He was wealthy, with property in London, Worcestershire and Surrey,[354] and associated with coal mining in Worcestershire and Staffordshire and the importing of coal to London (probably for iron-smelting).[355] He already had industrial interests in South Wales, owning a copperworks and collieries at Neath, in partnership with his brother-in-law Thomas Hill Cox,[356] and was an active proprietor of the Monmouthshire Canal.[357]

The first record of Raby's interest in the Llanelli Coalfield was c1792/93, when he considered taking a lease of Sir John Stepney's coal at Llanelli. The second phase of Llanelli's industrialisation was about to begin, but Raby was undecided about committing himself, William Hopkin writing to Sir John Stepney, in July 1793, "*I have heard nothing since from Mr Raby or Mr Cox, if you have I should be glad to know what they intend to do as by waiting for them you may lose an opportunity of letting to other people at Llanelly... am apt to think some of the lands about Llanelly may be let to advantage to them if Mr Raby has drop'd thought of this Country*".[358] Raby evidently took no action over Stepney's coal, which was leased to William Roderick in 1794.

He became involved in Llanelli through his association with John Givers and Thomas Ingman. They probably came to Llanelli on Raby's recommendation, as William Hopkin wrote, in February 1794, "*Mr Ingman is a gentleman I know nothing of further than that he and his partners have been very strongly recommended to Sir John* (Stepney) *by a Mr Raby*".[359] Raby possibly backed Givers and Ingman financially when they started to build an iron furnace* and foundry on Stradey Estate lands at Cwmddyche in 1793,[360] arranging to have their coal from Dame Mary Mansell's collieries in September.[361] They obtained a lease of iron ore under the tenement of Pontyetts and Ynisyetts, from William Owen Brigstocke, in July 1794[362] and it was reported, in November 1795, that they had greatly improved the Cwmddyche area around the iron works[363] (possibly Daniel Shewen's iron furnace mentioned in the footnote). Although they were apparently successful, Raby took over their concern in 1796.[364] D. Bowen says that they became insolvent and Raby, as their principal creditor, took over the ironworks,[365] but it is more likely that Raby decided to concentrate his efforts into developing the resources of the Stradey Estate, after the passing of the 1796 Act of Parliament, which allowed the granting of coal leases for 91 years. It is not known whether Raby foreclosed upon Givers and Ingman or whether he bought them out. The latter is probably correct as Givers was still involved in the concern in 1798.[366] (Plate 13)

Raby's primary interest was iron, but he also planned to mine coal at

---

* Givers and Ingman may have taken over an iron furnace previously worked by Daniel Shewen in the late 1750s, Shewen having written to his brother Joseph in 1762 "*We have had a bad time of it, the furnace being idle for 2 years has been £600 at least out of Sir Edward's pocket*" (Mansel Lewis London Collection 98, letter 14 Feb 1762).

Plate 13 — Remains of Alexander Raby's iron furnace at Cwmddyche. There were two furnaces on this site but the remains of only one is visible today. The village of Furnace takes its name from these furnaces.

Llanelli for his ironworks and the export trade. He approached Dame Mary Mansell, in 1796, for the lease of coal under Stradey Estate lands west of the River Lliedi.[367] The collieries were then idle[368] and, as Dame Mary doubtless welcomed the prospect of receiving substantial royalties and Raby had also lent her money during her stays in London,[369] there was little chance that she would turn down his request. In March 1797, Dame Mary, her children and the mortgagees and trustees of the Stradey Estate, leased coal in the Caerythen Vein (Bushy Vein), under Estate lands west of the River Lliedi, to Raby for 91 years, at a royalty of 2 shillings per wey. It was agreed that a waggonway or railway and a canal would be constructed from the furnace to the sea, that the lands occupied by Givers and Ingman would be granted to Raby, and that the minimum royalty payment per year would be for 1000 weys (5,500 tons) of coal raised.[370]

Raby was already working this coal, before the granting of the lease, to keep the furnace in operation,[371] but his long term mining plans were to exploit the Rosy, Fiery and Golden Veins west of the River Lliedi on a large scale. Negotiations for this coal took place between Raby and Mansell throughout 1797, but relationships between them deteriorated and the lease had not been granted by the end of the year. There were a number of reasons for this[372] but Raby's refusal to lend further money to Dame Mary seems to have been the primary cause. In January 1798 she wrote to him "*I never will (knowingly) saddle myself and Family in Concern with any unfriendly or capricious person — sooner would I prefer the idea of a prison — I have now to assure you that had you granted me the loan of the sum I mentioned to you it would be no more than what I might of had with my worthy friend Mr Pemberton on the same Terms and had I wanted a larger sum I am confident he would not have refused me he being as sensible (I hope) of my honest principles as I am that he is a perfect Gentleman*".[373] Dame Mary was now dependent on Raby to exploit her Estate's resources, however, and the lease was soon forthcoming, Raby being granted, on 20 January 1798, the coal in the Rosy, Fiery and Golden Veins under Stradey Estate lands west of the River Lliedi for 91 years, at a royalty of 3 shillings per wey. Raby agreed to erect a steam engine to take workings to 240 feet (73m) and to pay a minimum yearly royalty, after three years, equivalent to 3000 weys (16,500 tons) of coal raised.[374] Raby must have intended exporting this coal, because the lease was termed "*Lease of Shipping Colliery*". Some clauses were also clarified and written into a subsequent agreement dated August 1798.[375]

Raby soon began building his industrial complex to the north west of Llanelli. He obtained a 21 year lease of Caebad, Caemain, Caebach, Vauxhall, Kilvig and other Stradey Estate landshares on 6 January 1798[376] and started to build his railway, which ran over these lands and connected his ironworks with the collieries and with the shipping place he began forming at Llanelli Flats (See Chapters 5 and 6). He sank new pits at Caemain (44, 45),* one of which was equipped with the steam engine specified in the Shipping Colliery lease,[377] at Caebad (38) and Caerelms (36)[378] and he probably also opened a new level, termed the New Slip (24), into the Bushy Vein. He continued workings at some of his inherited pits, amongst them being the Old Slip (23) and Caerhalen (32),[379] and commenced the modernisation and expansion of the ironworks at Cwmddyche, providing steam engines and an extra furnace[380] (Fig. 9). He enlarged his coalfield, in 1799, by obtaining leases of coal under lands west of the River Lliedi which did not belong to the Stradey Estate — in April 1799 he was leased coal in the Rosy, Fiery, Golden and Bushy Veins under Cae Parson, Cae fair, Glascwrte fawr, Glascwrte fach, Park y Crach and a field belonging to the Swan Inn, by Morgan Thomas, of St Clement Danes, Middlesex, for 89 years at a royalty of 2/6 per wey[381] and, in June

* Formerly the site of Charles Gwyn's water-engine pit.

Plate 14 — Sites of Alexander Raby's collieries. (Top) Old and New Slip collieries. Both were adits into the outcrop of the Bushy Vein along the southern hillside of the glacial valley between the Dimpath and Furnace. The New Slip was on the left and the Old Slip on the far right of the hillside. (Bottom) Caerelms pit. The pit was located on the overgrown ground at the end of Stradey Park Avenue on the right.

Plate 15 — Ruins of Furnace House or The Dell. Situated opposite the iron furnaces at Cwmddyche and built by Alexander Raby, c1798.

1799, he was leased all the coal under Morfa Bach common by John Vaughan of Golden Grove for 21 years at the same royalty payment.[382]

During the last four years of the 18th century, Raby invested capital in Llanelli's metal and coal industries on a scale not experienced hitherto. He was convinced that the Coalfield would ensure rich dividends on his investment, because he concentrated his efforts and money into the Llanelli area, relinquishing most of his previous interests. He built *"Furnace House"* (or the *"Dell"*)[383] for himself, on surface lands leased from the Stradey Estate in January 1798, occupying it after its completion in the late 1790s/early 1800s. He also built *"Caereithin"* (or *"Cilfig House"*) for his manager or agent.[384] Alexander Raby was the first major capitalist to commit his entire fortune to the Llanelli Coalfield and his belief in the potential of the area undoubtedly encouraged other industrialists to come to Llanelli in the early 19th century.(Plates 14 & 15)

CHAPTER 3 — REFERENCES

1 It has been stated that metal smelting was drawn to the coalfield in the 18th century because the ratio of the weight of coals to the weight of ores for smelting purposes was upwards of 3:1. (See "Industrial Expansion in South Wales" by R. O. Roberts, contained in "Wales in the Eighteenth Century," ed. by D. Moore, 1976).
2 "Llanelly Parish Church" by A. Mee, p.lv.
3 Ibid.
4 Stepney Estate 1103, Bill in Chancery, Cawdor v Stepney, first filed in 1872.
5 Stepney Estate 59 (2 Aug 1705).
6 Nevill MS LVI, copy deed (30 Mar 1709).
7 "Cadets of Golden Grove" by Francis Jones, Trans. Hon. Soc. of Cymmrodorion 1971 Part II.
8 Sir Thomas's son John was baptised at Llanelli in 1693 ("Llanelly Parish Church" by A. Mee p. 3) and the Partition Deed of 1705 showed that Llanelly House, Llanelly Mills and Machynis were in Sir Thomas's possession at that time, although they were allocated to Anne Lloyd (Stepney Estate 59).
9 Francis Jones, op.cit.
10 Thomas Mainwaring, op.cit., quoting a letter no longer extant.
11 "The Stepneys of Prendergast", Hist. Soc. of West Wales Trans. Vol VII, 1917-18.
12 A. Mee, op.cit., p. 3.
13 Stepney Estate 1103 (1872) referring to an unseen deed poll dated 12 Oct 1705.
14 Francis Jones, op.cit.
15 Stepney Estate 1103 (1872).
16 Thomas Mainwaring, op.cit., quoting an unseen lease dated 23 Jul 1723.
17 Stepney Estate 1100 (1 Aug 1729).
18 Llanelli Public Library plan 9 (1814).
19 Nevill 41 (29 Sep 1815).
20 A. Mee, op.cit., p.1vi.
21 Ibid. His death is given as 1744 on p. 1vii and as 19 Jan 1745 on p. 96.
22 Ibid, p. 1vii.
23 "The Stepneys of Prendergast" op.cit.
24 "Old Llanelly" by J. Innes, p. 46 quoting an unidentified source. Innes gave the year of the lease to Morgan as 1747 but it is assumed that 1748, the year of Sir Thomas Stepney's inheritance, was meant.
25 Stepney Estate 1099 quoting an unseen lease dated 13 Dec 1749.
26 Nevill 19 (24 Oct 1752) quoting an unseen lease.
27 Stepney Estate 1116 (19 Nov 1748).
28 Stepney Estate 1101, copy letters (7 Sep, 27 Sep 1749).
29 Ibid, copy letter, undated but late 1749.
30 Cilymaenllwyd 208 (3 Jan 1750).
31 Stepney Estate 1101, copy letters between 1750 and 1754.
32 Nevill 19, 21 and 22 (all dated 24 Oct 1752).
33 Stepney Estate 1101, copy letter (22 Nov 1752).
34 Ibid, copy letter (20 Dec 1753).
35 Ibid, copy letter (10 Feb 1755).
36 Stepney Estate Office plan 72, "Map of Lower Pencoid" (1761) (CRO) and Llanelli Public Library plan 3 (Apr 1772).
37 Stepney Estate 1101, copy letter (10 Feb 1755).
38 Cilymaenllwyd 135 (28 Feb 1753) and Stepney Estate 1101, copy letter (22 Nov 1753).
39 Cilymaenllwyd 214 and 218 (29 Jan and 14 Apr 1752 respectively).
40 Stepney Estate 43 (21 Oct 1758).
41 Stepney Estate 1103 (1872).
42 "The Stepneys of Prendergast", op.cit.
43 Alltycadno Estate, correspondence of Alexander Scurlock (11 Jun 1759).
44 Thomas Mainwaring, op.cit., quoting an unseen lease dated 1759.
45 Ibid, quoting an unseen lease dated 1760.
46 Stepney Estate 1094 (28 Jul 1765).
47 Stepney Estate 1099, quoting an unseen lease dated 1 Jan 1765.
48 Ibid, quoting an unseen lease dated 21 Nov 1768.
49 Alltycadno Estate, correspondence of William Roderick, letter dated 9 Oct 1773 referring to an unseen lease; and letter dated 7 Jan 1774, stating that the lease was due for expiry in 1789/90.
50 Ibid, letters (2 Dec 1771 and 22 Aug 1772).
51 Cawdor (Vaughan) 63/6572, rental for the year ending Michaelmas 1772.
52 "The Stepneys of Prendergast", op.cit.
53 "Some Notices of the Stepney Family" by R. Harrison, private impression (1870) (LPL).

54 "The Stepneys of Prendergast", op.cit.
55 Alltycadno Estate, correspondence of William Roderick, letters dated 9 Oct 1773, 7 Jan and 26 Nov 1774.
56 R. Harrison, op.cit.
57 Nevill 27 (27 Jun 1785), 409 (6 Nov 1786).
58 Cawdor (Vaughan) 16/466 (1781), 112/8396 (1782).
59 Local Collection 1610 (1787).
60 Cilymaenllwyd 313 (18 Oct 1787).
61 Stepney Estate 1122 (1791).
62 Cawdor (Vaughan) 21/615 (1 Oct 1794); Mansel Lewis 731 (1 Apr 1800); Carmarthenshire lease (2 Oct 1802) (CCL).
63 "Calendar of the Diary of Lewis Weston Dillwyn", Vol I — 4 Apr 1819, (NLW) made reference to the fact that the collieries on the Llangennech Estate had been purchased from Sir John Stepney by Symmons although the date of the purchase was not given.
64 "The Old House of Stradey" by Francis Jones, Archaeologia Cambrensis Vol CXXII (1973).
65 Mansel Lewis London Collection 1, lease (27 Dec 1708).
66 Ibid, lease (5 Apr 1711).
67 Ibid, lease (1 Sep 1713).
68 Ibid, lease (12 Sep 1715).
69 "St Illtyd's Church, Pembrey" by E. Roberts and H. A. Pertwee (1898).
70 Francis Jones, op.cit.
71 Ibid.
72 Mansel Lewis 58 (21 Feb 1750).
73 Nevill 21 and 22 (both dated 24 Oct 1752).
74 Francis Jones, op.cit.
75 Mansel Lewis 110 (c1759) and 756 (30 Jul 1766).
76 Mansel Lewis 376 (1789).
77 A. Mee, op.cit., p.lix.
78 Ibid, p. 22.
79 Mansel Lewis London Collection 34 (1805).
80 Ibid, referring to unseen leases or agreements dated 21 May 1758 and 24 Nov 1759.
81 Mansel Lewis London Collection 98, letter (14 Feb 1762).
82 Mansel Lewis 57 (15 Nov 1814) and Thomas Mainwaring, op.cit.
83 Mansel Lewis London Collection 34 (1805).
84 Nevill 25 (2 Aug 1762).
85 Nevill 705 (2 Aug 1762).
86 A. Mee, op.cit., p. 97.
87 Trans. of Carmarthenshire Antiq. Soc. and Field Club Part LIX (1934).
88 Mansel Lewis London Collection 98, letters (13 Feb, 15 Feb 1763).
89 Ibid, letters (15 Feb, 27 Feb 1763).
90 Ibid, letter (13 Feb 1763).
91 Mansel Lewis London Collection 34 (1805). The opinion was expressed in this document that many of Mansell's debts were incurred on Daniel Shewen's behalf.
92 Ibid. and Mansel Lewis 376 (1789).
93 Mansel Lewis London Collection 98, letter (23 July 1763).
94 Ibid, the evidence of letters throughout 1764.
95 Mansel Lewis 376 (1789). Francis Jones, op.cit., gave a date of 1771 for Mansell's marriage to Mary Shewen but did not quote the authority for his statement.
96 Mansel Lewis London Collection 4 (1767).
97 Mansel Lewis 1811 (1796).
98 Mansel Lewis London Collection 98, letter (14 Apr 1769).
99 Mansel Lewis 378 (17 Dec 1802).
100 Mansel Lewis 1811 (1796).
101 Mansel Lewis London Collection 34 (1805).
102 Mansel Lewis London Collection 98, letters (20 Jul, 30 Jul 1763, 15 Jul 1765).
103 Nevill 702 and Mansel Lewis 127 (9 Oct 1776).
104 Mansel Lewis 1811 (1796).
105 Mansel Lewis London Collection 34 (1805).
106 Mansel Lewis 1811 (1796).
107 "St Illtyd's Church, Pembrey", by E. Roberts and H. A. Pertwee (1898).
108 Ibid.
109 Mansel Lewis 1811 (1796).
110 Ibid.
111 Mansel Lewis 12 (Nov 1795). The report was attributed to "W & J Black".
112 Mansel Lewis 1811 — "An Act for vesting certain detached parts of the Real Estate, late of Sir Edward Vaughan Mansell, Baronet, deceased, situate in the county of Carmarthen, in Trustees,

in Trust to be sold, and to apply the money to arise from such Sale in the reduction of the several Mortgages or other Incumbrances subsisting upon or affecting such Real Estate; and also for enabling such Trustees to demise the mines, veins or seams of coal lying under the Residence of such Real Estate in such Manner and with such content as therein is mentioned (36 Geo III, 1796)".

113 Mansel Lewis 1634, letter from Thomas Lewis to R. G. Thomas (28 July 1801). This letter said that no part of the Estate had then been sold.
114 Mansel Lewis 189 (1794).
115 "Hanes Llanelli" by David Bowen (1856), quoting an unidentified source. (LPL).
116 Mansel Lewis 480 (25 Mar 1797) and 49-50 (20 Jan 1798).
117 Francis Jones, op.cit.
118 Trans. of the Carmarthenshire Antiq. Soc. and Field Club Part LIX (1934).
119 D. Bowen, op.cit.
120 Llanelli Public Library plan 10 (undated but c1822).
121 Llanelli Public Library plan 9 (1814).
122 Cawdor (Vaughan) 5854 — plan of Morfa Bach (1787).
123 Mansel Lewis London Collection plan 60 (1800).
124 All the previous statements have been obtained from "The Vaughans of Golden Grove", Parts I, II and III by Francis Jones, contained in Trans. of the Hon. Soc. of Cymmrodorion 1962, 1963 and 1964.
125 Cawdor (Vaughan) 8057 quoting an unseen lease of 12 Dec 1718.
126 Cawdor (Vaughan) 13/372 (3 Sep 1733).
127 Francis Jones, op.cit.
128 Cawdor (Vaughan) 53/6124 — rental of the Estate of John Vaughan 1751-54, implying an unseen lease.
129 Cawdor (Vaughan) 21/623 (28 Jun 1754).
130 Cawdor (Vaughan) 102/8029, letters (22 Nov 1755, 6 Mar, May 1756).
131 Ibid, letters (22 Nov 1755, 27 Nov 1756).
132 Cawdor (Vaughan) 21/624 (22 Apr 1758).
133 Cawdor (Vaughan) 102/8028, rentals for 1758/59.
134 Cawdor (Vaughan) 102/8029, letter (9 Jul 1761).
135 Francis Jones, op.cit.
136 Ibid, and Stepney Estate 1103 (1872).
137 Cawdor (Vaughan) 14/414 (Sep 1765).
138 Cawdor (Vaughan) 53/6140 (July 1766).
139 Cawdor (Vaughan) 114/8430 (11 Jan 1768).
140 Alltycadno Estate, correspondence of William Roderick, 9 Oct 1773, referring to an unseen lease.
141 Francis Jones, op.cit.
142 Stepney Estate 1104 (1840) and 1103 (1872).
143 Cawdor (Vaughan) 103/8042 — "Accounts of Golden Grove Estate sold in 1783, 1784 and 1785".
144 Cawdor (Vaughan) 112/8396 (1782), 8038 (1783).
145 Nevill 27 (27 Jun 1785), 409 (6 Nov 1786).
146 Cawdor (Vaughan) 125/8649 (1790).
147 Cawdor (Vaughan) 41/5790 (21 Apr 1791) and 41/5795 (21 Nov 1795).
148 Cawdor (Vaughan) 5/118 (1 Sep 1794).
149 Cawdor (Vaughan) 8112 (1 Oct 1794).
150 Nevill 1738 (17 Jun 1799).
151 Cawdor (Vaughan) 41/5800 (16 May 1799).
152 Thomas Mainwaring's Commonplace Book (LPL), quoting an unseen letter from R. J. Nevill, to Messrs. Goodeve and Ranken dated 31 Aug 1819. Nevill stated in his letter that he obtained his information from an account contained in a *"large Box of Miscellaneous papers which I found at Trosserch"*.
153 Cawdor II — 2/187, "Abstract of the Title of Jn° Hughes Rees, Esq., to Kilymaenllwyd etc..." (18 Nov 1844).
154 Cawdor (Vaughan) 8057, quoting an unseen lease dated 12 Dec 1718. This document is a summary of leases previously granted by the Golden Grove Estate and retrospectively referred, in error, to Pawlett and his wife as the Duke and Duchess of Bolton, a title they would not hold until 1721/22.
155 Thomas Mainwaring, op.cit., quoting an unseen lease dated 23 Jul 1723. Mainwaring obtained his information from papers held by a Mr Brown (probably F. L. Brown, solicitor at Llanelli) which are not now extant.
156 Llanelli Public Library plan 3 (Apr 1772).
157 Stepney Estate 1100 (1 Aug 1729).
158 Cawdor (Vaughan) 13/372 (3 Sep 1733).

159 Stepney Estate 39, account book 1736 to 1771. The actual lease of Stepney's coal at "*Crofswthy*" has not been seen.
160 Cawdor (Vaughan) 659, accounts for 1749.
161 Cawdor (Vaughan) 21/623 (28 Jun 1754).
162 Cawdor (Vaughan) 102/8029, letter (15 Jan 1757).
163 Ibid, letter (9 Jul 1761).
164 "Hanes Llanelli" by D. Bowen (1856), quoting an unidentified source.
165 Mansel Lewis London Collection plans 69 and 87 (May 1846).
166 D. Bowen, op.cit.
167 Cawdor (Vaughan) 21/623 (28 Jun 1754) and 21/624 (22 Apr 1758).
168 The evidence of A. Mee, op.cit., and Stepney Estate 39 (1736-71).
169 Stepney Estate 1099, rent roll of Sir John Stepney's Estate quoting an unseen lease dated 1 Oct 1747.
170 Llanelli Public Library plan 3 (Apr 1772).
171 A. Mee, op.cit., p. 16.
172 Derwydd H 19, pedigree of Stepney and Popkin Families. (CRO).
173 "Old Llanelly" by J. Innes, p. 46, quoting an unidentified source. Innes gave the year of the lease as 1747 but it is assumed that 1748, the year that Sir Thomas Stepney inherited his father's estate, was meant.
174 Cawdor (Vaughan) 53/6124 — rentals 1751-54, implying an unseen lease.
175 "Cwmdwyfran Forge, 1697-1839" by M.C.S. Evans, The Carm. Ant. Vol XI 1975.
176 Cawdor (Vaughan) 53/6124 — rentals 1751-54.
177 Stepney Estate 1101, copy letter (10 Feb 1755).
178 Alltycadno Estate, letters (31 May and 11 Jun 1759) referring to an unseen lease or agreement.
179 Carmarthenshire document 118 (1 Oct 1759) (CCL). A "*Caellwyddes*" was shown near Brynsheffre lands on Llanelli Public Library plan 14 (c1825); this was probably the "*Cae yr Arglwddes*" mentioned in the lease.
180 Thomas Mainwaring, op.cit., quoting an unseen lease dated 1759.
181 Alltycadno Estate, letter (26 Oct 1761).
182 Cawdor (Vaughan) 102/8028, rentals 1758-61 and 49/5962, rental 1761/62.
183 "Carmarthen Tinplate Works 1800-1821" by T. James, Carm. Ant. Vol XII 1976.
184 Stepney Estate 1099, quoting an unseen lease dated 13 Dec 1749.
185 Mansel Lewis 58 (21 Feb 1750).
186 The evidence of Llanelli Public Library plan 3 (Apr 1772).
187 Thomas Mainwaring, op.cit.
188 Llanelli Public Library plan 1 (1751).
189 Cawdor (Vaughan) 102/8029, letters (22 Nov 1755, 6 Mar, - May 1756).
190 Cawdor (Vaughan) 112/8383 (1757), 44/5960 (1758), 102/8028 (1759-61).
191 Mansel Lewis 2179 (1762), quoting an unseen lease dated 1760.
192 Opinions expressed in Mansel Lewis London Collection 114, letter (10 Feb 1770) and Mansel Lewis London Collection 34 (1805).
193 Nevill 25 (2 Aug 1762).
194 Mansel Lewis London Collection 98, letter (3 Oct 1763). Mansell was confined in Fleet Prison at this time.
195 Ibid, letter (20 Feb 1765).
196 Llanelli Public Library plan 3 (Apr 1772).
197 Mansel Lewis London Collection 98, letter (26 Sep 1766).
198 Mansel Lewis London Collection 114, letter (25 Sep 1778).
199 Cawdor (Vaughan) 53/6135 (1779).
200 Cawdor (Vaughan) 112/8396 (1781/82).
201 Thomas Mainwaring, op.cit., stated that a pit in Cae Scyborddegwm had been originally opened by Dr Evans. Llanelli Public Library plan 3 and its associated schedule of lands (Apr 1772) shows that this pit was the Ox pit.
202 Penller'gaer B.18.64 — Case of the Llanelly Colliery 1817/18.
203 Nevill 19 (24 Oct 1752), quoting an unseen lease. The reference to Sir Thomas Stepney, 7th Bart., as one of the original lessors, confirms that the lease could not have pre-dated 1748 which was the year of his inheritance.
204 Cilymaenllwyd 135 (28 Feb 1753).
205 Llanelli Public Library plan 3 (Apr 1772).
206 Cawdor (Vaughan) 21/623 (28 Jun 1754).
207 Cawdor (Vaughan) 102/8029, letter (6 Mar 1756).
208 Cawdor (Vaughan) 112/8383 (Michaelmas 1757).
209 Cawdor (Vaughan) 21/624 (22 Apr 1758).
210 Mansel Lewis London Collection plans 69 and 87 (May 1846).
211 Cawdor (Vaughan) 41/5812, letter (8 May 1802).
212 "Old Llanelly" by J. Innes and the Hopkin Morgan Manuscript (LPL) both quoting unidentified sources.

213 Brodie Collection 69 (1757).
214 Cawdor (Vaughan) 49/5960 (1758), 102/8028 (1759-61).
215 Cawdor (Vaughan) 41/5812, letter (8 May 1802).
216 Mansel Lewis London Collection 98, letter (11 Jun 1763).
217 "History of Carmarthenshire", J. Lloyd, Vol II, p. 340, quoting an unidentified source.
218 Stepney Estate 1099 referring to an unseen lease dated 21 Nov 1768. Due to a tear in the document it is possible to interpret the described lands as Trostre, Llwyncyfarthwch and Cae Cefen y Maes or else as Trostre, Llwy...?, Llwyncyfarthwch and Cae Cefen y Maes. If the latter interpretation is correct then Llwynwhilwg would have been the most likely landshare.
219 Llanelli Public Library plan 3 (Apr 1772) referred to Bowen's activities in the Pencoed/Ffosfach region in the past tense.
220 Mansel Lewis London Collection 98, letter (14 Apr 1769). Sir Edward was Sir E. V. Mansell, 3rd Bart.
221 Cawdor (Vaughan) 63/6572 (1771/72), 16/466 (12 May 1781).
222 Stepney Estate 1099, rent rolls 1772, 73, 76, 80.
223 Llanelli Public Library plan 1 (1751). Information superimposed at a later date, probably by Rhys Jones of Loughor.
224 Cawdor (Vaughan) 5854 (1785) and Llanelli Public Library plan 10 (c1822).
225 Cawdor (Vaughan) 16/466 (12 May 1781).
226 Cawdor (Vaughan) 112/8396 (1781/82), 8038 (1782/83), 8039 (1787/88).
227 *The Cambrian* (14 May 1809) reported Thomas Bowen's death at the age of 94 years.
228 Mansel Lewis 2620 (30 Jan 1822).
229 Llanelli Public Library plan 1 — "Map of Lands near and about Berwick Chappel" by Wm. Jones, 1751.
230 Nevill 21 (24 Oct 1752).
231 Nevill 22 (24 Oct 1752).
232 Nevill 19 (24 Oct 1752).
233 Nevill 709 (20 Nov 1752).
234 "Canals of South Wales and the Border" by C. Hadfield, p. 31.
235 Llanelli Public Library plan 1, information superimposed at a later date.
236 Brodie Collection 69 (1757).
237 A. Mee, op.cit., p. 97.
238 Nevill 25 (2 Aug 1762).
239 Mansel Lewis London Collection 98, letters (11 Jun, 20 Jul 1763).
240 Ibid, letter (15 Jul 1765).
241 "The Smelting of Copper in the Swansea District" by G. Grant Francis, p. 117 and "Chauncey Townsend's Waggonway" by P. R. Reynolds, Morgannwg, Vol XXI, 1977.
242 "Some Aspects of the History of Carmarthenshire in the late 18th Century" by W. J. Lewis, Carmarthenshire Antiquary 1963; "Lead Mining in Wales" by W. J. Lewis (1967).
243 "Economic History of Wales prior to 1800" by D. J. Davies, p. 133 (1933).
244 Mansel Lewis London Collection 98, letter (30 Jul 1763).
245 Mansel Lewis London Collection 34 (1805).
246 Cawdor (Vaughan) 14/414 (Sep 1765), 53/6140 (July 1766).
247 Nevill 5 (11 Apr 1766).
248 "Report from Committee on Petition of the Owners of Collieries in South Wales 1810". Evidence of Henry Smith, M.P., given on 16 May 1810.
249 This pit was still in use as the pumping pit to the Genwen Colliery in 1907.
250 Llanelli Public Library plan 3 (Apr 1772).
251 Mansel Lewis London Collection 114, letter (5 Sep 1768).
252 Mansel Lewis London Collection 98, letter (14 Apr 1769).
253 Thomas Mainwaring, op.cit., quoting an unidentified source.
254 Mansel Lewis 756 (30 Jul 1766).
255 Mansel Lewis 382 (undated but c1806).
256 Mansel Lewis London Collection 34 (1805).
257 Ibid.
258 Llanelli Public Library plan 3 (Apr 1772) entitled "A Map of Colliery Lands to the East of Llwynhendy Fire Engine upon lease to the Executors of the late Chauncey Townsend, Esq."
259 Mansel Lewis 2082, letter (8 Apr 1775).
260 Mansel Lewis London Collection 114, letter (10 Sep 1775).
261 Mansel Lewis 382 (un-dated but c1806).
262 Mansel Lewis London Collection 114, copy letter (10 Feb 1776).
263 Nevill 702 and Mansel Lewis 127 (9 Oct 1776).
264 Mansel Lewis London Collection 114, letter (25 Sep 1778).
265 Ibid, letter (16 May 1780).
266 Mansel Lewis 1866 (24 Oct 1781).
267 Nevill 27 (27 Jun 1785).

268 Nevill 23 (7 Jan 1786).
269 Nevill 409 (6 Nov 1786).
270 The evidence of Llanelli Public Library plan 3 (1772) and 8 (1812) which both showed a 240 feet deep Engine Pit.
271 Llanelli Public Library plan 1 (1751). Information superimposed at a later date, probably by Rhys Jones of Loughor.
272 Mansel Lewis London Collection 34 (1805).
273 Mansel Lewis London Collection 10, document entitled "Stradey Collieries and Binie — Queries for Mr Rees Jones dated February 1828".
274 Report from Committee on Petition of the Owners of Collieries in South Wales 1810. Evidence of Henry Smith, M.P.
275 Mansel Lewis London Collection 34 (1805).
276 Cawdor (Vaughan) 16/466 (1781), 112/8396 (1782), 103/8038 (1783), 103/8039 (1788), 125/8649 (1790), 103/8040 (1795) and Stepney Estate 1099, rental for 1797.
277 Glamorganshire deed MS-35.32 (11 Dec 1799) (CCL).
278 Nevill 702 (9 Oct 1776).
279 Report from the Committee on Petition etc . . . 1810.
280 Nevill 24 (11 Dec 1799) and Glamorganshire deed MS-35.32 (11 Dec 1799) (CCL).
281 Cawdor (Vaughan) 102/8029, letter (22 Nov 1755). Although John Rees was referred to as of "*Killmaenllwyd*" in this letter, Hector Rees was referred to as of "*Tuwyn*" in a plan of 1751 (Llanelli Public Library plan 1).
282 Ibid, letter (27 Nov 1756).
283 Brodie Collection 69 (1757).
284 Cawdor II, 2/187 — "Abstract of the Title of Jn° Hughes Rees, Esq., to Kilymaenllwyd etc . . ." (18 Nov 1844).
285 "St Illtyd's Parish Church" by E. Roberts and H. A. Pertwee (1898).
286 Cawdor II, 2/187 (18 Nov 1844).
287 Roberts and Pertwee, op.cit.
288 Stepney Estate 1094 (28 Jul 1765).
289 Ibid, memorandum dated 21 Nov 1765 written on the reverse of the lease.
290 Cawdor (Vaughan) 114/8430 (11 Jan 1768).
291 Stepney Estate 1094 (28 Jul 1765) referring to an unseen lease or agreement.
292 Thomas Mainwaring, op.cit., quoting an unidentified source.
293 Ibid.
294 Cawdor (Vaughan) 112/8396 (Michaelmas 1782).
295 Thomas Mainwaring, op.cit., quoting an unidentified source.
296 Cawdor (Vaughan) 102/8028 (1758-61), 49/5962 (1761/62), 112/8394 (1762/63).
297 Cawdor (Vaughan) 102/8029, letter (23 Oct 1756).
298 Carmarthenshire deed 118 (1 Oct 1759) (CCL).
299 Thomas Mainwaring, op.cit., — "Minutes of Old Leases granted by the Stepneys (from Mr Brown's papers)". Mainwaring worked for F. L. Brown, a Llanelli solicitor, and must have seen a lease which has not survived.
300 Stepney Estate 1099 — rent rolls, accounts and schedules of the Stepney Estate for certain years between 1741 and 1802.
301 Nevill 19, 21, 22 (all dated 24 Oct 1752).
302 John Innes, op.cit., referring to an unidentified source.
303 Mansel Lewis London Collection 34 (1805).
304 Mansel Lewis London Collection 98, letter (30 Apr 1763).
305 Ibid, letter (11 Jun 1763).
306 Mansel Lewis London Collection 10, document entitled "Stradey Collieries and Binie — queries for Mr Rees Jones dated February 1828".
307 Mansel Lewis 127 (9 Oct 1776) and Mansel Lewis London Collection 114, letter (10 Sep 1775).
308 Mansel Lewis London Collection 10 — "Stradey Collieries and Binie etc."
309 Cawdor (Vaughan) 5854 — "Plan of Morfa Bach 1787".
310 Stepney Estate 1099 — rent roll of Sir John Stepney's Estate for 1773, quoting an unseen lease or agreement.
311 Stepney Estate 1094 (28 Jul 1765).
312 Stepney Estate 1099 — rent rolls of Sir John Stepney's estate for 1773 and 1776.
313 Mansel Lewis London Collection 98, letter (14 Apr 1769).
314 Alltycadno Estate, correspondence of William Roderick (14 Aug 1773).
315 Ibid, letters (9 Oct 1773, 7 Jan 1774).
316 Ibid, letter (26 Nov 1774).
317 Carmarthenshire deed 140 (20 Oct 1774) (CCL).
318 Alltycadno and Gwylodymaes Estate rentals 1769 to 1785, MS4.830.
319 Stepney Estate 1099 (1741-1802).
320 Local Collection 465 (1773 to 1792).

321 Alltycadno and Gwylodymaes Estate rentals 1769 to 1785, MS4.830.
322 "Cadets of Golden Grove" by Francis Jones , op.cit.
323 Mansel Lewis 154 (20 Nov 1776).
324 Mansel Lewis 2590 (Jan 1822).
325 Local Collection 465 entitled "Mary Hopkin's Book 1790" but containing accounts between 1773 and 1792. An entry stated that William Hopkin and Mary Davies, both of the Parish of Llangunllo in the Deanery of Brecon in the County of Radnor were married on 24 November 1772.
326 Mansel Lewis 2620 (30 Jan 1822).
327 Cawdor (Vaughan) 103/8039, rental (1787/88).
328 Cawdor (Vaughan) 92/7635 (1 Jan 1789).
329 Statement contained in Carmarthenshire deeds 126, 127 (20 Feb 1793) (CCL).
330 Local Collection 465 (1773 to 1792).
331 Carmarthenshire deeds 126, 127 (20 Feb 1793) (CCL).
332 Stepney Estate 1088 (evidence of letters throughout 1793) and Cilymaenllwyd 314 (4 Feb 1794).
333 Stepney Estate 1088, letters (30 Mar, 24 Jul, 12 Nov 1793).
334 Cawdor (Vaughan) 87/7350 (1 Oct 1794/95).
335 Carmarthenshire deed 81 (28 Jan 1805) (CCL).
336 *The Cambrian* (3 Oct 1807, 8 Sep 1810).
337 Nevill 363 (7 Oct 1793).
338 Cawdor (Vaughan) 5/118 (1 Sep 1794).
339 Ibid.
340 The evidence of Local Collection 37 (11 May 1824) and Thomas Mainwaring, op.cit., quoting an unidentified source.
341 Report on the Baccas Colliery by Rees Jones, undated but c1824, contained in Local Collection 37 (11 May 1824).
342 Thomas Mainwaring, op.cit., quoting an unidentified source.
343 Stepney Estate 1099 — "Sir John Stepney's Rental for the year 1797".
344 Alltycadno Estate correspondence from 1770 onwards.
345 Cawdor (Vaughan) 21/615, copy lease (1 Oct 1794).
346 Cawdor (Vaughan) 8112, draft lease (1 Oct 1794). There is evidence that this lease was never fully executed although coal was worked. Apparently Roderick signed the lease on 1 Oct 1794 and delivered it to John Vaughan's agent (Rev. Thomas Beynon) who retained it, with the result that Vaughan never signed it. (Cawdor 2/133 — Brief of a Bill in Chancery, R. A. Daniell and others against Lord Cawdor and others, 1814).
347 Cawdor 2/133 (1814).
348 Nevill 411 (29 Apr 1825) quoting an unseen assignment dated 20 Mar 1800.
349 Llanelli Public Library plan 7 (Jul 1806) showed the pit to be abandoned at a depth of 201 feet at that time.
350 Cawdor (Vaughan) 43/5847 — "Mr Roderick's proposal", undated but c1803.
351 Llanelli Public Library plans 4 — 7 (July 1806).
352 "Hanes Llanelli" by D. Bowen (1856) quoting an unidentified source.
353 Cawdor (Vaughan) 103/8049 (25 Mar 1797 to 29 Sep 1800) and 103/8050 (1 Jan 1795 to 29 Sep 1800).
354 Mansel Lewis London Collection 3, draft assignment (10 Jun 1807).
355 Report on the State of the Coal Trade (23 Jun 1800), evidence of Alexander Raby given on 7 May 1800.
356 Thomas Mainwaring, op.cit., quoting an unidentified source.
357 Document G 779 (CCL), miscellaneous accounts and papers bound in a volume dated 1811.
358 Stepney Estate 1088, letter (14 Jul 1793).
359 Cilymaenllwyd 314, copy letter (4 Feb 1794).
360 Ibid, which stated that the furnace and foundry had been erected by Feb 1794. David Bowen, op.cit., stated that the furnace was erected in 1791.
361 Mansel Lewis 189 (1794), quoting an unseen agreement dated September 1793.
362 Brigstock of Blaen Pant MSS, Box VIII, lease (14 Jul 1794) (NLW). I am indebted to Mr M. C. S. Evans for bringing this reference to my attention.
363 Mansel Lewis 12 (28 Nov 1795).
364 Mansel Lewis London Collection 34, Mr Bell's opinion (1805).
365 "Hanes Llanelli" by D. Bowen, quoting an unidentified source.
366 Local Collection 1586 — Stradey Furnace Accounts, May 1797 to July 1798.
367 Mansel Lewis London Collection 34, Mr Bell's opinion (1805).
368 Mansel Lewis 12 (28 Nov 1795).
369 Mansel Lewis London Collection 112, draft letter entitled "Exparte Mrs Mansel" c1825.
370 Mansel Lewis 480 (25 Mar 1797).
371 Ibid.
372 See Mansel Lewis London Collection 97, letters (4 Nov and 16 Dec 1797).

373 Ibid, letter (Jan 1798).
374 Mansel Lewis 49-50 (20 Jan 1798).
375 Mansel Lewis 247 (3 Aug 1798).
376 Mansel Lewis 244 (6 Jan 1798).
377 Mansel Lewis London Collection 98, letter (1 Mar 1799).
378 "Hanes Llanelli" by D. Bowen (1856).
379 Mansel Lewis London Collection 97 — document entitled "Account of Coals worked and landed by Mr Raby from Cae eithin and Cae Hallen in 1797 and 1798."
380 D. Bowen, op.cit. and Thomas Mainwaring, op.cit.
381 Nevill 155 (3 Apr 1799). This coal had previously been leased to Dame Mary Mansell and Raby had agreed to take the coal from her. (See Mansel Lewis London Collection 97, letter dated 3 Aug 1798).
382 Nevill 1738 (17 Jun 1799).
383 "Llanelly Parish Church" by A. Mee, p.xxxiii.
384 Mansel Lewis 1966 — William Rees's Statement (27 Sep 1825).

CHAPTER 4

COAL MINING 1800 TO 1829

INTRODUCTION

The first 30 years of the 19th century saw unparalleled industrial growth throughout much of Britain and events at Llanelli reflected the national trend, as the area emerged from the period of industrial stagnation experienced between c1770 and 1800. Alexander Raby undoubtedly gave the initial impetus to this second phase of Llanelli's industrialisation by committing most of his money to the area in the late 1790s / early 1800s. His action persuaded other English developers to invest in Llanelli and John Symmons, General George Warde, Charles and Richard Janion Nevill (acting for John Guest and William Savill) and the Pemberton Family, all arrived between 1800 and 1804. Their efforts, and those of local developers such as Roderick, Bowen and Griffiths, resulted in coal exports increasing 5-fold between 1800 and 1829, and ensured the continuation of Llanelli's industrial expansion of the late 1790s, with the establishment of ironworks, copperworks and leadworks, the formation of railways and the construction of docks capable of taking large vessels.

This rapid industrial expansion significantly affected the area's character and Llanelli must have experienced most radical changes during the first 30 years of the 19th century. The population increased[1] and many of the newcomers, especially the overseers and skilled workers, were brought in from other parts of Wales and the United Kingdom.* These newcomers would have brought the customs of their own areas with them and many of Llanelli's traditional outlooks and values must have been eroded at this particular time. Additionally, all three of the main land-owning families, who had dominated most aspects of Llanelli life up to

* Insufficient evidence is available to assess the origins of the new population but individual references show that people were recruited from well outside Llanelli. When the Llanelly Copperworks Company took over the Wern Colliery in 1807, James Morrison of Cardiff was appointed as superintending engineer (Thomas Mainwaring's Commonplace Book) and "*men from Shropshire*" were employed to expand the existing workings (Nevill MS XLIX).

1800, relinquished their local interests between 1804 and 1811: John Vaughan left the Golden Grove Estate to John Campbell of Stackpole Court; Mary Anne Mansell left the Stradey Estate to her solicitor Thomas Lewis of Llandilo; and Sir John Stepney, 8th Baronet, virtually disinherited his family by leaving the Stepney Estate to a succession of named English friends and their male line. These three estates, therefore, passed to strangers and a new estate, the Llangennech Estate, was created by John Symmons of London, through land purchase. All these changes in such a short time led to the creation of a new Llanelli, with a character which owed more to the events of the early 1800s than it did to the traditions of the previous centuries. Llanelli's present character undoubtedly has many of its origins firmly rooted in the years between 1800 and 1829, when Raby, Roderick, Bowen and Griffiths, Symmons, Warde, the Nevills, the Pembertons and others laid the foundations to the modern industrial township we know today.

Except for the Nevills, acting for the Llanelly Copperworks Company, few of the new industrialists made high profits, and many lost money in the Coalfield. Roderick, Bowen and Griffiths sold out to the Pembertons and the Llanelly Copperworks Company, in 1807, to avoid bankruptcy; Alexander Raby and General Warde, after many financial crises, ended up so deeply in debt to Richard Janion Nevill and the Llanelly Copperworks Company that they lost all their industrial interests; the Pembertons, after the expensive but abortive sinking of the Llwyncyfarthwch pit, withdrew from the area, selling their interests to George Bruin. The Earl of Warwick, John Vancouver, Richard Davenport, George Morris and William Rees all lost money, and Symmons made little profit, exploiting coal in the Llangennech area. By 1830, only Richard Janion Nevill was left and he, and the flourishing Llanelly Copperworks Company, had gained virtual control of the Coalfield. Only the newly-constituted Llangennech Coal Company could present any challenge to his monopoly, but the problems experienced in the sinking of the St David's pit effectively reduced the Company to a subsidiary role in Llanelli's coal trade at that time. The best years lay ahead for the Llangennech Coal Company but it would never quite rival the coal empire that R. J. Nevill and his son C. W. Nevill, acting for the Llanelly Copperworks Company, built after 1829.

A substantial amount of information, including plans, is available for 1800 to 1829. Fuller accounts of the actions and possible motives of the principal participants in the industry can, therefore, be given and many of the collieries, and their allied transportation systems and shipping places, can be located. Additionally, a valuable source of contemporary account and opinion is provided by *The Cambrian,* which was the first newspaper to report events at Llanelli. A strict chronological narrative will, again, not be adopted, as the main themes would be obscured by detail, although the individuals and partnerships will be considered, sequentially, in terms of their first appearance in the industry. Appendices C, D and E should be consulted for a summary of the ownership and devolution of the Stepney, Stradey and Golden Grove Estates.

(I) THE MAIN ESTATES AND THEIR OWNERS

(1) *The Stepney Estate — Sir John Stepney, 8th Baronet; the Earl of Cholmondeley and Richard Henry Alexander Bennett; William Chambers*

As we have seen, Sir John Stepney, 8th Bart., had little interest in his inheritance apart from the income it produced. He had tried to sell it since 1787 but had found no takers, apart from John Symmons of Paddington Green, who had made an offer for lands in the Llangennech region.[2] In Stepney's absence, William Hopkin, his agent since c1785, handled all matters concerning his coal interests, which produced small royalty income in 1800. This was mainly because most of the coal was under lease to Chauncey Townsend's inactive inheritors and also because Roderick and Bowen ran into difficulties in mining Stepney's coal in the Llanelli Town region. Matters quickly changed as Roderick and Bowen met with success and as the surrender of leases, held by Chauncey Townsend's grandsons, Charles and Henry Smith, was negotiated. The surrenders were arranged by October 1802 and Stepney, together with the other inheritors of the Llanelli Vaughans' original Estate, re-leased this coal, together with coal under Pencoed and Ffosfach, to George Warde, Francis Morgan and James Morgan for 99 years.[3] In two years, therefore, the Estate's mining situation was transformed, with virtually all of its coal coming under active lease.

Sir John drew up his will at this time, and, for reasons which are not apparent, virtually disinherited his family. The will's provisions were complicated and, as they indirectly affected the Llanelli coal industry, especially a clause restricting the granting of coal leases to a period of 21 years, a brief summary of its provisions is necessary.[4] Stepney appointed four trustees — Lord Robert Seymour, Roger Wilbraham, William Owen Brigstocke and Anthony Goodeve — who were entrusted to convey his estate, on his death, to "*his three dear friends*", Henry, Duke of Beaufort, Rt. Hon. George James the Earl of Cholmondeley and Richard Henry Alexander Bennett, or their survivors. After their deaths, the estate was to pass, in turn, to a number of named friends and to their survivors in *"tail male"*, i.e. to the eldest sons in perpetuity. In order of inheritance, Stepney named Lt. Col. Orlando John Williams and his survivors and, in default of such male issue, Lt. Richard Falkland and his survivors and then William Chambers and his survivors. Stepney's own family came after all these other people, with his brother Thomas, and his issue in tail male, being the first of his family named in the will. If Thomas had no son, the estate was to pass, in succession, to the following and their survivors in tail male — Eliza Gulston (daughter of Sir John's younger sister Elizabeth Bridgetta), John Stepney Cowell (son of Sir John's youngest sister Maria Justina), Frances Head (Maria Justina's daughter by a first marriage), Richard Henry Alexander Bennett (son of the previously named Richard Henry Alexander Bennett) and Joseph Gulston (Elizabeth Bridgetta's son). There is no doubt that the 8th Baronet was consciously disinheriting his family by his will.[5] His reasons are not apparent but he was in debt, trying, unsuccessfully, to sell lands in the region of Llanelli Town in 1805[6] and might, therefore, have borrowed from his friends and had pledged his estate to them.

Sir John Stepney went to live abroad again in 1803/04,[7] and William Hopkin continued to manage his estate.[8] George Warde's successful exploitation of coal previously leased to Townsend and Smith, and Roderick and Bowen's success at mining in the Town area, brought a significant increase in royalties paid to the Stepney Estate. The 8th Baronet did not enjoy this increased income for long, however, because he died on 3 October 1811, aged 68, at Turgau near Temesar in Hungary. His brother Thomas became the 9th Baronet[9] but not, under Sir John's will, the owner of the Estate and its income. When the will was proved by the trustees, in the Prerogative Court of the Archbishop of Canterbury, they conveyed the Estate into the control or trust of a Charles Ranken, who was empowered to pay the income to Stepney's inheritors. Henry Duke of Beaufort had died in October 1803, leaving the Earl of Cholmondeley and Bennett as co-heirs and they jointly succeeded to the Estate. Neither lived at Llanelli and Bennett died in March 1814, leaving the Earl of Cholmondeley in full possession of the Estate.[10] No information relating to Cholmondeley's involvement in Llanelli's life has been seen, apart from a specially commissioned version of John Howell's 1814 plan of Llanelli, which was entitled, *"A Map of an Estate situate in and near the Town of Llanelly, Carmarthenshire, Belonging to the Right Hon. the Earl of Cholmondeley — surveyed and mapped by John Howell, Llanelly 1814"*.[11] The Estate's coal was already fully leased, and Cholmondeley probably took no interest in his inheritance, apart from enjoying its rents and its steadily increasing royalty payments. Cholmondeley died in April 1827 and, as Lt. Col. Orlando John Williams and Lt. Richard Falkland had both died without male issue, in 1812 and 1824 respectively, the ownership of the Estate passed to the next in line, William Chambers of Bicknor, Kent.[12]

William Chambers took up residence at Llanelli shortly after Cholmondeley's death, because the Vicar of Llanelli (Ebenezer Morris) was married before him on 27 July 1827[13] and, in 1828, he was the High Sheriff of Carmarthenshire[14] and also involved in a dispute over royalties with the Pembertons.[15] So, for the first time in 55 years, the Estate's affairs were in the control of an owner who lived in Llanelli. This aspect of the management of the Estate had always been important — an agent, no matter how competent and conscientious, is seldom an adequate substitute for an active, interested owner — and was particularly so at this time of industrial change and development. (Coincidentally with Chambers's arrival in Llanelli, Richard Janion Nevill took over Raby's and Warde's collieries and the Pembertons ceased to be involved in mining). The condition in which Chambers found the one-time Stepney Mansion, Llanelly House — in disrepair and let out as lodgings and offices[16] — must have reflected that of much of the Estate. After 1830, Chambers renovated, and lived in, Llanelly House, ensured that the Estate's mineral resources were fully exploited and, until his death, at 81 years, in 1855,[17] played a significant part in the industrial development of Llanelli.

(2) *The Stradey Estate — Dame Mary Mansell; Mary Martha Ann Margaret Mansell and Edward William Richard Shewen (Mansell); Thomas Lewis; David Lewis*

When Dame Mary Mansell died c1800,[18] her daughter, Mary Martha Ann Margaret Shewen, assumed full control of the Estate and its revenue, with her husband Edward William Richard Shewen. Shortly afterwards, Shewen adopted the surname and arms of the Mansell Family and became known as E. W. R. Mansell.[19] Raby's royalty payments in the early 1800s provided a steady income to the Estate and, with the sale of detached lands at Cwmlliedi, Brynygroes, Trebeddod, Bryn, Ffosfach, Draenenbicca, Maesyrddaven, Llwynhendy, Heolhene, Drawsdre, Duffrin, Tyrbaccas and other locations — some of which Symmons had agreed to buy in 1802,[20] and which were finally purchased for some £19,000 by Symmons, Thomas Williams, John Rees and others in 1805[21] — the Mansells were able to pay off most of the family's debts. In 1806, then, the family's financial situation was considerably improved but E.W.R. Mansell did not live to benefit from it. He died suddenly, on 22 October 1806, aged 28, and was buried in the Mansells' family vault, at St Illtyd's Church, Pembrey.[22]

His widow was apparently inconsolable, confining herself to the upper chambers of Stradey Mansion and refusing to enter the dining and drawing rooms.[23] She drew up her will, on 5 February 1807, only three and a half months after her husband's death and, apart from certain family bequests, quite unexpectedly left the entire reversion and residue of the Stradey Estate to her solicitor, Thomas Lewis of Llandilo,[24] who had managed her affairs, and her husband's, since about 1804.[25] She died, of an apoplectic stroke, at the Mackworth Hotel, Swansea, on 18 January 1808, aged 40,[26] less than two years after her husband's death. Thomas Lewis proved her will on 8 March 1808[27] and took possesion of the Estate. The remaining family members of Mansell/Shewen blood unsuccessfully contested the validity of the will in 1810,[28] and the Mansell Family's association with Stradey and Llanelli, which had lasted more than 150 years, ended.

Thomas Lewis inherited an estate which was reasonably sound financially, but he did not live at Stradey, choosing to remain at Llandilo.[29] The legal disputes he had entered into on Mary Mansell's behalf now became matters of personal interest. He questioned the validity of the coal leases, granted to Townsend and then to Warde, on the grounds that Sir E. V. Mansell, 3rd Bart., was not fully responsible for his actions when he signed the leases, but failed to make a case, although he sought legal opinions on the matter as late as 1825.[30] He also took proceedings, without success, to recover royalties and rent from Alexander Raby, who was in financial trouble. As most of his coal was leased to Raby, Lewis never received the royalties he could have expected on his inheritance. The dispute was still not settled when Lewis died, on 10 March 1829. His son, David, inherited the Estate[31] and later reached an amicable settlement with the then holder of Warde's and Raby's leases, the Llanelly Copperworks Company. David Lewis also became personally involved in mining his Estate's coal after 1837.

(3) *The Golden Grove Estate — John Vaughan; John Campbell the first Lord Cawdor; John Frederick Campbell the second Lord Cawdor and first Earl Cawdor*

When John Vaughan inherited in 1781, the Estate was saddled with a mortgage of £23,000 which had its origins in the mortgage of £20,000, raised by Anne Vaughan in 1724 to pay off the Duke of Bolton's debts. John, too, lived beyond his means and, although he sold off land to pay his creditors,[32] his financial situation did not improve. The industrial situation in Llanelli in 1800 brought a prospect of increased income, as he still owned land in Llanelli and held the manorial rights. Vaughan's agent, the Rev. Thomas Beynon, and his solicitor, Thomas Lewis of Llandilo, negotiated with the industrialists to obtain the best terms for the lease of coal under Estate lands,[33] Vaughan being one of the lessors of coal (previously leased to Townsend and Smith) to George Warde in 1802.[34] But his financial situation did not improve and he died on 26 January 1804, still owing some £41,000, despite having sold 9,000 acres (3,645 hectares) of land in the previous 20 years.[35] His will, drawn up in 1786, stipulated that his close friend John Campbell, the first Lord Cawdor, was to inherit the Estate, although members of his family were still living.[36]

John Campbell of Stackpole Court, Pembrokeshire, created the first Lord Cawdor in 1796,* was a landowner and an industrialist who had been interested in the industrial potential of the Llanelli area as early as 1793, when he had considered moving his lead smelting operations from Carmarthen to the disused Pencoed Lead Works, to reduce coal transport costs.[37] Within two months of taking control of the Golden Grove Estate, he leased part of Morfa Bach Common to Alexander Raby to erect an iron forge.[38] In 1807 he supported the Llanelli Burgesses in their application for an Act of Parliament to drain, embank and lease some 600 acres (243 hectares) of marsh and common land within the Borough boundaries, although, by doing so, he would lose manorial rights over substantial lands, for which he would be compensated by being allocated one-fourteenth of the enclosed lands. His support ensured a smooth passage for what was Carmarthenshire's first Enclosure Act, in August 1807,[39] by which considerable areas of lands in and around Llanelli were reclaimed and improved and rents were raised which could be applied, by the Burgesses, to improve facilities in the Town and Port of Llanelli. He became directly involved in Llanelli's industry in 1812, when he dismantled his disused Carmarthen leadworks and had it re-erected, under Charles Nevill's supervision, at Llanelli, where coal was immediately available.[40] Campbell commenced shipping lead ore from Carmarthen to Llanelli on 19 January 1813, and smelting at the "*Llanelly Lead Works*" (at the southern end of New Dock Road) probably commenced shortly after this time.[41] The first Lord Cawdor, therefore, became more involved than any of the Estate's previous owners in Llanelli's industrial development, but his ownership of lands in the area was steadily diminishing. A second Enclosure Act of 1812, relating to lands outside the Borough boundaries, further reduced his holding.[42] He died in 1821 and his son, John Frederick Campbell, became the second Lord Cawdor.[43]

*John Campbell's grandfather, Sir Alexander Campbell of Cawdor, married into the Lort family of Stackpole Court, near Milford Haven, and his father, John Campbell, added to the family estate by marriage into the Pryse family of South Cardiganshire.

There is little record of his involvement in the Llanelli area between 1821 and 1829. He was created the first Earl Cawdor in 1827[44] and, in the same year, he supported the newly-formed Llangennech Coal Company when he leased the coal under his lands at Penllwyngwyn and Penprys to the owner of the Llangennech Estate (E. R. Tunno), who sub-leased the same coal to the Company.[45] He also granted Tunno permission to construct a railway over the foreshore, between Llangennech Quay and Pencoed, so that the Llangennech Coal Company could transport their coal down to Spitty Bank.[46] Unlike most others with vested interests in Llanelli's coal industry, he supported the directors of the Company when they applied, in 1828, for an Act of Parliament to construct a railway, from their coalfield in the Bryn/Llangennech region to a projected floating dock at Machynis Pool (later the New Dock). Although he required the directors to enter into various covenants binding them to carry out the measures specified in the Act,[47] there is little doubt that his strong support helped the Company in their early years, their eventual success in the increasingly important steam coal market contributing significantly to Llanelli's later economic growth.

(4) *The Llangennech Estate — John Symmons; the Earl of Warwick and John Vancouver; Edward Rose Tunno*

The Llangennech Estate was formed between the end of the 18th century and the beginning of the 19th century* when John Symmons of Paddington House, London, purchased lands from Sir John Stepney, Mary Martha Ann Margaret Mansell and possibly others.[48] Plans, schedules and rentals for 1819 to c1825[49] show that the Estate covered almost 4000 acres (1620 hectares), with an annual rent roll of £2,722. It extended from Mynydd Sylen in the north, to Spitty and Tyr Baccas in the south, most of the lands being concentrated in the Llangennech region, although substantial areas north west of Felinfoel were also held.[50]

Symmons purchased land in the Llanelli area to acquire a coalfield on which to establish metalworks. It seems that he originally intended to become involved personally in Llanelli's industry, as he purchased the working collieries and railroads on Stepney's Llangennech lands for £6,300[51] and was one of the original proprietors named in the Carmarthenshire Railway Act of 1802.[52] Also, in 1803, he tried to persuade John Vaughan and Sir John Stepney to transfer William Roderick's

---

* The Stepneys' lands at Llangennech (probably gained by the 5th Baronet's marriage to Eleanor, daughter of John Lloyd of Llangennech) were sometimes referred to, in the 18th century, as the Llangennech Estate. Symmons's purchased estate, although very much larger, must have taken its name from the Stepneys' original estate.

coal leases to him and, at the same time, secretly approached Roderick's partner, Thomas Bowen, and offered to buy him out.[53] Symmons failed to gain Roderick's coal leases but he continued working his own collieries in the Llangennech region. Symmons remained domiciled in London, however, and, by late 1803/early 1804, decided against retaining a personal interest in Llanelli's coal industry. He engaged Edward Martin of Morriston to assess the Llangennech Estate, who reported, in September 1803, that there were 10 million tons of unworked coal, which could be exploited for the outlay of a small amount of capital.[54] Symmons soon advertised the Llangennech Estate, for sale or letting, saying that the collieries were being worked on a small scale but could yield a "*net amount*" (profit?) of up to £15,000 per annum for an initial investment of £6—10,000; that there was a suitable site on the Estate for "*smelting houses*"; and that iron-stone was available within ten miles.[55]

Symmons's advertising soon yielded results, the Earl of Warwick and John Vancouver purchasing the Llangennech Estate before September 1804.[56] Vancouver, who put up part of the purchase price, was the managing partner, later living at Llangennech Mansion, and Warwick the financial partner, who agreed to provide most of the £70,000, the reputed price of the Estate.[57] *The Cambrian* reported in October 1804 "*considerable coal and copper works are immediately to be set on foot on the Llangennech Estate*"[58] and Vancouver soon commenced the development of the Estate's collieries and the construction of a transport system and shipping place. He obtained additional coal holdings in the Llangennech area, in January 1805, when Sir William Clayton leased him coal under Brynsheffre and Talyclyn for 21 years at a royalty of 5 shillings per wey.[59] The main colliery developments were probably the Clyngwernen pit (133), the Brynsheffre Level (142), the Glanmwrwg Level and pit (144, 143), perhaps the Gelligele pits (152, 153) and possibly other collieries on the Estate. Railways were built from these workings to the shipping place opposite the tenement of Box — previously used by William Hopkin and John Jenkin — which was enlarged and formed into Llangennech Quay (Fig. 16), Charles Nevill writing, in April 1805, that the railway system was within six weeks of completion and that Vancouver was about ready to ship coal.[60]

In addition to mining coal at Llangennech, Warwick and Vancouver also planned to become involved in the copper industry, seeking Richard Janion Nevill's opinion on the suitability of Glanmwrwg Coal (Swansea Four Feet Vein) to smelt copper. Nevill reported favourably, in May 1806,[61] and Warwick and Vancouver must have been on the point of investing even more heavily in the Llanelli area. Their appearance of

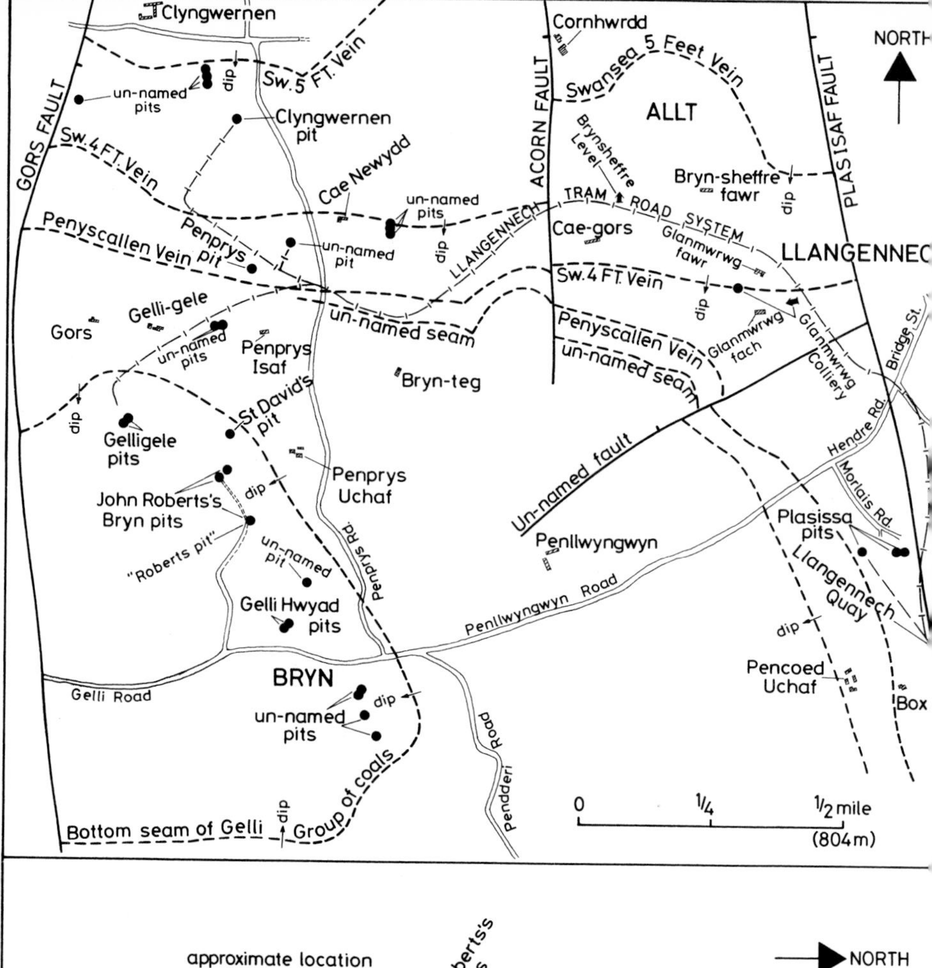

KNOWN MINING ACTIVITIES IN THE BRYN/LLANGENNECH AREA 1800 TO 1829: JOHN SYMMONS; JOHN VANCOUVER AND THE EARL OF WARWICK; DAVENPORT, MORRIS AND REES; LLANGENNECH COAL COMPANY. (ALSO MARTYN JOHN ROBERTS)

Fig. 16

prosperity was entirely illusory, however. Warwick was unable to provide the promised capital and, before the close of 1806, they had to surrender the Llangennech Estate back to Symmons.[62] Symmons later wrote "*£70,000 was bona fide given by the late Earl of Warwick for the property in question, who would have turned it to good account had not certain of his creditors covered the whole with executions and at once swept away all means of converting it to any advantage beyond that of the actual rents*".[63] The surrender was effected about November 1806.[64] It was followed by disputes, with Vancouver resisting an attempt by Symmons to sell Llangennech Park, (purchased with Vancouver's own money), and taking action against the High Sheriff of Carmarthenshire for compensation for goods, which had been sold as a result of a writ issued by the Sheriff, and which, Vancouver claimed, belonged to him and not the Earl of Warwick.[65] They lost considerable amounts of money during their two year possession of the Llangennech Estate but they commenced the large scale exploitation of the steam coal resources of the Llangennech area, which would be successfully worked in later years.

Symmons repossessed an Estate which had been greatly improved, in an industrial sense, but he obviously still did not wish to be personally involved in its industrial exploitation, because an advertisement he issued, in April 1807, stressed the Estate's potential for the establishment of "*Iron, Copper or other works of that description*". He was also careful to stress that his collieries were in full work, with coal ready for shipping at his new dock (Llangennech Quay).[66] As a result, the collieries were taken over by Richard Davenport, George Morris and William Rees, who also built a copperworks at Spitty Bank, a detached landshare forming part of the Estate, where there was an established shipping place.[67] John Symmons lent Rees, who was his agent,[68] £5,000 but Symmons was not personally involved in the partnership.[69] Davenport, Morris and Rees started to build the Spitty Copperworks in 1807/08[70] and probably formed additional shipping places there. They probably developed the collieries and may have extended the Llangennech Tramroad system down as far as Bryngwyn Bach, to convey coal overland to Llanelli.[71] Their enterprise failed and John Symmons issued a writ of bankruptcy against George Morris and William Rees in late 1813/early 1814. A sale of their one-half share of the stock-in-trade of the "*late Firm of Davenport, Morris and Rees*" was advertised in February 1814, with Richard Davenport's remaining one-half share to be sold three months later.[72]

Symmons was once again in sole possession. Immediately, but unsuccessfully, he advertised the Spitty Copperworks for letting,[73] but kept the collieries in operation, because their coal was advertised in

1815.[74] In 1817, a proposal, undoubtedly initiated by Symmons, to raise £100,000 capital by subscription to form a new company, the Cambrian Copper Company, to smelt copper ore at Spitty Bank, was published.[75] This proposal also failed and Symmons, still left in control of his Estate's industrial interests, which he did not want, decided once more to try to sell the Estate. In March 1819 he negotiated with Lewis Weston Dillwyn, owner of the Penllergaer Estate, who already possessed property in the Llanelli area and who had expressed interest in gaining extra lands. Dillwyn, however, was interested only in land purchase. He offered £48,000, which excluded his purchasing the Spitty Copperworks and Spitty Bank, and was conditional on Symmons leasing the collieries at Llangennech from him.[76] Symmons was tempted by Dillwyn's offer but was reluctant to retain any involvement in the Llanelli area, as he made plain when he wrote to Dillwyn "*my object in selling this property is principally to get rid of the trouble, perplexity and vexatious delays attendant on the distant management of complicated concerns in coal, copper etc. Encreasing in years, I have found these difficulties become less and less supportable and it was therefore that I was induced to make the sacrifices which I have made in the proposal submitted to you*".[77] Dillwyn discovered, by June 1819, that the Estate was mortgaged and that this would complicate the purchase[78] and, having come to distrust Symmons, later withdrew his offer. Dillwyn wrote of him "*His well known character added to some suspicious circumstances which I detected rendered extreme caution necessary*".[79] Dillwyn also wrote to his solicitor "*Having anything to do with Symmons is like fishing in dirty water and unless you are positively sure that he cannot put you to any expense or inconvenience respecting the Papers I should advise you to give them up. I will of course pay all the costs and have done with the old rogue for ever*".[80]

The sale to Dillwyn having fallen through, Symmons again advertised the Estate for sale, in December 1820.[81] This time he was successful, and Edward Rose Tunno's purchase of the Estate, between 1821 and May 1824,[82] ended Symmons's 20 to 25 year involvement in Llanelli's industrial life. He was mainly a speculator in industrial property, although he may have initially intended becoming personally involved in coal mining and metal smelting. If Dillwyn's assessment of Symmons was correct, then the Earl of Warwick, Vancouver, Davenport, Morris and Rees had all experienced the problem of dealing with an unscrupulous man, who foreclosed on them as soon as they ran into difficulty.

Edward Rose Tunno of Upper Brook Street, Grosvenor Square, London, like Symmons before him, remained in London and tried to attract entrepreneurs or industrialists to develop the Llangennech Estate.

He commissioned Symmons's estate manager, Richard Cort, to write a report on the advantages of "*smelting and manufacturing copper ore on commission at Spitty Bank works situated on the Burry River, on the Llangennech Estate, with coal very considerably cheaper than at any other work in Wales*",[83] which quickly produced results, Daniel Tower Shears and James Henry Shears taking over the Spitty Copperworks in 1825.[84] At the same time, with Thomas Margrave and William Ellwand, they formed "*The Llangennech Coal Company*" to exploit the Llangennech Estate's coal resources, and commenced a major sinking of a pit (154) to the Swansea Four Feet Vein on Gelli-gille Farm lands (later to be called the St David's Pit).[85] To enable the Llangennech Coal Company to complete their coalfield, Tunno obtained a lease of the coal under Penllwyn-gwyn and Penprys from John Frederick Campbell (Lord Cawdor), in June 1827, and sub-let it to them.[86] Tunno gave the Company strong backing in their efforts to develop his Estate's resources and, in 1827/28, was involved, on their behalf, in a dispute regarding a railroad constructed from Llangennech Quay to Spitty Bank.[87] He also supported the proprietors of the Company in 1828 when they, with others, sought Cawdor's backing for an Act of Parliament to construct a railway from the St David's Pit to Machynis Pool, where a floating dock was to be constructed.[88] Tunno's support in their early years helped ensure that the Company met with success in the 1830s. Llanelli benefited from this success in terms of the shipping trade it attracted and Tunno benefited in royalty payments for coal worked.

(II) THE INDIVIDUALS AND PARTNERSHIPS

(1) *William Roderick, Thomas Bowen and Margaret Griffiths (Eaton)*

As we have seen, in 1800 Roderick, Bowen and Griffiths held all the coal, extending from Llanerch to the Wern, previously granted to Roderick in 1794, and had, unsuccessfully, tried to sink a deep pit at the Bres, which they abandoned in favour of shallower mining at the Wern. Their capital outlay had been considerable but their bad start, with consequent low coal production, produced little return on their investment in the first five years, when money was scarce.

Their financial situation worsened, despite a significant increase in coal production between 1800 and 1803, when they raised 11,440 tons of coal leased by John Vaughan.[89] They were committed, as we have seen, to minimum royalty payments to Vaughan and Sir John Stepney and the money from the sale of this coal — with any money realised by selling an unknown amount of coal raised from Stepney lands — was obviously not enough to meet royalty payments, sleeping rents and to keep the collieries in production. To survive, they used the money primarily to meet running costs, paying part of the sleeping rent and royalties due with the money that was left. Vaughan's agent was prepared to postpone the payment of sleeping rent but insisted on the royalties being paid, his solicitor applying to Roderick, in November 1803, for payment of overdue royalties amounting to £444.5.3.[90] With Sir John Stepney's agent also showing increasing concern over payments, they could have failed at this time, were it not for newly-arrived English industrialists, intent on gaining the best situated coalfields for the erection of metalworks and for shipping.

As Roderick held the leases of a valuable and proven coalfield near the seashore, it was to his advantage to come to some financial arrangement with one of these new industrialists, to save him from bankruptcy. In 1803 he offered to sub-lease the "*North Crop*" (i.e. the coal from the Bres to Llanerch on the northern limb of the Llanelli Syncline) to Raby if he would agree to erect a steam pumping engine of sufficient power to dewater both the North and South Crop coal. The coal offered lay adjacent to Raby's existing coalfield and would have been a valuable acquisition, but Raby wished to gain control of all of Roderick's collieries and asked to be made a co-partner with Roderick, Bowen and Griffiths. They rejected his request and the proposal fell through, Roderick later explaining "*we were fearful of embarking in a concern with a person of so speculative a turn of mind*".[91] Symmons also hoped to gain Roderick's leases but, true to his devious nature, he did not deal with Roderick directly. Instead, he approached John Vaughan and William Hopkin, Stepney's agent, in December 1803, requesting that Roderick's leases should be transferred to him because of the continuing royalty default. Symmons said he would give Roderick and his partners £1,000 plus a half share in the South Crop coal, which would be dewatered by a steam engine he intended erecting at the Bres, adding in a letter "*In fact they will have nothing to do but land their coal and sell it at a large profit, as a remuneration for their having originally undertaken a concern far above their abilities to support, an error, to call it by no harsher term, that yearly lodges numberless individuals in prisons for life*".[92] Symmons also approached Thomas Bowen secretly and offered to purchase his interest (and probably his daughter's) in the concern.[93] Roderick, meanwhile, was considering a proposal by another new industrialist, John Pemberton, who offered to erect a powerful steam engine at the Bres, pay the arrears of sleeping rent and provide a lump sum of £1,000, in return for a sub-lease of North Crop coal. During these negotiations, Roderick became aware of Symmons's surreptitious attempts to gain his leases and immediately wrote to John Vaughan, seeking clarification on the position of the accumulating sleeping rent and royalty payments. He expressed his hope that Vaughan would postpone taking action until an agreement had been finalised with Pemberton.[94] This was hurriedly drawn up and signed, on 6 January 1804,[95] nullifying Symmons's threat to the partnership.

Encouraged by Pemberton's undertakings, the partnership re-commenced full scale operations and advertised, in June 1804, that an extension to their works had allowed a reduction in price of their coal.[96] Their new-found freedom from the demands of creditors was short-lived, however, because Pemberton failed to meet his obligations. The arrears of rent and the £1,000 were not paid and the promised steam engine was not installed. Vaughan's and Stepney's agents once more became anxious about the accumulating royalty payments and, before the close of 1804, legal action was taken against Roderick for the recovery of £3,040, owed to Stepney for sleeping rent.[97] Roderick and his partners were again in deep financial trouble, although they were to some extent sustained by the hope that John Pemberton would honour his agreement.

A fourth new industrialist now became involved in the partnership's affairs. Charles Nevill, acting on behalf of John Guest and William Savill, came to Llanelli to find a site and a suitable supply of coal for a copperworks. By November 1804, Nevill had decided to site the works in the Seaside area and negotiated with Roderick to supply coal. Roderick found that he could not guarantee the quantities required, unless Pemberton went ahead with his agreed developments,[98] which made Nevill doubt the wisdom of finalising terms, although strongly influenced by the fact that the South Crop colliery lay close to the proposed works. He was not at all impressed by Roderick and his partners, nor by the Pembertons, and explained that his uncertainty stemmed from "*the Disposition of the Proprietors* (i.e. Roderick and Bowen) *neither of whom are Men of business and the people with whom they are connected, the Pembertons, everyday appearing more in character as men of small property and merely acting in speculation*".[99] Nevill very soon realised that the partnership's financial troubles could lead to Roderick having to surrender his coal leases, and he and his son, Richard Janion Nevill, took steps to ensure that they would gain these leases, if this occurred. They offered to lend money to Roderick and his partners, to expand their concern to produce the quantities of coal required by the copperworks, and drew up an agreement for the supply of 200,000 tons of Rosy Vein coal over 25 years.[100] Edward Martin of Morriston was engaged to survey the South Crop colliery and estimate the cost of developing it on a large scale.[101]

The Nevills were not acting philanthropically, however, their actions being motivated by the prospect of gaining Rodericks' leases. They approached Sir John Stepney's agent, William Hopkin, and obtained an assurance that they would be granted Roderick's lease if he was forced to surrender it. R. J. Nevill's letter to the secretary of the proposed Copperworks Company, on 12 December 1804, leaves little doubt that a take-over of Roderick, Bowen and Griffiths's interests was planned: "*the Colliery of Roderick and Company is under lease from Sir John Stepney and it is stipulated that should any circumstances occur to dispossess them that in such case we shall have liberty to take a lease of the said colliery ... It affords me very considerable Gratification to learn that you approve of the Agreement made with Roderick and Bowen and that our opinion respecting the advantages of having the Ground for our buildings totally distinct from their Colliery and the probability of R & Co's lease falling in our hands so fully meet your ideas*".[102] The agreement for the supply of Rosy Vein coal was finalised on 26 January 1805, £500 being lent to Roderick and his partners, to undertake repairs and improvements in their colliery, sink a new pit and construct a railway from the Wern to the Copperworks.[103] The agreement also stipulated that Nevill would have the right to take possession of Roderick's collieries if the agreed coal supply to the copperworks fell into regular default. In March 1805, William Hopkin, who kept the Nevills informed of Roderick's affairs, expressed the opinion that the partnership would be unable to provide the quantities of coal specified,[104] thus providing an early opportunity for dispossession. This prospect disappeared when Edward Martin completed his survey and reported, in April 1805, that there was sufficient coal available for the contract to be met.[105] The Nevills realised that their take-over of Roderick's interests would not be the quick affair they had envisaged, so they concentrated on erecting and commissioning their copperworks, which began smelting in September 1805.[106] The partnership was now committed to supply the coal specified in the agreement and default would provide the Nevills with the excuse to exert further pressure.

The partnership's debts were steadily mounting, as they failed to meet sleeping rent and royalty payments and, by October 1805, they owed Sir John Stepney £1,800, of which payment of £1,000 was to be immediately enforced; a similar sum was owed to Lord Cawdor. Charles Nevill observed that a split was apparent in the partnership, with Thomas

Bowen and Margaret Griffiths (by now Margaret Eaton)* wishing to dispose of their one-half share.[107] By January 1806 Charles Nevill was confident that the Llanelly Copperworks Company was about to gain Roderick's leases and wrote "*Roderick & Co are going on in the same way, rather worse than better, there cannot be a doubt but that 'ere long they must be compelled to relinquish their undertaking and then we shall have everything we could wish for*".[108] Roderick and his partners made one last attempt to remain solvent, in May 1806, when they offered to sell a one-third share in their collieries to Nevill, but they were approaching the wrong man and he declined.[109] Bowen and his daughter decided to advertise their one-half share in the colliery for sale, in August 1806,[110] Bowen informing Nevill that it had been placed to discover whether Roderick (whom Bowen described as "*a very needy man*") had involved partnership property in his private affairs. He also assured Nevill that the Llanelly Copperworks Company would be given the first chance to purchase his and his daughter's share if a sale took place, although Alexander Raby, John Pemberton and another un-named person had already approached him.[111]

Bowen, with 50 years experience in the industry, now showed that he had as much business acumen as any English industrialist. He realised that the Copperworks Company constituted the greatest threat to any chance he and his partners might have of emerging from their ill-fated venture without suffering bankruptcy, and, while apparently co-operating with Nevill, he was actually intending to sell to others, who were not so intent on driving him out of business. He asked Nevill a very high price for his and his daughter's half share, which Nevill, as Bowen had calculated, refused to pay, deciding instead to wait for the reduction in price, which he felt must come, as the partnership's position worsened.[112] Despite his own questionable conduct in the events leading up to the sale, Nevill was indignant about the whole affair and wrote disparagingly of Bowen "*he has not behaved with honour or policy in the Business having given the promise of a decided preference and then asking a higher price than what he at the same time gave to others*".[113] Although the partnership was unable to pay their workmen by November 1806,[114] Bowen had bought sufficient time to ensure his escape with money in his hand. He sold his and his daughter's one-half share in the South Crop colliery to John Pemberton, in January 1807, for £2,500 which, after deduction of liabilities, left £500 for him and his daughter.[115] Roderick was probably involved in the deception of Nevill because, just one month after the sale,

---

* Margaret Griffiths, Bowen's widowed daughter, was married, a second time, to Henry Eaton on 20 September 1805. (Llanelly Parish Church by Arthur Mee, p. 54).

he combined with Bowen and his daughter in an agreement with Pemberton, which modified and clarified their previous agreement of 1804.[116] This agreement, combined with Bowen's sale, gave John Pemberton (who represented the Pemberton Family of the north of England) a sub-lease of all the North Crop coal and the collieries upon it, a half-share in the collieries on the South Crop coal and an interest in the Wern Canal, the Wern Railway and the partnership's dock. The Pembertons immediately began to install a large steam engine at the Bres pit and started to exploit the North Crop coal.[117]

Roderick was left with the South Crop coal and a half-share in the collieries upon it. It is feasible that he could have remained in business, because his relationship with the Pembertons was good and his workings were soon to be dewatered by their steam engine, but the Copperworks Company took swift action to ensure that everything would not be lost, by offering to buy Roderick's remaining interests. Although heavily in debt, Roderick was able to dictate terms, because the Company was afraid of losing out completely to the Pembertons. He named a fair price but also stipulated that the Company should pay the arrears of sleeping rent and royalties due to Stepney and Cawdor. His terms were accepted and Roderick, with Bowen and his daughter, who must have still been financially involved with him, sub-leased the South Crop coal to the Company, and assigned them the collieries, and their associated transport systems and shipping place, in two agreements of 11 May and 21 May 1807.[118] The collieries, transport systems and dock were valued at £4,003, the "*present interest*" of Roderick and his partners in the coal leases at £2,500, and the rents owed to Stepney and Cawdor came to £1,847.[119] Roderick, Bowen and Eaton thus ceased to be associated with Llanelli's coal mining industry and must have been greatly relieved to escape with so much in hand, having faced bankruptcy a few months previously. Roderick was still in debt, due to his involvement in other failed enterprises, and was declared a bankrupt on 24 May 1808.[120] He was not ruined, however, because he continued to live at Bradbury Hall (a large house once situated near the York Hotel) until his death, on 10 May 1823, aged 85 or 87.[121] He also retained control of the leases from Stepney and Vaughan till he died, and it is quite possible that the Pembertons and the Copperworks Company helped to maintain him after his bankruptcy, to avoid the risk of the leases being surrendered and regranted to other industrialists. Bowen died at Llanelli, on 7 May 1809, aged 94,[122] having been associated with Llanelli's coal industry for over 60 years. He had unrivalled knowledge of the Coalfield, and was an important link between the first real industrialists of the 1750s/1760s and the new industrialists of the late 1790s/early 1800s.

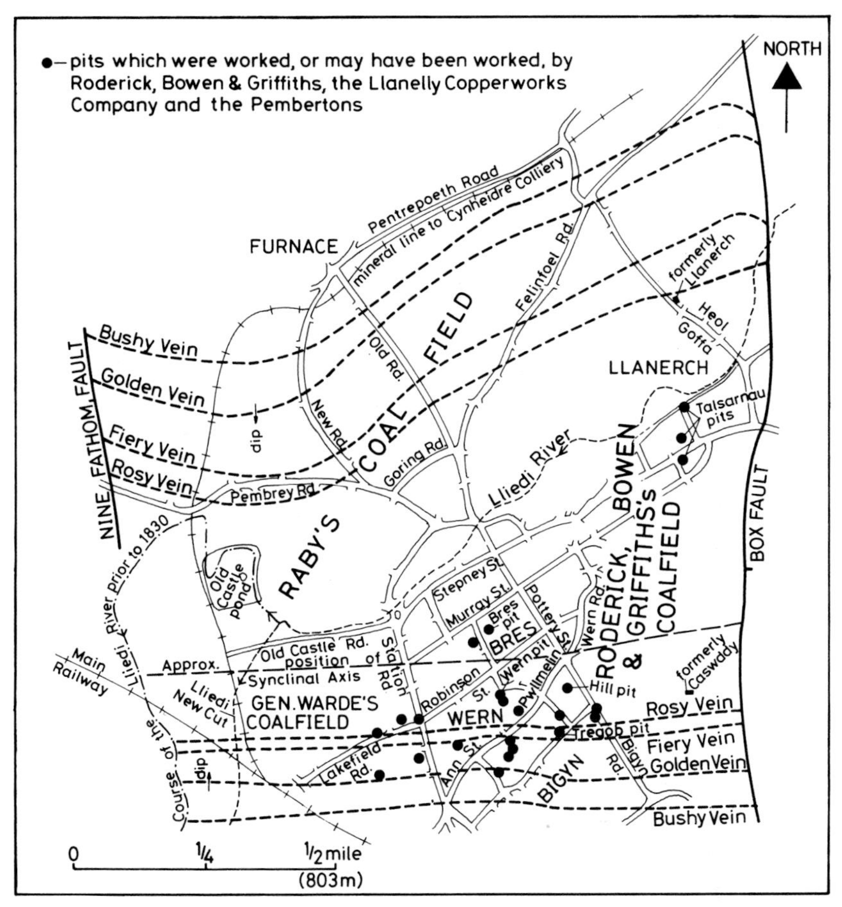

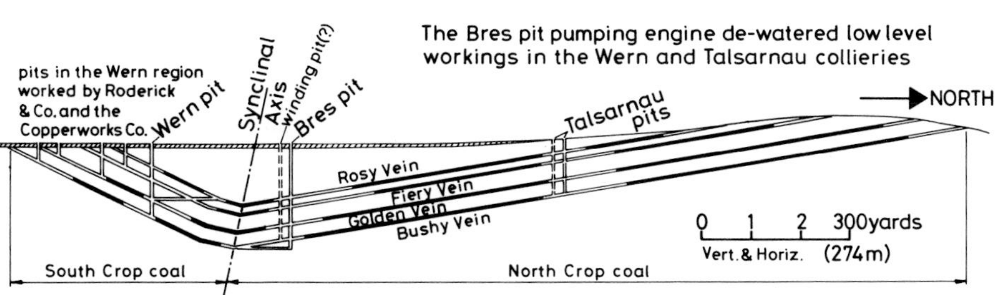

MINING ACTIVITIES OF RODERICK, BOWEN & GRIFFITHS, THE LLANELLY COPPERWORKS COMPANY AND THE PEMBERTONS TO THE WEST OF THE BOX FAULT 1800–1829

17

In 13 years, Roderick, Bowen and Griffiths sank a number of new pits, and re-opened others previously worked by John Allen and perhaps by Thomas Bowen himself (Fig. 17). It is known that they commenced the sinking of the Bres Fawr (or Bres) pit (49), sank the Wern (or Engine) pit (73), the Hill pit (82) and probably the Tregob (or Bigyn) pit (80). They either sank or re-opened Pwll Cae Pulpud (the Cae Pulpud pit) (72), Pwll-ycornel (the Corner pit) (71?), and Pwll Melyn (the Yellow pit) (75). They also sank the Meadow pit (not located) and could have worked a number of the un-named pits in the Wern/Bigyn region, along the southern outcrop of the Rosy, Fiery, Golden and Bushy Veins, to the west of the Box Fault, shown on the plan in Appendix I. (Plate 16)

They never recovered from the losses incurred when trying to sink the Bres pit to coal, and their financial problems attracted the predatory attention of the new industrialists. They emerged from their difficulties relatively unscathed because of Bowen's shrewdness, their case history providing a well-documented example of the attitudes and outlooks which prevailed in early 19th century Llanelli, when money, rivalry and intrigue overshadowed most ethical considerations.

(2) *Alexander and Arthur Raby*

Intent on the establishment of his iron and coal empire on Stradey Estate lands, Alexander Raby rapidly expanded his operations and, in 1800, he was enlarging Givers and Ingman's iron works at Cwmddyche, sinking major new pits and constructing transport systems and a dock. By this time, he had undoubtedly committed the bulk of his considerable capital to the development of the Llanelli Coalfield. Raby's coal holdings consisted of the north crop of the Rosy, Fiery, Golden and Bushy Veins, west of the Lliedi River. Almost all of this coal had been leased to him by Dame Mary Mansell, the Bushy Vein coal being called the Caereithin Colliery, and the Rosy, Fiery and Golden Vein coal being called the Shipping Colliery.

Raby's mining complex depended mainly on a large pumping engine at the Caemain Colliery (41 to 45). It could dewater all the Shipping Colliery workings back up to the northern outcrop of the seams, and first operated in late 1799/early 1800.[123] He initially concentrated his efforts into mining the coal granted by the Shipping Colliery lease, raising some 23,000 tons from the Fiery and Golden Veins[124] between Christmas 1799 and Christmas 1802. Little coal was extracted from the Bushy Vein in this period.[125] Although the income from the sale of this coal would have exceeded the minimum agreed royalty payments, Raby could have made no profit in the first six years of operations. This probably did not worry

Plate 16 — Sites of Roderick, Bowen and Griffiths's collieries. (Top) Wern Colliery. The site is now occupied by Waddle Engineering and Fan Co. Ltd. with the Engine or Wern pit located inside the works compound. (Bottom) Bigyn or Tregob pit. The grassed mound is the recently landscaped tip to the colliery. The pit was located just above, in Bigyn Park Terrace.

him unduly. He was still implementing his plans and he was also a principal participant, probably the prime mover, in the establishment of the Carmarthenshire Railway Company, which was empowered by an Act of Parliament, in 1802, to build a railway, from Llanelli Flats to Castell-y-Garreg near Llandybie, to develop the mineral resources of the Carmarthenshire hinterland.[126] Raby's influence was clearly evident in the Act, which stipulated that the proprietors of the Railway were to purchase his main railway line, from the Stradey Furnaces to Llanelli Flats, and his shipping place, which was to be enlarged into a major dock. Additionally, the railway was to run to an area with an abundance of ironstone and limestone which Raby, primarily an ironmaster, required for his furnaces.

Raby was, therefore, expanding, engrossing all the industrial interests which became available in the Llanelli area. He quickly took control of the Carmarthenshire Railway Company's development at Llanelli by securing 138 of its 250 shares by personal holdings, and by proxy holdings whose votes he could control, when it only required 126 shares to constitute a General Assembly of the Company, which could take decisions.[127] He tried, unsuccessfully, to gain Roderick, Bowen and Griffiths's collieries in 1803[128] but took steps to expand his iron interests, in May 1804, when he leased part of Morfa Bach Common from Lord Cawdor to build a new iron forge.[129] In late 1805/early 1806 he negotiated for a major new coalfield, east of the Box Fault, and finalised the transaction, in September 1806, when he was leased the coal under "*Llanlliedy, Westva, Wainllan yr Avon late Badgers, Llanlliedy Vach, Bryngwyn Mawr, Bryngwyn Fach and Box*", by Lucinda Hayton for 80 years.[130] This coalfield, extending from Felinfoel/Dafen to the Box, was of comparable size to his field west of the Box Fault. Raby immediately commenced sinking the first Box Colliery pits (97, 98) and the construction of a railroad, just over 1 mile (1.6km) long, from the pits to the Carmarthenshire Railway[131] (Fig. 18). Raby's existing coal and iron interests on the Stradey Estate were now beginning to repay his heavy investment. Armaments and ammunition for the Napoleonic wars had kept his iron furnaces at work day and night,[132] and he had raised 66,000 tons of coal in the three years between Christmas 1802 and Christmas 1805,[133] so money from sales would have exceeded sleeping rents and running costs.

If this improvement had been maintained, he would have been successful, but problems of his own making, from the effects of which he never fully recovered, caught up with him in late 1806. Raby controlled the Carmarthenshire Railway Company's development but misused this control to further his own interest, rather than that of the Company, which was

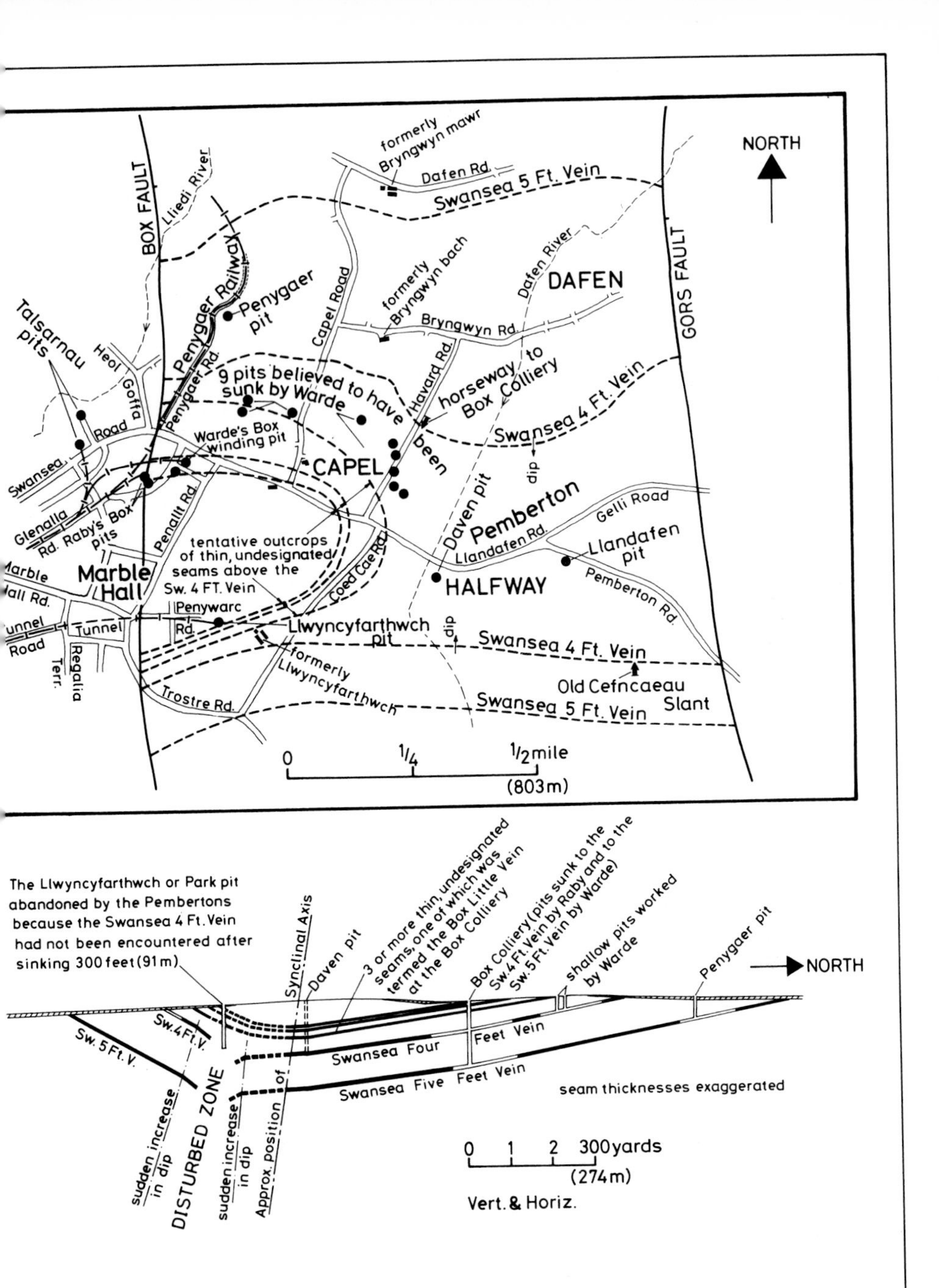

COAL MINING ACTIVITIES OF ALEXANDER RABY, GENERAL WARDE AND THE PEMBERTONS BETWEEN THE BOX AND GORS FAULTS, 1800 TO 1829

8

the development of the Carmarthenshire hinterland. He had constructed unauthorised branch lines to all his works and interests and had avoided payment of tolls, with the result that all the share and loan capital had been used up before the line reached Castell-y-Garreg. No dividend had been paid, and dissatisfied shareholders convened a General Assembly of the Company, in August 1806, which appointed a sub-committee to survey the railway and its branch lines.[134] The sub-committee's findings exposed Raby's malpractices[135] and precipitated a crisis in his affairs. He had borrowed extensively, on the security of his lands and properties,* to finance his industrial activities. His creditors, fearing the loss of their money, began to press for repayment and Thomas Lewis, solicitor to the Stradey Estate, also took action to recover £1,000 owed for rents.[136] Faced with these demands, Raby had no alternative but to convene a meeting of his creditors, to avoid bankruptcy. He was compelled to assign his personal estates to his principal creditors, on 10 June 1807, so that they could be sold to reduce his debts and also had to transfer his leasehold coal and metal interests at Llanelli, Swansea and Neath to trustees for sale, if and when it became necessary. The trustees agreed to pay workmen's wages up to specified limits, and to maintain the ironworks and collieries, but allowed Raby to keep control of his industrial interests.[137]

Raby had survived his first crisis at Llanelli but there was to be little respite from his creditors. Thomas Lewis served Raby's trustees with a notice of ejection from the Stradey Estate in September 1807,[138] Raby counterclaiming that his coal output was low because the seams in the Shipping Colliery were not as thick as those described when he took the 1798 lease.[139] He developed the Box Colliery, however, raising coal in November 1808.[140] The pits were sunk to the nine feet (2.7m) thick Swansea Four Feet Vein and he must have had great hopes of high profits or he would not have continued the colliery's development throughout 1807 and 1808, years of grave financial crisis for him. Any optimism on Raby's part soon disappeared. The sale of his personal estates provided insufficient money to meet his debts and, as a result, the trustees advertised his industrial and leasehold interests at Llanelli, Neath and Briton Ferry for sale by auction, at Garraway's Coffee House, Cornhill, London, on 16 November 1809. The Llanelli interests listed in the sale were the Box Colliery, the Llanelly Colliery (Raby's Stradey Estate collieries), two iron furnaces, a "*newly erected Forge and Rolling Mill*", managers' and

---

* Mansel Lewis London Collection 3 (10 June 1807) listed Raby's estates in that year as:- properties in Bush Lane and Cannon Street, London; Old Swinford in Worcestershire (11 acres); Castle Moreton in Worcestershire; 8 messuages in Cobham and lands at Chertsey, Surrey. Mee says that Raby sold Cobham Park in Surrey to raise capital for his developments at Llanelli. (Llanelly Parish Church by A. Mee p.xxxi).

Plate 17 — The Box Colliery. (Top) The Box Colliery c1820 (from a painting by an unknown artist). The painting depicts Alexander Raby's original colliery with its closely situated pumping and winding pits. (Bottom) The present-day site of the Box Colliery — until recently the Myrtle Hill Dairy.

workmens' houses, iron railways, steam engines, and the leasehold of the farms of "*Kille and Kilveg, Cae Mawr and Trebeddod*".[141] All Raby's possessions at Llanelli were being sold and he was again on the verge of bankruptcy. (Plate 17)

This time, he was saved by Charles and Richard Janion Nevill of the Llanelly Copperworks Company. The Nevills advanced Raby a loan of £5,000, against a mortgage on the Box Colliery lease, and agreed to provide him with an income of £200 a year until the £5,000 plus interest had been repaid, when Raby would regain a one-third interest in the concern. To disassociate themselves and the Company from Raby's general debts, the Nevills used Thomas Hill Cox and Stephen Jones as intermediaries and it was agreed, in April 1810, to assign the mortgage on the lease to Cox and Jones in return for a loan of £5,000, supplied by Messrs Waters, bankers at Carmarthen, acting under instructions from Nevill.[142] As Cox was Raby's wife's brother, and had been involved with Raby in a copperworks on the Neath River in the early 1790s, the arrangement was obviously of mutual benefit to all concerned. The assignment was completed in October 1810 between Raby's trustees, Raby, Cox and Jones and Rowland Wimburn (probably a financier).[143]

Raby's industrial and leasehold interests on Stradey Estate lands remained unsold and were re-advertised in December 1810, and again in March 1811.[144] The latter advertisement showed that the trustees were determined to end the long drawn-out affair of Raby's debts, because it stated that the collieries, ironworks and farms were to be "*Peremptorily Sold by Auction*" on 10 April 1811, but once again Raby was saved at the last moment. On 4 May 1811, the creditors and debtors of his Llanelli concerns were invited to contact Stephen Jones "*in consequence of an arrangement about to take place between Mr Alexander Raby of Llanelly, in the County of Carmarthen, and his Trustees, for the Sale of the Llanelly Works, Colliery and Property*" and it was also announced that "*the extensive Iron-works at Llanelly will be immediately resumed and carried on by Alexander Raby, Esq*".[145] This sudden change in Raby's fortunes was undoubtedly due to intervention by R. J. Nevill, although no documentary confirmation of this has been seen.[146] In July 1811, Raby informed coal dealers and ships' masters that he was once more actively engaged in Llanelli's coal trade and, for the first time, reference was made to his son, Arthur Turnour Raby, being in partnership with him.[147] The Rabys enjoyed only a brief respite from their financial troubles, because a D. Morgan of Neath announced, in March 1812, that a meeting of Alexander Raby's creditors was to be held at the Mackworth Arms, Swansea, on 13 April 1812.[148] Raby successfully challenged the legality of this meeting[149] but ran into further problems, in July 1812, when David Lewis (Thomas Lewis's son) took legal action at Hereford Spring Assizes for the recovery of arrears of rent and wayleave payments, owed to the Stradey Estate.[150] The Rabys must have had the backing of the trustees and that of R.J. Nevill and the Copperworks Company, because no crisis resulted, and Alexander Raby even subscribed £300 towards the new Kidwelly and Llanelly Canal Company, set up under an Act of Parliament of June 1812.[151] The Copperworks Company's association with the Rabys was openly confirmed by an advertisement of January 1813, which invited colliers to apply for employment at the Wern Colliery (owned by the Copperworks Company) and the Caemain Colliery (owned by the Rabys),[152] and agreements for the Rabys to supply coal to the Copperworks were reached, in May 1814 and March 1815.[153] But, as in the case of Roderick, Bowen and Griffiths, the Company was not acting philanthropically: there is little doubt that R.J. Nevill was preparing the ground to take the Rabys' affairs over when the time seemed right.

In the meantime, Alexander Raby was in dispute over the Box Colliery and applied to the Court of Chancery, in November 1814, for an injunction to stop it working. Raby maintained that he had not been paid his £200 a year, although the colliery had been run at a profit since Cox and Jones took it over in 1810. Furthermore, his right to regain a one-third share in the concern had been extinguished when Cox and Jones sold the colliery to General Warde in 1813,[154] and Raby wanted compensation. He obtained his injunction and the Box Colliery was stopped. It did not re-open until July 1815, when agreement was reached.[155]

The Rabys were fighting a rearguard action, however, and Nevill's intervention had only warded off imminent bankruptcy. Their iron-works had been failing before 1808[156] and the furnaces were apparently blown-out in 1815, although the forge still operated.[157] Their collieries on Stradey Estate lands were also in a depressed state by 1815[158] and the proprietors of the Copperworks Company, aware of the Rabys' situation, instructed Nevill, in May 1815, to reduce the debts the Rabys owed to them "*as speedily as may to him appear advisable*".[159] The Rabys' trustees also realised that there was little prospect of creditors being paid out of profits and arranged a sale of their effects. This did not suit Nevill, who wished to acquire the Rabys' industrial interests, and he persuaded the trustees to postpone the sale, to get a higher price. On 27 January 1816, the parties agreed that the Rabys would transfer overall control of the collieries to appointed trustees acting for the creditors, with Alexander and Arthur being retained at annual salaries of £500 and £200 respectively.[160] Having bought time, Nevill acted swiftly to secure the Rabys' concerns and, within one month of the agreement, purchased the Rabys' collieries and goods through a writ of fieri facias,* executed by the High Sheriff of Carmarthenshire.[161] The day the purchase was finalised, Nevill agreed that the Rabys should continue to control their collieries.[162] Nevill undoubtedly deceived the Rabys' trustees to gain the Stradey Estate collieries, and allowed the Rabys to retain control of their interests because they still held the Caereithin and Shipping Colliery leases, and because Nevill was careful to dis-associate himself and the Copperworks Company from the Rabys' debts.

From 1816, then, the Rabys were in a curious situation. They still owed money to various creditors and their collieries were owned by, or mortgaged to, the Llanelly Copperworks Company, but they still held the 1797 and 1798 coal leases and were in charge of all the mining operations. The Company also acted as a bulwark against their creditors and put up capital to improve the run-down Stradey Estate collieries. Most of the investment was probably put into the Caemain Colliery, with the result that Fiery Vein coal, which had not been worked there since 1813, was advertised for sale in January 1819.[163] Alexander Raby took new heart from his changed circumstances and sought to enlarge his coalfield, in May 1819, when the Commissioners of His Majesty's Woods, Forests and Land Revenues agreed to lease him the Rosy, Fiery, Golden and Bushy Veins, under the seashore west of Llanelli between high and low water marks.[164] Raby was required to sink a steam engine pit to cut all four seams in the region of the synclinal axis, within 15 years of the start of the agreement, and must have intended extending his existing coalfield southwards.** The Copperworks Company must have supported Raby in this agreement, as he subsequently confirmed that the seashore area, occupied by the copperworks, would be assigned to the Company if he were successful in his application for the lease.[165]

But the Rabys' financial troubles continued. In February 1821 Thomas Lewis, owner of the Stradey Estate, had the plant and machinery at the Caemain Colliery valued under a distress for rent,[166] and the Rabys' workmen obtained a magistrate's order for the recovery of unpaid wages.[167] This new crisis caused Nevill to draw up a formal agreement, in March 1821, by which Nevill took over the superintendence of the collieries for 3 years, at a personal stipend of £200 per annum. Nevill was to have "*the exclusive management and control of the whole of the Property, Foundry, Cottages etc*" and sole discretion in using the colliery profits, to reduce

* A writ commanding the Sheriff to execute the judgement by raising the sum payable out of the goods and chattels of the defendant.

** Raby did not commence this development but it was implemented by the Copperworks Company in 1839 with the sinking of the Old Castle Colliery pits.

outstanding debts in his chosen order of priority. Nevill agreed to advance money to satisfy immediate creditors and to pay the Rabys whatever allowance he thought fit (not to exceed £500 per year), maintain the two houses occupied by them (the Dell and Caemawr Cottage), supply coal for their fires and provide for the keep of one horse.[168] Arthur Raby decided to leave Llanelli at this time, because a sale of his effects at Caemawr Cottage was advertised, in March 1821, when it was stated that he intended leaving Wales,[169] but he changed his mind and continued in partnership with his father.

Afforded temporary respite yet again, the Rabys continued working their collieries (under R. J. Nevill's management) and some 22,000 tons of coal were raised under the Caereithin and Shipping leases in 1821.[170] This output fell to 17,000 tons in 1822[171] and Thomas Lewis, dis-satisfied with his royalties, authorised a surveyor to enter the Caemain Colliery in September 1822 to "*ascertain the way the said coal or culm is carrying on*".[172] The Rabys also found it difficult to sell coal in a depressed economic period[173] and bad weather during the 1823 coal shipping season further affected sales, by impeding the coal trade vessels.[174] The Copperworks Company agreed to take extra coal to reduce their stocks, but at 2 shillings per wey below the previous agreed price, and the Rabys were unable to meet royalty payments, due to Thomas Lewis on Ladyday 1823.[175] As 1823 ended, the Rabys owed £13,000[176] and the 77 year old Alexander decided, or else was prevailed upon, to relinquish his interests at Llanelli in favour of Arthur. By an agreement of 1 December 1823, Alexander assigned his leases and interests to Arthur for a payment of £200, plus £25 per month for his life, or that of his wife, if she survived him.[177] Nevill undoubtedly arranged this transaction because, on the same day that Alexander and Arthur signed their agreement, Nevill was involved in two other complementary agreements with Charles Druce (the Copperworks Company's London solicitor) and George Haynes (a Swansea banker). Firstly, George Haynes agreed to provide immediate financial aid to the Rabys, with Druce and Nevill agreeing that the Rabys' debts could be discharged over a period of time by supplying coal to the copperworks at reduced prices (the Rabys owed Druce £1,897, Nevill £9,762, Druce and Nevill together £800 and Messrs. Waters and Co., bankers at Carmarthen, £878).[178] Secondly, George Haynes agreed to provide continuing financial assistance, to enable Arthur Raby to carry on the collieries being assigned to him by his father. It was stipulated that Arthur would have to operate subject to the advice of Nevill and Haynes, but would be paid a salary of £250 per year, in monthly instalments, plus the money he had agreed to pay his father.[179] Alexander Raby's 27 year involvement in the iron and coal industries at Llanelli, therefore, came to an end on 30 November 1823. He subsequently left the area and lived at Burcott House, near Wells, Somerset, where he died in February or March 1835, aged 88 or 89.[180]

Arthur Raby continued to mine the coal granted by the Caereithin and Shipping Colliery leases, and even sank a new pit on lands belonging to Kille Farm, in 1824.[181] His father's forge had been idle since about 1820[182] and he leased the freehold lands known as "*Forge*" (behind Park View Terrace and Raby Street, where the forge was once situated) to R. J. Nevill and John Roberts the younger, in February 1825, and agreed to supply Bushy Vein coal to the forge, which Nevill and Roberts intended re-opening and expanding. Raby agreed to provide the coal from the Pwllydrisiad pit (not located), £1,000 being placed at his disposal to sink a new pit or to drive a new level to achieve a greater winning of coal from the Bushy Vein.[183] He also obtained a sub-lease of the Golden Vein, under Old Castle Farm lands near the Iron Bridge, from General Warde in March 1825[184] and must have intended extending the workings from the Caebad Colliery.

Arthur Raby was, therefore, attempting to expand his coal interests but any ambitions he may have had of rebuilding his father's industrial empire and paying its debts soon disappeared, when R. J. Nevill and Charles Druce decided that the time was right for a complete take-over of his collieries. On 19 May 1826, Nevill and Druce agreed that a writ of fieri facias, relating to debts owed to Nevill, be used to direct the High Sheriff of Carmarthenshire to assign Raby's collieries to James Guthrie (manager of Raby's collieries on the Stradey Estate) and Nevill Broom (R. J. Nevill's partner in a timber business at Llanelli), to hold in trust for Nevill and Druce, with Nevill paying off the concern's debts and undertaking the management of the collieries.[185] Raby could not resist their proposals and probably co-operated willingly to escape bankruptcy. His property at Llanelli was advertised for sale, on 17 June 1826, when it was stated that he was going to live on the Continent[186] and, on 8 July 1826, Raby empowered Guthrie and Broom to manage his affairs "*during his absence*".[187] The writ of fieri facias was executed and the collieries and effects of Alexander and Arthur Raby were sold to Guthrie and Broom, on 11 July 1826, for £2,995.[188] Arthur Raby cancelled the agreement for the lease of Forge on 14 July 1826,[189] and must have left Llanelli shortly after, because Caemawr Cottage was advertised for letting, on 23 September 1826, when it was described as "*late the residence of Arthur T. Raby*".[190] By October 1826, James Guthrie was requesting that receipts be written out to "*James Guthrie and Co., late A. & A. Raby*"[191] and the Rabys' direct interest in Llanelli's industry had come to an end. Arthur Raby did not leave for the Continent immediately, because Nevill wrote to him, in March and May 1827, regarding his intention to go abroad to avoid his creditors,[192] but he did finally leave Britain, because he was living in Germany in 1856.[193] The complicated procedures adopted by Nevill in the final take-over were due to his fears of becoming involved in the Rabys' general debts. His caution was justified because, as late as 1849, unpaid creditors were trying to prove that he was responsible for certain of the debts, through his ownership and management of the Rabys' collieries.[194]

Between them, Alexander and Arthur Raby had been involved in Llanelli's iron and coal industries for 30 years and had invested, and lost, a fortune, perhaps £175,000,[195] exploiting the area's resources. Alexander Raby's belief in the financial potential of the Llanelli Coalfield and the commitment of his entire wealth to its development, undoubtedly attracted many of the early 19th century entrepreneurs and capitalists. In this respect, the traditionally held view that Raby laid the foundations of modern industrial Llanelli[196] is, to some extent, correct. Raby remains an enigmatic figure, however, and he, and his son Arthur, only survived after 1809 because of Richard Janion Nevill's welcome, but predatory, intervention. By any standards, the Rabys were unsuccessful and their numerous financial crises inhibited the development of the Stradey Estate coal, with a resulting detrimental effect upon Llanelli's coal industry. Additionally, Alexander Raby's questionable use of the Carmarthenshire Railway Company's funds, in the early years of its existence, must have contributed to the venture's lack of success, with the result that the planned exploitation of the mineral resources of the Carmarthenshire hinterland was delayed for many years.

It is not possible to locate all the pits and adits worked by the Rabys because working plans of their collieries have apparently not survived. Nevertheless, it is known that they inherited and worked a number of Dame Mary Mansell's collieries (See Chapter 3), although positive proof of this exists only for the Caerhalen pit (32), the Old Slip (23) and the water-engine pit at Caemain (46), which was incorporated into the Caemain Colliery complex. They also opened a number of new pits and adits, some of which were provided with steam engines.* These new sinkings were:- the Caemain Colliery (one or more of entries 41 to 46), Caebad pit (38), Caerelms pit (36), un-named pit near Dulais Mill (4), New Slip (probably 24), Box Colliery original pits (97, 98), Field pit (not located)

* At least 5 steam engines were provided at their collieries — one at Caemain, one at Caebad, two at the Box and one at an undetermined site but possibly Engine Fach (35). One steam engine was also provided at the Furnaces and two at the Forge.[197]

and the Pwllydrisiad pits (not located). They also used a horseway or adit (34), probably to take horses into the Caemain Colliery workings, and many unidentified pits or adits on the northern limb of the Llanelli Syncline, to the west of the Box Fault, may have been worked by them.

(3) *Charles and Henry Smith; General George Warde, Francis and James Morgan*

In 1800 Charles and Henry Smith held the extensive coalfield at Llanelli, previously leased to their grandfather Chauncey Townsend and their father John Smith, who had worked it to a minor extent only. From 1797, Charles and Henry Smith were inactive at their inherited collieries, thus sterilising a large part of the Llanelli Coalfield, and the local landowners were eager to obtain a surrender of the leases and to grant the coal to the new industrialists, who were ready to develop the area's resources.

Negotiations relating to the Smiths' leases must have been under way by 1801/02, at the latest, because George Warde of Bradfield, Berkshire and Francis and James Morgan of Whitechapel, were granted all the Smiths' coalfield, with other areas, by a series of leases and assignments between October and December 1802. On 2 October, Sir John Stepney, John Vaughan and Jane Langdon (inheritors of the Llanelli Vaughans' original Estate) granted all the coal under their lands, previously leased to Chauncey Townsend and John Smith in 1752 and 1785, to Warde and the Morgans, for 99 years. Also included were the Great and Fiery Veins (Swansea Four and Five Feet Veins) under Ffosfach, which had not previously been leased.[198] The same day, Sir John Stepney also granted to Warde and the Morgans, for 99 years, coal under Pencoed and Ffosfach, which he personally owned, and which was not included in the coal and wood commonality clause of the 1705 Partition Deed, and the Old Lead House at Pencoed.[199] These leases set out fully their developments and their undertaking, within two years, to erect a steam engine near Pencoed for draining their workings, to construct a copper or lead furnace at Pencoed and to commence a canal from Penrhyngwyn to a pit at "*Daven*" (the present Halfway). Both grants were made subject to the surrender or assignment of the original leases, which were made on 15 and 17 December 1802.[200] The leases they gained were encumbered, as some of Chauncey Townsend's and Daniel Shewen's descendants still held interests, which had to be acknowledged. Accordingly, on 17 December 1802, it was agreed to pay the Rev. Joseph Townsend and Thomas and Sarah Biddulph 8 pence per wey for coal raised in the parishes of Llanelli and Pembrey,[201] and to pay Anna Maria, Martha and Mary Shewen and Rose Thomas 1s 3d per wey for coal raised, which had previously been leased to Daniel Shewen in 1758 and 1759.[202]

They engaged Rhys Jones of Loughor as their colliery agent and quickly set to work to develop their extensive coalfield, initially concentrating on opening inherited pits or levels, and working coal above the water table, while work proceeded on the major task of installing steam engines and sinking new pits. Coal was being worked at Ffosfach by October 1803, and at Pencoed and Genwen before the close of 1804.[203] Warde, a Colonel in the Army and Inspecting Field Officer of the Severn District, was promoted Brigadier General in June 1804,[204] and, subsequently, was invariably referred to as General Warde in all contemporary documents and correspondence. He became the sole holder of the coal leases, in November 1804, when the Morgans, who had been financial sleeping partners, assigned their interests to Warde.[205]

Although he began to exploit his coalfield with the re-opening of some small pits and levels, Warde's initial strategy was based on large scale developments at Dafen Bridge (Halfway), Genwen and Erwfawr, which were near convenient shipping places. He planned: 1) to sink the Daven pit (119) at Halfway (started by John Smith c1786) to the Swansea Four Feet Vein, install a steam engine and construct a canal to a new shipping place at Penrhyngwyn; 2) to replace Chauncey Townsend's steam engine at Genwen with a powerful engine, which would dewater the Swansea Five Feet Vein in the vicinity through Townsend's original pit (182), with coal winnings effected through inherited pits — the 27 fathom pit (183), Cae Gwyn pit (not located), 21 fathom pit (210), Ox pit (209), and through a major new sinking called the Apple pit (208). Winnings in the Swansea Four Feet Vein, at Bynea, were to be effected through inherited or newly-driven adits. The existing canal would be used to ship coal at Townsend's Pill but a major railroad was to be constructed, from the Apple pit to Spitty Bite, where shipping places would be formed; 3) to re-open old pits and sink new pits on the southern limb of the Llanelli Syncline, west of Roderick, Bowen and Griffiths' coalfield, on Erwfawr lands, to work the Rosy, Fiery and Golden Veins. These pits would be connected to the Carmarthenshire Railway by a branch railroad.

Warde's schemes proceeded slowly before 1805. Charles Nevill, who was negotiating with Warde for coal for the proposed copperworks, commented on this, in late 1804, when he wrote "*I rode to Gen. Warde's colliery to learn how they were going on and found they were come to the coal but pursuing their discoveries in a very awkward manner*" and "*it appears plainly to me that no permanent Engagement could be made with him unless his collieries were opened, he has had them two years and has not made one trial of any consequence*".[206] Nevill's words implied inexperience on the part of Warde and Rhys Jones but Warde's absence on military duty could also have been a contributory factor.[207] Warde extended his coalfield in the Llwynhendy/Pencoed area, in December 1805, when he obtained a 40 year lease of coal under John Llewellyn's lands[208] and developments at Dafen Bridge and Genwen proceeded apace. A small steam engine became operative at the pit at Dafen Bridge (called the Llandaven or Daven Colliery) before the end of 1806,[209] and a start was made upon a new shipping place in the Dafen Pill/Penrhyngwyn region.[210] A large single-acting Boulton and Watt steam engine, with a 52 inch diameter cylinder, was installed into a new engine house, erected on the site of Chauncey Townsend's old engine house at Genwen[211] and the old workings were being dewatered by early 1808.[212] Despite some initial trouble at the Genwen Colliery,[213] the large engine proved adequate and a number of pits, previously sunk by Townsend and perhaps by John Smith, were re-opened in 1808. The railroad from the coal raising pits to the shipping places at Spitty Bite was also commenced in 1806/07 and was nearing completion by the Summer of 1808.[214] Meanwhile, the Daven Colliery was not producing as anticipated, because the engine was too small to cope with the water in the workings[215] and the coalfield west of the Bres/Wern region was not started.[216]

By the Summer of 1808, Warde had held his leases for almost six years and was understandably impatient to start recouping his heavy capital investment, his main hope being the Genwen colliery. The coal shipping season ended in October and Warde was anxious to establish his reputation as a major supplier before this, so that ships' masters would return for his coal the following Spring. Accordingly, he formally launched his enterprise at Llanelli in August 1808, as soon as the Apple pit railroad and the shipping places at Spitty Bite were completed, although the Apple pit itself was still being sunk.[217] Warde called his concern "*The Carmarthenshire Collieries*"[218] and advertised that coal from the Swansea Four and Five Feet Veins was being shipped at Spitty.[219] The master of the first vessel loaded with the "*new*" coal declined the customary offer of a gold laced cap to mark the occasion and opted instead for a payment of £1.13.0.[220] Perhaps he needed money, but he could also have considered the occasion to be hardly memorable, as Warde had shipped the same coal for five years. Warde ran into trouble, immediately after the opening of the Carmarthenshire Collieries, when Customs Officers began seizing vessels in breach of an Act regulating the length of bowsprits[221] and, by 17 September 1808, at least six vessels, intending to load with coal in the Burry River, had been seized. Warde, convinced that rival interests at Swansea were trying to suppress Llanelli's coal trade, advertised the affair in *The Cambrian* newspaper, inferring that the Customs Officers at

Swansea were acting under instructions or had accepted bribes, because the Swansea trade was not hampered. Public correspondence ensued between Warde, the officers involved and the Collector of Customs at Swansea[222] and, as a result, seizures ceased and vessels returned to ship his coal. Developments at Genwen continued and the Ox pit, previously worked by Squire, Evans and Beynon, was re-opened, with celebrations, in December 1808.[223] A temporary set-back was experienced in January 1809 when an inrush of water occurred,[224] but this was soon overcome and the major coal-winning Apple pit became operational by October 1809.[225]

Warde now realised that the Daven Colliery would not be viable, so he abandoned it[226] and exploited the more limited coalfield on the southern limb of the Llanelli Syncline, west of the Bres/Wern area. This area had seen previous workings by John Allen, Thomas Bowen and, perhaps, by others before them and many old pits were available for re-opening. Warde sited his colliery, which he first called the Llanelly pit, on Erwfawr lands (between Queen Victoria Road and Lakefield Road near the laundry) and commenced developments there by December 1809.[227] The Llanelly Pit was actually three separate pits (63, 64, 65), at least one of which was a new sinking to the Fiery Vein. One pit was provided with a steam or water engine and served as a pumping pit, coal being raised from the other two. A railroad was formed from the colliery to the Carmarthenshire Railway to transport coal to the docks at Llanelli.[228] When the colliery was about to be opened, Warde decided to change its name from Llanelly Pit to Old Castle Colliery, the notice of the opening in April 1811 explaining that the name had been chosen "*from the antiquity of the situation*".[229]

During the first ten years Warde had successfully implemented his main plans, although Daven had been a failure. He had also become involved in Llanelli's copper industry, probably on a small scale, at some undetermined location[230] (perhaps at the Old Pencoed Lead Works from the terms of the 1802 lease from Sir John Stepney). In 1812 he had at least three steam engines and a number of water and horse engines at work[231] and was the area's major active coalmaster. Like all the other new industrialists at Llanelli, he adopted a policy of engrossment, and, by 1812/13, as has been seen, he had negotiated with Cox and Jones to purchase the Box Colliery. This would give him an uninterrupted coalfield, extending over Box, Llandafen, Capel, Penygaer, Bryngwyn, Llanlliedi and Westfa and he was understandably eager to acquire it.

Warde was short of money by 1812, however, and defaulted on royalties, having to enter a bond for £800 in the Court of Great Sessions, Wales, for money owed to the Penllergaer Estate, for coal worked under the 1805 lease from John Llewellyn.[232] Further investment was also urgently required at the Genwen Colliery, to ensure that production continued in the Apple pit[233] and the Old Castle Colliery required additional development.[234] Warde could not raise money for all these purposes, and so decided to purchase the Box Colliery at the expense of developing his other collieries. The scheme to prolong the working life of the Apple pit (by deepening the Genwen pumping pit) was not implemented[235] and the Old Castle Colliery was sub-let to George Walker.[236] Released from these financial commitments, Warde purchased the Box Colliery in April 1813[237] and, by May, advertised that he was shipping coal from the "*Box*" and "*Pinygare*" Veins (Swansea Four and Five Feet Veins),[238] confirming that he was in working control. As Richard Janion Nevill was the real owner of the Box Colliery, Warde must have arranged the purchase with him, although no bill of sale has been seen. Warde's financial relationship with Nevill, which was to last to the end of his life, probably began at this time.

The Box Colliery purchase must have added to Warde's financial problems and, to raise capital, he borrowed money from a Major Rohde, in May 1814, pledging certain of his freehold lands and leasehold interests at Llanelli as security.[239] Alexander Raby added to his problems, in November 1814, when he obtained an injunction from the Court of Chancery to stop the Box Colliery operating,[240] eight months elapsing before a settlement was reached and working re-commenced.[241] On the re-opening, Warde advertised that "*Box Coal*" was available at reduced prices but that only money payments would be accepted,[242] suggesting that he was short of ready money. R. J. Nevill helped, in August 1815, by purchasing coal for use at the copperworks[243] but Warde's position deteriorated to such an extent by October 1815 that he could not meet wages and running costs at the Box and Genwen Collieries. To safeguard his financial interest in the Box Colliery, Nevill agreed to pay these for six months, from 24 October 1815,[244] and, apparently, advanced £2,000 to Warde.[245] A further agreement was drawn up for a supply of coal to the copperworks (at reduced prices) and Nevill was given a Warrant of Attorney respecting the "*Box and Chapel Collieries*", as security for money already advanced and to be provided later.[246] As a further safeguard to his investment, Nevill took out a policy on Warde's life, with the Pelican Life Assurance Society, in February 1817.[247] Warde's relationship with Nevill was fast becoming similar to that which already existed between the Rabys and Nevill, and there is little doubt that a take-over of Warde's interests at the right time was the primary stimulus for Nevill's actions.

Meanwhile, Warde's failure to improve the Genwen Colliery led to its closure, when available coal was worked out. The date on which workings ceased is not known but it was probably in late 1815/early 1816.[248] With the owner of the Penllergaer Estate (L. W. Dillwyn) also threatening legal action for the recovery of royalty money, which had not been paid despite Warde's entry of a bond in the Court of Great Sessions in 1812,[249] Nevill sensed an imminent crisis in Warde's affairs and decided to take personal control of the running of Warde's collieries, although he was careful to dis-associate himself from Warde's general debts. Warde (and Rohde) agreed that Nevill should manage Warde's collieries, for 13 years from 1 January 1818, at a salary of 3 shillings per wey for all coal sold.[250] Nevill sacked Rhys Jones, who had been Warde's colliery agent since 1802, and appointed his own agent, John Furnaess.[251] He re-opened the Genwen Colliery (the steam engine at the Genwen pumping pit had been re-started on 24 November 1817)[252] and formulated plans to develop the Box Colliery, which entailed high capital investment. Nevill did not wish to proceed with this expansion until the legal obstacle posed by Alexander Raby's claim, that he was entitled to regain a one-third share in the concern, was removed. Raby was as dependent on Nevill's financial backing as Warde by this time, however, and, in return for certain financial considerations, agreed to withdraw the Chancery Suit still pending,[253] Warde thus gaining possession, unencumbered by any legal ties, of the Colliery in August 1819. Nevill then began the planned expansion, installing new steam engines and sinking deeper pits 450 feet (137m) to the Swansea Five Feet Vein. (Fig. 18).[254]

Warde's status, then, had changed significantly between 1815 and 1819. Initially, he was glad to accept Nevill's help but, in the end, Nevill took all decisions. Nevertheless, he still held the coal leases and, because Nevill was careful not to become involved in Warde's debts, all transactions were negotiated in Warde's name and a balance of power probably existed between them. Nevill's salary as manager was altered from 3 shillings per wey to £750 per annum, on 1 June 1820[255] and, two days later, royalties owed to the Penllergaer Estate were paid (almost certainly by Nevill).[256] The developments continued at the Box Colliery, a scheme was proposed to extend Warde's Yspitty Canal westwards to Maesarddafen,[257] and Warde obtained a re-lease, from William Chute Hayton, of the coal contained in the original Box Colliery lease to Raby, for 65 years with an option to purchase the lands and mineral rights for £6,500, payable by 29 September 1824.[258] All this required money, and Nevill, with or without the support of the Copperworks Company, was undoubtedly providing it. The Box Colliery expansion was completed in

June 1823 and the first shipping of coal from the Swansea Five Feet Vein, on 7 June 1823, was marked by great celebrations. *The Cambrian,* reporting the event, also confirmed Warde's and Nevill's relationship, saying *"We have pleasure in recording all occurrences of prosperity in the Principality; and now have that of the completion of the costly and laborious pursuit of an untried deep vein of coal on General Warde's Collieries which under the judicious management of R. J. Nevill and the operative skill of Mr Furnease has been successfully effected by two pits of 150 yards depth each, a work of four years, unattended by any accident".*[259] The Box Colliery pits completed, Warde and Nevill commenced sinking the Penygare pit (29) and extended the Box Railroad to it.[260] All efforts were probably concentrated into the Box/Penygare Colliery complex after this, with workings at the Genwen Colliery being stopped.

Warde was now heavily in debt to Nevill and so, on 1 January 1824, they agreed repayment terms. Coal would be supplied to the Copperworks at a reduced price for seven years and Nevill's salary, as colliery manager, would be increased to £1,000 per annum as soon as profits reached £3,000 per year, after deducting all expenses.[261] Nevill, in turn, agreed to provide the £6,500, required to purchase the lands and mineral rights comprising the Box Colliery lease from William Chute Hayton, on the condition that they would pass to Nevill if Warde had not repaid him by 1829.[262] Warde's position was steadily deteriorating as his debts accumulated and the profit from his only money-making asset, the Box and Penygare Collieries, was insufficient to pay all his creditors. Nevill had to request that any claims upon Warde should be postponed and made secondary to his own claims, to achieve greater security for the Box Colliery[263] and Warde had to sell or mortgage some of his leasehold interests to raise £2,000, in March and April 1825.[264] His future prospects depended entirely upon the successful exploitation of the Swansea Four and Five Feet Veins in the Box, Capel, Bryngwyn and Penygaer regions and he obtained a lease of the coal in Warde's Fiery Vein (Swansea Five Feet Vein), under the glebe lands at Capel Uchaf and Capel Isaf, from the Vicar of Llanelli, in November 1826,[265] and, in the same month, Nevill paid Edward Rose Tunno £210 for coal under Pencilogi lands which may have lain within reach of the Box Colliery workings.*[266] By June 1828, Warde realised that he could not repay Nevill the £6,500, advanced to purchase the lands comprising the Box Colliery lease, and agreed that they should be transferred to Nevill.[267] Warde's position was becoming

---

* Warde and Nevill possibly commenced sinking the Llandafen pit (124) at this time (Fig. 18) and may have intended working northwards to Pencilogi. The date the Llandafen pit was started is not known, although it was working before 1836. (Nevill MS XVIII, Jan 1836).

hopeless. In October 1828, he mortgaged Genwen Farm to Nevill for £200 and informed the Vicar of Llanelli that he had to abandon workings under the glebe lands at Capel.[268] He must have still hoped that his fortunes would improve, because he objected, in January 1829, to a proposed bridge from Loughor to Spitty, because it would ruin his plan to establish "*Manufactories*" above Spitty,[269] but it turned out to be only a token protest.

He owed R. J. Nevill close on £30,000 by this time,[270] industry was generally in a depressed state and a general strike of the colliers at Llanelli, in June 1829,[271] may have dealt the final blow to his ailing concern. He either withdrew from further involvement at Llanelli or Nevill decided to foreclose on him. An agreement was drawn up between them, on 24 December 1829, to transfer all Warde's estates and collieries in Carmarthenshire to Nevill in lieu of payment of Warde's debts to him, Nevill also agreeing to pay Warde an annuity of £400 while Nevill held possession of the assigned property.[272] Warde left his home, Woodland Castle (now Clyne Castle, between Swansea and Mumbles), before March 1830[273] to live at Charlton Kings, Cheltenham. He died, on 20 June 1830, aged 70.[274]

Although he failed financially, Warde's contribution to the development of Llanelli's coal industry was significant between 1802 and 1830. He exploited the large coalfield, mostly neglected by the Townsends and the Smiths, on a wide geographic scale, proving many of the main coal seams. He over-committed himself financially and, like his contemporaries Alexander and Arthur Raby, found that Richard Janion Nevill's readiness to advance money led rapidly to Nevill controlling and finally possessing his interests. The extent of Warde's activity is evidenced by the number of pits and adits he is known to have worked, alone, or with R. J. Nevill as his manager. He re-opened and worked pits at Ffosfach and Pencoed, the Ray pit, Cae Gwyn pit, a pit or adit at Penygaer, Horse pit, Horse Engine pit, Old Fiery Vein level, Llwynhendy level, Pit F, Pit G, Engine Top pit, a pit at Bryngwyn (all not located), Genwen Engine pit (182), 27 fathom pit (183), 21 fathom pit (210), Ox pit (209) and a pit or pits at Erwfawr (one or more of 63 to 65). He finished the sinking of the Daven pit (119) and sank at least one new pit at the first Old Castle Colliery on Erwfawr lands (one or more of 63 to 65). After purchasing the Box Colliery, he sank at least one new pit (100) and either deepened the existing engine pit (98) or sank a new engine pit (99). To complement the workings from the Box Colliery, he opened the Penygare pit (29), the Bryngwyn Fiery Vein pit (not located) and, perhaps, the Llandafen pit

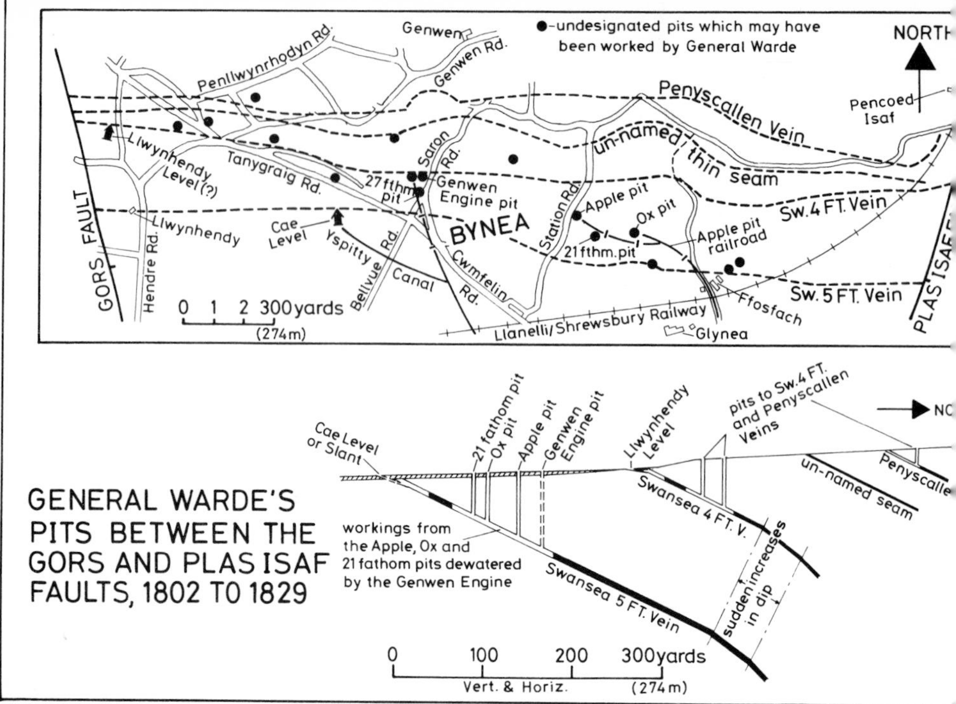

Fig. 19

(124) (Figs. 18 & 19). He also probably re-opened or sank a number of the undesignated pits in the Box, Penygaer, Capel, Bryngwyn, Llandafen, Maesarddafen and Bynea regions, shown on the plan in Appendix I, but his major efforts were concentrated into the Genwen, Daven, Old Castle, Box and Penygare collieries.

With all these colliery developments, his construction of supporting railways and shipping places and his involvement in the copper industry, Warde has to be recognised as one of the major figures in the early 19th century industrialisation of Llanelli.

(4) *Charles Nevill, Richard Janion Nevill and the Llanelly Copperworks Company**

Charles Nevill came to South Wales c1795, when he took over the management of a copperworks at Swansea. Before that he had managed a copperworks and button manufactory in Birmingham and had also been a mayor of that city.[275] South Wales produced some 90% of Britain's copper

* The Company was known by a number of names in the years between 1805 and 1830. It originally began as "*Daniell, Savill, Guest and Nevill*" in 1805, became "*Daniell, Savill, Sons and Nevill*" in 1817, "*Daniell, Son and Nevill*" in 1819 and "*Daniell, Nevill and Company*" in 1824. Additionally the Company was, in these years, known as the "*Llanelly Copperworks Company*", the "*Llanelly Copper Smelting Company*", the "*Copperworks Company*" or just as the "*Llanelly Copper Works*". To avoid confusion, "*Llanelly Copperworks Company*" or "*Copperworks Company*" is used.

in 1800, with most of the smelting works sited between Swansea and Neath.[276] The heavy demand for coal in this region led to uncertainty in supply, and fluctuations in price, and so Charles Nevill and his son Richard Janion Nevill — acting on behalf of John Guest (a Birmingham merchant and industrialist) and William Savill (a London copper merchant) — looked to Llanelli, where coal was plentiful, as a potentially desirable location for a new copperworks. They reconnoitred the Llanelli area, observing the activities of the new industrialists and, by November 1804, were convinced that the proposition made economic sense. The availability of suitable coal was the deciding factor, Charles Nevill writing to Guest and Savill "*The result of our enquiries are that it will be most to the Interest of the Co. to erect our Works at Llanelly providing a Contract for Coal can be made at a price not exceeding 57/6 per Wey and Ground secured for the Erection of Works. Our reasons for this Determination were the very superior quality of Coal in that Neighbourhood, the abundant Recourses which surround it and the very great difficulty and uncertainty that would attend a Treaty with the proprietors of the Surrounding Collieries*"[277] (i.e. the collieries around Swansea).

The Nevills approached Roderick, Bowen and Griffiths, Alexander Raby, John Vancouver and George Warde, in October and November 1804, regarding a supply of coal to the proposed copperworks, and also considered Penrhyngwyn, Ffosfach Marsh and Spitty Bank as alternative sitings. Initially, they were convinced that Penrhyngwyn was the ideal site but George Warde would not accept their price for coal nor grant them a sub-lease of Penrhyngwyn. They turned instead to Roderick, Bowen and Griffiths for their coal and decided to site the copperworks on the Flats or Beach near Penrhos (in the Seaside area). Their decision was strongly influenced by their assessment that Roderick and his partners would soon face bankruptcy and, before committing themselves, they obtained assurances that Roderick's leases would pass to the proposed copperworks company, if he were dispossessed.[278] Arrangements were made for the company to use Roderick, Bowen and Griffiths's dock to import copper ore and export smelted copper,[279] and a legal contract was drawn up, on 26 January 1805, which committed Roderick and his partners to the supply of 200,000 tons of coal over a 25 year period, and contained the stipulation that failure to meet the agreed annual quantities would result in the transfer of Roderick's leases.[280] The construction of the copperworks and its associated dock commenced in February 1805[281] (although the leases for the land were not signed until later in the year)[282] and, on 1 March 1805, the resolutions and bye-laws relating to "*the newly formed Llanelly Copperworks*" were agreed. 120 tons of ore were to be smelted each week, of which 60 tons would be purchased by the company at the public ticketings, with the other 60 tons processed at agreed rates for other purchasers. It was estimated that £22,614 initial capital was required for a predicted annual profit of over £7,000.[283] Four days later, on 5 March 1805, a formal deed of co-partnership[284] was signed between Ralph Allen Daniell (a Truro merchant and banker), William Savill, John Guest and Charles Nevill, and the company, "*Messrs. Daniell, Savill, Guest and Nevill, coppersmelters*", came into being. Although deeply involved in the formation of the Company, Richard Janion Nevill was not made a partner but he was legally assigned a share in his father's co-partnership, with the approval of the other partners, in January 1807.[285]

The Nevills came to live in Llanelli, superintending the construction and commissioning of the copperworks. As the import of copper ore and the export of smelted copper by sea was an integral part of their enterprise, they were also involved, throughout 1805, in early attempts to effect improvements in the safety of, and ease of access into, the Burry Estuary.[286] Smelting commenced at the copperworks on 20 September 1805,[287] and the copperworks dock became operative in late 1805/early 1806.[288] The enterprise had, therefore, got under way within a year or so of its inception, and the only aspect of the Nevills' plans which had not been realised was the acquisition of William Roderick's coalfield in the Bres/Wern region.

Roderick, Bowen and Griffiths supplied the agreed quantities of coal throughout 1806, but the steady deterioration in their finances led the Nevills to believe that their failure, and the subsequent acquisition of Roderick's leases by the Llanelly Copperworks Company, was imminent.[289] As we have seen, the Pembertons' intervention dashed the Nevills' hopes and the Company managed to gain only a sub-lease of the South Crop Wern Colliery, and a share in its collieries, railways, canal and dock, when Roderick, Bowen and Griffiths sold out in 1807. Nevertheless, despite the limited realisation of the Nevills' expectations, the Copperworks Company also became involved in coal mining at Llanelli, in May 1807,[290] an involvement which was to last for exactly one century.* The Company took possession of the Engine (or Wern) pit (73), which had a steam engine, the Hill pit (82), the Meadow pit (not located), the Pwllmelin pit (75), the Tregob pit (80) and possibly other smaller pits (Fig. 17). Charles Nevill advertised for a coal agent in July 1807[291] and John Furnaess was appointed to manage the South Crop Wern Colliery,[292] which was expanded and re-named the Wern Railway

* The Llanelly Copperworks Company, operating under the name of "*Nevill, Druce and Co,*" ceased to be involved in Llanelli's coal industry on 8 April 1907, when the Pencoed Colliery was abandoned.

Colliery.[293] Over 10,000 tons of coal were raised from the Rosy, Fiery and Golden Veins between 30 September 1807 and 30 June 1808,[294] and, as less than half this quantity was required for smelting,[295] it seems that the Copperworks Company entered the coal export trade at the first opportunity. The primary expansion of the colliery was completed by August 1808, when it was announced that "*the extensive collieries of Messrs. Daniell and Co., at Llanelly (lately belonging to Messrs. Roderick & Bowen) are completely opened and in full work*".[296]

The vessels which used the Company's dock usually brought in copper ore and took out coal. The partners realised that, to take full advantage of this trade, they must have their own coal available for export from the dock, but the Wern Railway Colliery had a limited life, its seams being steeply inclined and most of the easily-mined coal worked out before 1807.[297] The Nevills, therefore, took immediate action to gain control of the Box Colliery in 1809, when Alexander Raby's financial position deteriorated to such an extent that his trustees put his industrial interests up for sale. The Nevills displayed the careful business acumen that distinguished them from many of Llanelli's other early 19th century entrepreneurs, and took the greatest care to ensure that they, and their Company, could not be held responsible for debts incurred by people whose interests they took over. Accordingly, the Box Colliery was purchased in October 1810, using Cox and Jones as intermediaries[298] who, apparently, took charge of operations. William Hopkin, agent to the Stepney Estate, was appointed mining agent to the Box Colliery,[299] John Furnaess not being involved in its day to day running. The Nevills took further action to extend the Company's involvement in Llanelli's coal industry in 1811, when they provided Alexander and Arthur Raby with the necessary money to allow the Rabys to keep control of their industrial interests.[300] Their financial help was certainly given in the hope that the Stradey Collieries would come into their possession if the Rabys were forced to relinquish their interests, the Nevills, thus, re-enacting the process gone through with Roderick, Bowen and Griffiths, some five years earlier. Having gained a foothold in the extensive Stradey Collieries, the Nevills may have felt that the Company's future involvement in the coal trade no longer depended on the Box Colliery, which they sold to General Warde in April 1813.[301] In view of the Company's obvious expansionary policy, this sale is puzzling and no satisfactory explanation has been encountered for it. They had no thoughts of withdrawing from active participation in coal mining because, throughout 1813, the Nevills unsuccessfully tried to sink a pit at Machynis (89) to coal.[302] No agreement or bill of sale of the transaction has been seen, however, and subsequent evidence implied that Warde did not purchase the colliery outright. The Company may have over-committed itself financially and agreed that Warde should undertake the cost of running and developing the Box Colliery, with the Copperworks Company retaining a holding interest.

The Company was certainly expanding on other fronts, the Nevills being recognised as two of the area's leading industrialists and Charles being commissioned, by Lord Cawdor, to supervise the re-erection of the Carmarthen leadworks near the copperworks,[303] which meant increased trade for the copperworks dock through the import of lead ore. Dock facilities were improved in 1812/13[304] and it was intended to make it available for public use.[305] In 1813 it was proposed that an Act of Parliament be obtained to construct a canal or railway from the dock to the Llanedy region,[306] probably because it was feared that the newly-formed Kidwelly and Llanelly Canal Company would take trade away from the copperworks dock. The Nevills initiated all these activities and proposals but, in November 1813, as some of them were being implemented, Charles Nevill died at his home (Vauxhall House), at the early age of 60.[307] Richard Janion Nevill became the Company's main representative at Llanelli but was not admitted to a full partnership on his father's death.

Plate 18 — **Richard Janion Nevill.**

The 28 year old Nevill[308] succeeded to the local control of a flourishing enterprise. The copperworks was expanding and revenue from its dock was boosted by copper and lead ore imports and coal exports. The Company's involvement in the coal industry was well established, with its colliery at the Wern, a foothold in the Rabys' collieries (which guaranteed a coal supply to the copperworks, should the need arise, and the promise of the future acquisition of a large coalfield) and a probable part interest in the Box Colliery. Nevill soon capitalised on his financial association with the Rabys, and they agreed, in 1814 and 1815, to supply coal to the copperworks.[309] These agreements had undoubtedly been precipitated by a growing dispute between the Pembertons and the Copperworks Company, which could have stopped production at the South Crop Wern Colliery. The dispute had arisen because the Pembertons had diminished the flow of a water course from the River Lliedi by diverting it to their water-wheels at the Bres, with the result that the coal raising water-wheels at the South Crop Wern Colliery pits, lower down the water course, were not receiving sufficient water to impel them.* The Company also alleged that the Pembertons were not working their North Crop Colliery satisfactorily, resulting in water accumulating in the deep workings. The Copperworks Company intended entering a Bill in Chancery to this effect, in 1814, and a counterclaim was filed against them, alleging that they themselves were working the South Crop Wern Colliery inefficiently.[310] The Company had been disappointed by acquiring the sub-lease of only the South Crop coal in 1807, and this, with the disruption produced by the dispute, led Nevill to re-assess the value to the Company of the Wern Colliery, which he decided to offer for sale to the Pembertons in May 1815.[311] They declined to purchase and the Company kept the colliery but Nevill decided to abandon it when he could gain control of an alternative supply of coal for the copperworks.

Nevill soon acquired his coalfield. Both Warde and the Rabys experienced financial crises, in late 1815/early 1816, and the trade depression after the Napoleonic Wars precluded any possibility that they could stay solvent without further borrowing. In August 1815, Nevill agreed to purchase Warde's coal for the copperworks and, in October, advanced money to him and agreed to meet the running costs at the Box and Genwen Collieries.[312] As security Nevill was given a Warrant of Attorney respecting the Box Colliery[313] and this meant that the Llanelly Copperworks Company had taken a significant step towards acquiring

* Innes says that the water-wheels at the Bres had to be up to 12 feet (3.7m) wide to provide the necessary power because of the low velocity of flow of the water courses in this region. ("Old Llanelly" by J. Innes, p. 48). A decrease in this already restricted water supply would have adversely affected the Llanelly Copperworks Company's water-wheels at the Wern.

Warde's collieries. In February 1816, Nevill purchased the Rabys' collieries, by means of a writ of fieri facias issued by the High Sheriff of Carmarthenshire, although he agreed to allow the Rabys to remain in charge.[314] This transaction gave the Company actual possession of the Stradey Estate collieries (although Alexander Raby still held the coal leases) but, as was the case with Warde, the Rabys were left in control in order that the Llanelly Copperworks Company would not become liable for, or involved in, the concern's existing and future debts. Having obtained a guaranteed coal supply for the copperworks, Nevill proceeded to abandon the South Crop Wern Colliery, work ceasing in late 1816/early 1817.[315] At this time the Copperworks Company apparently withdrew from active participation in Llanelli's coal industry, but this is not so, because the Company virtually controlled the future development of the large coalfields leased to Warde and the Rabys.

The constitution of the Llanelly Copperworks Company was changed, in 1816 and 1817, R. J. Nevill being made a full partner and given legal power to transact business on behalf of the other partners. Daniell, Savill and Guest provided Nevill with letters of attorney empowering him to execute leases and grants, in June 1816,[316] and, in December, Guest sold his interest to the other partners.[317] A new co-partnership was formed, on 29 March 1817, between Ralph Allen Daniell, William Savill, Edwin and Joseph Savill (William Savill's sons), William Thomas Daniell (R. A. Daniell's son) and Richard Janion Nevill,[318] which traded under the name of "*Messrs. Daniell, Savill, Sons and Nevill*".[319] Thereafter, many transactions were undertaken in R. J. Nevill's name alone, and it is sometimes difficult to distinguish between his personal enterprises and those undertaken on behalf of, or backed by the finances of, the Llanelly Copperworks Company.*

R. J. Nevill now completely controlled the Company's local affairs and became personally and openly involved in the coal industry again. He took over the management of General Warde's collieries on 1 January 1818,[320] sacked Rhys Jones, Warde's colliery agent since 1802, and appointed the Copperworks Company's agent, John Furnaess, in his place.[321] He persuaded or induced Raby to drop the Chancery Suit pending over the ownership of the Box Colliery[322] and began major redevelopment there, providing new steam engines and sinking deeper pits to the Swansea Five Feet Vein.[323] This high cost expansion would take several years to complete and Nevill undoubtedly intended that the Box

* It seems that Nevill allowed his personal interests and those of the Company to overlap in some instances. The situation regarding R. J. Nevill's and the Llanelly Copperworks Company's ownership of coal leases and property would become the subject of dispute and subsequent clarification in 1836. (Nevill MS VII, loose letter dated 29 Mar 1836).

Colliery should replace the South Crop Wern Colliery. He may also have considered expanding the Genwen Colliery, a scheme being proposed to extend the Yspitty Canal westwards, probably to convey coal from the Bynea region to the copperworks dock.[324] Further changes took place in the composition of the Company's partnership in 1819 but these did not affect Nevill's control at Llanelli, although new buyers had to be found for the smelted copper. The Savills, who were copper merchants, trading separately under the name "*William Savill and Sons*", ceased operations and withdrew from the partnership in February 1819.[325] On their withdrawal, the Llanelly Copperworks Company altered its title to "*Daniell, Son and Nevill*" and arrangements were made with other London merchants for the sale of copper.[326] In March 1821 Nevill took over the management of the Stradey Estate collieries, although the Rabys were allowed to retain an interest.[327] The Llanelly Copperworks Company was now virtually in charge of most of the Llanelli Coalfield, although no colliery was actually worked in its name.

The expansion or improvement of the copperworks and the dock, and the financial liability incurred in taking over the running of Warde's and the Rabys' coal interests, had involved substantial capital, £10,000 being borrowed on long-term loan before 1821.[328] To raise more money, a new co-partnership with an increased share capital of £50,000 was formed, in August 1821, when R. A. Daniell, Thomas Daniell, Joseph Savill and Richard Janion Nevill agreed to continue operating under the name "*Daniell, Son and Nevill*",[329] although Charles Druce (the Company's London solicitor) may have been given, or was promised, a partnership at that time.[330] The new partnership pressed ahead with existing and new schemes, including a plan to convert the copperworks dock into the first floating dock equipped with gates in Wales,[331] and, before the close of 1823, the Company's loan capital equalled its share capital, because of sums borrowed on permanent loan from Messrs. Esdaile and Co. (the Company's London bankers) and from Charles Druce.[332] R. A. Daniell, the last of the Company's original partners, died in March 1823,[333] and a new co-partnership was formed, in April 1824, when Thomas Daniell, Joseph Savill, Richard Janion Nevill and Alexander Druce (Charles Druce's son) operated under the title "*Daniell, Nevill and Co*".[334]

By this time the Box Colliery developments had been completed[335] and Alexander Raby had retired, leaving his interests with his son Arthur Raby, who discharged his debts to the Company by supplying coal at reduced prices to the copperworks.[336] In May 1826, however, Nevill foreclosed on Raby and purchased his remaining interest in the Stradey Estate collieries. Nevill acted through James Guthrie and Nevill Broom, who were empowered to manage Arthur Raby's affairs during his absence (he went to live abroad to avoid payment of his debts).[337] At this time, the extensive Stradey Estate collieries, west of the River Lliedi, passed fully into the possession of the Llanelly Copperworks Company, although the pretence that they still belonged to the Rabys was maintained in order that no liability would be incurred for their debts.

These transactions, and an involvement in the Landore Copperworks,[338] severely taxed the resources of the Company. By the end of 1825, they owed Messrs. Esdaile and Co. over £79,000 and further loans were provided by the Bank of England in 1826 and 1827 (a temporary loan of £10,000 to be repaid within 9 months) and by the Miners' Bank of Truro (to purchase copper ore).[339] Although the depressed economic climate was against expansion, General Warde's financial plight became so serious that Nevill had no option but to foreclose and take over his interests. Warde owed Nevill and the Company nearly £30,000 by 1829, and it was agreed, in December, that all his estates and collieries in Carmarthenshire would be conveyed to Nevill, to settle his debts.[340]

The Copperworks Company, therefore, emerged as outright victors of the conflict between Llanelli's early 19th century industrialists who had fought and intrigued to acquire the most desirable coal areas. There is little doubt that the Company's success, during a period which saw the failure of most of the other new industrialists, was due to Charles Nevill's and Richard Janion Nevill's careful, efficient management, which was in direct contrast to that of their main rivals. Although the Copperworks Company had only worked under its own name, between 1807 and 1817, at the small South Crop Wern Colliery, the outright acquisition of Raby's and Warde's colliery interests had set the scene for the Company's emergence as the largest single coal mining concern ever to operate in the Llanelli Coalfield. Richard Janion Nevill had gained considerable coal mining experience at the Wern Colliery and as manager of Warde's and Rabys' collieries. After 1830, he and his son, Charles William Nevill, built a coal mining empire which reached its peak output in 1870 and still existed at the start of the 20th century.

(5) *The Pemberton Family**

The Pembertons, from the north of England, probably first became aware of the industrial potential of the Llanelli area, c1796/97,

* The Pembertons' records do not appear to be extant but it seems that the family worked as an industrial partnership. By 1813 at least seven members of the family — Francis Pemberton, Christopher Pemberton, Richard Pemberton, John Pemberton, Ralph Stephen Pemberton, Richard Pemberton, junior, and Thomas Pemberton — were involved in developments in the Llanelli/Pembrey area. (Cawdor [Vaughan] Box 228, "*Llanelly Harbour Bill*" 31 Mar 1813).

when a "*Mr Pemberton*" established good relationships with Dame Mary Mansell and offered to lend her money.[341] These overtures were no doubt motivated by the hope of being granted the Stradey Estate coal, but Alexander Raby gained the leases and the Pembertons temporarily lost interest in the area. By 1803, they must have become convinced that investment in the Llanelli Coalfield would yield high profits and John Pemberton, a barrister, came to Llanelli and generally advertised the fact that he had £20,000 available for industrial development.[342]

Pemberton formed an amicable relationship with the financially-troubled William Roderick[343] and offered to pay Roderick's arrears of sleeping rent, give him a lump sum of £1,000, and erect a powerful steam engine, capable of dewatering workings each side of the synclinal axis, for a sub-lease of the North Crop coal. An agreement was finalised on 6 January 1804,[344] the Pembertons taking possession of the unfinished Bres pit (49) where the steam engine was to be erected, possibly other pits near Talsarnau (95, 96) and being given permission to use Roderick, Bowen and Griffiths's Wern Canal and their shipping place. Ralph Stephen Pemberton came to Llanelli about this time, living in a large house, on the site of Llanelli Public Library, which became known as the "*Pemberton Mansion*".[345]

But John Pemberton had mis-represented the family's financial situation. The money he had mentioned was not immediately available[346] and so they could not carry out the agreement with Roderick, although the purchase of a steam engine for the Bres pit was put in hand.[347] Charles Nevill expressed the opinion, in November 1804, that the Pembertons were merely speculating and would be unable to meet their commitments[348] and, by March 1805, William Hopkin questioned the legality of their sub-lease, because the North Crop coal was not being worked and the Stepney Estate was not receiving the anticipated royalty payments.[349]

The Pembertons did not work coal in 1805 or 1806[350] but a situation rapidly developed which forced them to take action. As already seen, William Roderick had contracted to supply the Copperworks Company with coal, the Company being empowered to possess Roderick, Bowen and Griffiths' collieries if the supply were regularly in default.[351] The Nevills held great hopes that this would occur, and also felt that agreement could be reached with the Pembertons over the 1804 sub-lease of the North Crop coal.[352] The Pembertons did not relish the prospect of the Copperworks Company gaining Roderick, Bowen and Griffiths' collieries any more than Roderick and his partners did, because their sub-lease of the North Crop coal could be rendered valueless. They, therefore, negotiated with Thomas Bowen and his daughter and purchased their one half share in the partnership for £2,500, in January 1807.[353] William Roderick must have been involved in this, because he and Bowen and his daughter agreed, in February, to modify and finalise the 1804 agreement with the Pembertons.[354] Charles Nevill wrote at the time "*Roderick & Co. have I understand finally closed with the Pembertons who have taken possession of the North Crop and talk of getting to work immediately*".[355] The Copperworks Company gained a sub-lease of the South Crop coal in May 1807[356] and Roderick's coalfield was, therefore, shared between the Pembertons and the Copperworks Company, with the lion's share passing to the Pembertons. The Pembertons had undoubtedly been forced to act, before they were really ready, by the fear that the Copperworks Company would gain Roderick's leases.

Having taken this action, the Pembertons committed themselves whole-heartedly to the development of the North Crop coal. A 45 inch diameter cylinder, double-acting steam engine was installed at the Bres pit and became operative on 9 January 1808.[357] By May 1808, the pit had been sunk from its inherited depth of 201 feet (61m) to 216 feet (66m)[358] but the first coal seam (the Rosy Vein) lay another 100 feet (30m) further down. The Pembertons planned ultimately to sink the pit to the Bushy Vein, at about 580 feet (177m), but this would take a number of years. For the present, they were content to reach and exploit the Rosy and Fiery Veins. In conjunction with this sinking, they also commenced at least one new pit at Talsarnau (94) and possibly re-opened and deepened an existing pit (95) (Fig. 17). No detail of these developments appears to have survived but the Bres pit had been sunk some 370 feet (113m) to the Fiery Vein by 1813,[359] and the pit at Talsarnau (94), also to the Fiery Vein, was termed "*Messrs. Pembertons' New Pit*" in 1814.[360] These developments probably took some years to complete after 1808, and coal may not have been worked until c1811/12. As we have seen, the Pembertons diminished the water supply to the Copperworks Company's water-wheels at the Wern and a dispute resulted.[361] The Copperworks Company, realising the problems of working the South Crop Wern Colliery in isolation from the North Crop coal at the Bres, offered to sell it to the Pembertons in May 1815.[362] The Pembertons, heavily committed to their own developments and to land purchases in the area, declined and the Wern Colliery was run down and abandoned in 1816/17.[363]

By 1817, the Pembertons had been at work in the Llanelli Coalfield for 10 years. They had working pits at the Bres and Talsarnau, an interest in the railway from the Wern to the Seaside, and possessed Roderick, Bowen and Griffiths' original dock, now enlarged and known as "*Pembertons' Dock*". They also acquired land, purchasing most of the lands comprising

Dorothy Parker's one-twelfth share in the Llanelli Vaughan's original estate, from Lucinda and William Hayton between 1807 and 1809* (Appendix C) and lands at Llwyncyfarthwch, from the Haytons and from John Rees of Cilymaenllwyd between 1812 and 1816.[364] Their enterprise was well established, they were issuing their own bank notes from Pemberton Mansion[365] and, in September 1817, they advertised that their coal trade required 5,000 tons of shipping.[366]

The Pembertons undoubtedly wanted to expand at this time and sought to enlarge their coalfield by sinking a new pit on their own lands at Llwyncyfarthwch. This pit (93), called the Llwyncyfarthwch or Park pit,[367] (later named on plan the Pemberton pit)[368] was east of the Box Fault and south of the Box Colliery. The Pembertons intended exploiting the Swansea 4 feet Vein, which had been proved to be 9 feet (2.7m) thick at the nearby Box Colliery, and anticipated reaching the seam at about 240 feet (73m). Sinking commenced c1820/22 and the railway system required for the transportation of coal to the Pembertons' Dock at about the same time.[369] This could follow a natural downhill route towards Trostre Farm and then to Seaside (approximately 1½ to 2 miles; 2.4 to 3.2km), or an artificial downhill route could be formed by tunnelling through high ground at Marble Hall, between the pit and the Wern (approximately ½ mile; 0.8km). They chose the latter and constructed a 300 yards (274m) long tunnel. Colliery enterprises of this magnitude required considerable ongoing capital outlay, and the Pembertons were evidently satisfied that their involvement in Llanelli's coal industry would continue for many years.

The Copperworks Company, and Richard Janion Nevill in particular, did not share the Pembertons' view and openly re-commenced the old conflict between them. Through R. J. Nevill's control of Warde's and the Rabys' collieries, the Copperworks Company held most of the Llanelli Coalfield outside of Llangennech, and the Pembertons were the only barrier to their achieving a monopoly. Disputes over the payment of royalties to the Pembertons (because of their ownership of one-twelfth of the Llanelli Vaughans' original estate) had led to Court of Chancery proceedings in 1822 and 1824[370] but, in 1825, the Copperworks Company directly challenged their right to the North Crop coal. Roderick had died in 1823 and his assignees, with Margaret Eaton (formerly Griffiths), sole executrix of Thomas Bowen's will, combined with the Copperworks Company against the Pembertons. They surrendered the lease of Cawdor's coal under the Wern, to allow its re-lease to the Copperworks

* The lands purchased from the Haytons included Llandafen. This has given rise to the present name of Pemberton for the area.

Company,[371] and also assigned the Company part of the North Crop coal.[372] This assignment was made on the basis that the 1807 agreement between Roderick, Bowen and Griffiths and John Pemberton was "*supposed not performed*" but it was anticipated that the Pembertons would take legal action against the involved parties.[373]

This challenge could not have come at a worse time for the Pembertons. They had taken the Llwyncyfarthwch pit down to 300 feet (91m)[374] but had not reached the Swansea Four Feet Vein because the coal seams were disturbed in that region (Fig. 18). Faced with the probability that they would not find the Vein, they stopped sinking. They were still working the Bres and Talsarnau collieries[375] but the assignment of part of the North Crop coal to the Copperworks Company was an added serious psychological blow to the Pembertons' confidence, especially as there were signs of the trade recession of the late 1820s/early 1830s. They abandoned the Llwyncyfarthwch pit before any coal had been worked* and, to recoup some of their investment, tried to persuade the Llangennech Coal Company to link its St David's pit (154), which was being sunk, to the tunnel railway.[376] The Pembertons would have received significant wayleave payments if the Company had accepted the offer, but its proprietors had other plans. In 1828, they promoted a Bill to construct a separate railway, from the St David's pit to a proposed floating dock at Machynis Pool. The Pembertons opposed the Bill[377] but it received Royal Assent on 19 June 1828.[378]

This was the Pembertons' last attempt to retain a personal involvement in the Llanelli Coalfield. Although agreement was reached, in November 1828, over royalties, leases of the North Crop coal and the boundaries of Llanelli's various coalfields,[379] the Pembertons decided to withdraw. They advertised the "*Llanelly Colliery*", described as having two working pits drained by a powerful steam engine (i.e. the Bres and Talsarnau pits), for sale or letting in February 1829, stating that particulars could be obtained from "*Ralph S. Pemberton, Llanelly, R. Pemberton, Holborn, London or Thomas Pemberton, Sunderland*".[380] They quickly found a taker and George Bruin of Hermes Hill, Pentonville, was sub-leased all the Pembertons' industrial interests at Llanelli, for 20 years from November 1829,[381] when Ralph Stephen Pemberton probably left the area, ending the Family's 25 year direct involvement in Llanelli's coal industry.

Although they met with only limited success, the Pembertons contributed significantly to Llanelli's early 19th century development. They

---

* Coal was never subsequently worked from this pit.

Plate 19 — Site of the Bres Pit. The old structure incorporated into the newer buildings on the left may be the remains of the Bres engine house because the pit was located immediately in front of it. The derelict building must have been part of the colliery.

sank the Bres pit to over 580 feet (177m) — then by far the deepest pit in the Llanelli Coalfield — and extensively exploited the Rosy, Fiery, Golden and Bushy Veins on the northern limb of the syncline, west of the Box Fault, from the Bres and Talsarnau pits. They also enlarged Roderick, Bowen and Griffiths' original shipping place into a large tidal dock, and their pits and dock were used long after they left the area. The abortive sinking of the Llwyncyfarthwch pit, and the building of its allied railway, may have strained their financial resources and affected their confidence at a critical time and these, seemingly, were the main reasons for their decision to withdraw from Llanelli. (Plate 19)

Because records have apparently not survived, it is difficult to be certain which members of the Family were directly involved in Llanelli's coal industry and even contemporary writers seemed uncertain, referring in letters to "*Mr Pemberton*" or "*The Pembertons*". Nevertheless, it seems that John Pemberton was involved in the initial negotiations leading up to their involvement at Llanelli. Richard Pemberton, senior (died 23 March 1838, aged 94, in County Durham)[382] and his sons Thomas (died 27 February 1839), Ralph Stephen (died 22 February 1847) and Richard, junior (died 3 November 1843),[383] were also involved at an early stage, but their relationship to John Pemberton is not known. Ralph Stephen was the Pemberton who lived at Llanelli. The involvement of Francis and Christopher Pemberton (proposed as Harbour Commissioners in 1813) in Llanelli is not known but both worked to promote the Kidwelly and Llanelly Canal Company. The family left Llanelli 150 years ago but they, and their work, live on in present names — "*Pemberton*" for the Llandafen area and "*Tunnel Road*", along which their railway from the Tunnel to the Wern once ran.

(6) *Richard Williams*

Alexander Raby mined coal under Stradey Estate lands at Stradey Woods, c1800,[384] as a temporary measure until his main developments got under way. Richard Williams, of Moreb, also mined coal under Stepney Estate lands in the area about this time,[385] and he would have wanted to acquire Raby's collieries when working ceased. Raby probably abandoned these collieries in 1804 because, in that year, E. W. R. Mansell agreed that Williams could extract coal under Stradey Estate lands, in the Stradey Woods and Penywern regions, at a royalty of 3 shillings per wey, with Williams having the use of the existing road from the pits, past the cottage called Yard House, to the beach. It was also agreed that, if coal was discovered at Penywern, Williams could construct a railroad over Kille Farm lands to the Carmarthenshire Railway.[386]

Williams was mining there by July 1804,[387] the lease being signed in May 1805.[388] He paid royalties, in November 1805, for 322 weys (1,610 tons) of coal "*Shipt and Sold from Stradey Wood Works,*" between 1 July 1804 and 1 October 1805,[389] but no further reference to his activities has been seen.

The actual locations of Richard Williams's collieries are not known, although it is suspected that he worked the pit adjacent to Dulais Mill (4), shown on a plan of 1805.[390] One or more of the many unidentified pits and levels at Stradey Woods could have been worked by him, but it is unlikely that he undertook any development at Penywern, because the branch railroad to the Carmarthenshire Railway was never constructed. Evidence has survived to suggest that he supplied coking coal to the Kidwelly Tinplate Works.[391] He would have transported this coal by horse and cart from Stradey Woods to the beach, where the Stradey or Yard river formed a pill navigable to small ships.[392]

(7) *George Walker*

George Walker was agent to Alexander Raby and Thomas Hill Cox, at Neath, in the early 1790s[393] but came to Llanelli, c1796, to help establish Raby's industrial empire on Stradey Estate lands.[394] Although an employed industrial agent, his activities suggest a higher status. He was one of the proprietors named in the Carmarthenshire Railway Act in June 1802,[395] was the Carmarthenshire Railway Company's first secretary, between its inception and June 1803, when E. W. R. Shewen (Mansell) replaced him,[396] and was appointed a trustee to execute the provisions of the 1807 Enclosure Act.[397]

Walker must have had sufficient personal capital to mine in his own right and, when General Warde decided, in 1812/13, to purchase the Box Colliery at the expense of other interests, he obtained a sub-lease of the original Old Castle Colliery (63 to 65). He worked for a limited period only, ceasing operations there c1817.[398] It is not known if this was due to Walker's lack of capital or whether R. J. Nevill, who was about to take control of Warde's interests, wished to regain the Old Castle Colliery. Whatever the true reason, George Walker's role as one of Llanelli's early 19th century coalmasters lasted for only 5 or 6 years, and his contribution to the development of Llanelli's coal industry has virtually been forgotten.

(8) *Martyn John Roberts**

Martyn John Roberts first became involved in Llanelli's coal industry about 1820, although he was previously involved financially with Alexander Raby,[399] and may have held a part interest in the Brondini Colliery (near Cynheidre).[400] His first known direct involvement began when he obtained a 21 year lease of coal, under M. Wall's lands at the Bryn, on 10 July 1820.[401] He sank "*Roberts*" pit (157), probably equipped with a steam engine, working it with two other pits (155, 156) which were new sinkings or re-openings of old pits. Roberts called this three-pit complex the Bryn Colliery[402] (Fig. 16) and it soon became operative, because he exported some 2,000 tons of "*Llangennech Vein*" coal** to France in 1823/24.[403]

Roberts's first venture in the Coalfield was successful and he soon became involved in other industrial activities in the area. He became, or was already, associated with Richard Janion Nevill[404] in developments west of Llanelli, where there was renewed interest in the unbuilt transport link between Pembrey and the Carmarthenshire Railway, authorised by the 1812 Act of Parliament to the Kidwelly and Llanelly Canal Company[405] (See Chapter 5). In 1825, Roberts and Nevill obtained a joint

---

* Also referred to as John Roberts, John Roberts, junior or John Roberts the younger in the correspondence and documentation of the period. His father was also named John Roberts, but no problem of identity exists because the father did not become involved in Llanelli's coal industry.

** The "*Llangennech Vein*" worked by Roberts at the Bryn Colliery was most probably the Gelli Vein.

lease of the Forge land and buildings from Arthur Raby, to erect new furnaces and re-start the forge[406] and, probably in 1825, Roberts re-opened the Pool Colliery (one or more of entries 1 to 3) (Fig. 14), idle for many years, but which was on the route of the proposed railway link to Pembrey.[407] Nevill was not officially involved in the Pool Colliery re-opening but backed Roberts financially.[408] The plan to re-start the forge fell through by 1826[409] but, on Arthur Raby's departure from Llanelli in that year, Roberts and James Guthrie managed the Rabys' Stradey Estate collieries,[410] owned and controlled by R. J. Nevill. The railway was constructed to the Pool Colliery by 1826/27[411] and coal was being raised and transported to the docks at Llanelli by 1827.[412]

Roberts was apparently successful during his first ten years in Llanelli's coal industry. His Pool and Bryn collieries were working, the former linked to the docks at Llanelli, and the latter sited close to the proposed route of the newly-formed Llanelly Railway and Dock Company's railway; he was involved in the management of the Stradey Estate collieries and had R. J. Nevill's financial backing and co-operation. His future in Llanelli's industrial life, therefore, seemed assured but by 1837/38 he was no longer mining[413] and left Llanelli owing large sums of money to many creditors.[414]

(9) *The Llangennech Coal Company (and the Llanelly Railway and Dock Company)*

Edward Rose Tunno gained some 4,000 acres (1,620 hectares) of land (mostly at Llangennech), a partly-developed coalfield and the Spitty Bank copperworks, when he purchased the Llangennech Estate from John Symmons, in the early 1820s. Tunno, like Symmons, was a speculator. He advertised the potential of his newly-acquired estate, in May 1824, to attract developers[415] and, as "*Llangennech Coal*" was now well known in the London markets,* he soon found takers, Daniel Tower Shears, James Henry Shears, Thomas Margrave and William Ellwand the younger (all London merchants) taking over the copperworks and the working collieries in 1825. They formed a company — "*The Llangennech Coal Company*" — to work the coal but Margrave and Ellwand were not involved in the copperworks, taken over by Daniel and James Shears only.[416] No formal lease was then drawn up and, initially, they must have been working under an agreement.

---

* Prior to 1824 a Mr Wiburn of Goswell Street, London, a master smith, had stated that he would "*as soon pay 50s a chaldron for such coal as he would for the best Tanfield Moor coal which is the most esteemed for smith use known in the Metropolis, because the Llangennech Coal possessed a property the other had not, by being entirely destitute of sulphur and dirt*". (Local Collection 37, 11 May 1824).

The Company inherited working collieries and a railway system linking the pits and adits to Llangennech Quay. Working plans have apparently been lost,* so one cannot be precise about which collieries were inherited, but it is likely that the Company took over the Gelligille pits (152, 153), the Clyngwernen pit (133), the Gelli Hwyad pits (159, 160), the Brynsheffre level (142), the Glanmwrwg level and pit (144, 143), an un-named pit near Cae-Newydd (148), the Plasissa pits (165 to 167), perhaps the first Penrys pit (149) and probably others in the Bryn/Llangennech region, a number of seams being worked** and at least one of the pits having a steam engine.[417] The Company planned to exploit the thicker Swansea Four Feet Vein on a large scale for the coal export market, with a major new pit and a powerful steam engine on Gelli-gele Farm lands (later the St David's pit), from which initial winnings would be effected. When operating, this pit would be the main pumping pit, with workings from inherited collieries and new sinkings linked at the first opportunity. The Company based its plans on Edward Martin's 1803 report, which said that the "*Glanmurrog Vein*" (Swansea Four Feet) was not more than 360 feet (110m) deep at its synclinal axis[418] and, as the pit site was on the northern limb of the Syncline, the seam should have been reached at about 200 to 250 feet (61 to 76m). Unfortunately, Martin was wrong, the seam lying at 660 feet (201m). Without knowing, the Company had started to sink what was, then, by far the deepest pit in Wales.*** (Plate 20)

The 12 feet (3.7m) diameter shaft was started in 1825,[419] when the Company also partly reorganised its unsatisfactory shipping system. Llangennech Quay could only take small vessels and, to transport coal to large vessels at Spitty Bank and Llanelli Docks (also to the Spitty Copperworks), it had to be loaded into barges at the Quay, towed by steam-tug to its destination, and off-loaded.[420] This double-handling was expensive. To avoid it, it was decided to extend the railway from Llangennech Quay to Spitty Bank but owners along the proposed route refused to give permission to pass over their lands, fearing permanent loss of access to the

---

* A number of working plans of the Llangennech Estate collieries, dated 1824, (probably prepared by Tunno for his advertisements) were still extant in 1930 but did not pass into the hands of the National Coal Board on nationalisation.

** The collieries on the Llangennech Estate worked the Gelli, Penyscallen, Swansea Four Feet and Swansea Five Feet Veins and "*Llangennech Coal*" was the product of all these seams. Because of its greater thickness, the Llangennech Coal Company concentrated on mining the Swansea Four Feet Vein, and so it became equated to "*Llangennech Coal*" in the 1830s.

*** Martin's opinion was due to the misconception — still held in 1825 — that the Penyscallen and Swansea Four Feet Veins were one seam. This led him to predict a dip less than that which exists (Fig. 16).

Plate 20 — Sites of the Llangennech Coal Company's collieries. (Top) Remains of the St. David's pit. (Bottom) Remains of the Penprys pit.

River Loughor. The Company, through Tunno, then obtained permission from Lord Cawdor, holder of the manorial rights, to route the railway over commonlands and the foreshore.[421] Warde also agreed to its passage over lands he owned near Ffosfach.[422] The railway was constructed, probably operating by 1827, but landowners were still unhappy about the possible permanent loss of access to the River Loughor and continued to oppose it.[423]

The agreement with Tunno was formalised by a 40 year lease of all coal under the Llangennech Estate on 1 January 1827[424] and, some 6 months later, Tunno also sub-leased to the Company coal under Penllwyngwyn and Penprys, which Cawdor had granted to him.[425] The Company, obviously well satisfied with progress in the first two years and completely unaware of the problems that lay ahead in sinking the St David's pit, made ambitious plans to transport and ship the large quantities of Llangennech coal the colliery would produce when operational. The Company had probably planned to re-organise its shipping arrangements completely since 1825, but had not acted until certain that the probable market for its coal justified the high capital cost. £14,000 was needed for the scheme, which involved building a railway 2 miles (3.2 km) long from the St David's pit to Machynis Pool (at the southern end of New Dock Road), where a floating dock to take vessels of 300 tons was to be built. To raise the capital, the proprietors of the Llangennech Coal Company joined with seven others* to form the Llanelly Railway and Dock Company and permission to proceed with the scheme was sought, in 1828, through an Act of Parliament.

The Bill was vigorously opposed by parties whose interests could be affected by the scheme (the Burry Harbour Commissioners, the Carmarthenshire Railway Company, Thomas Lewis of Stradey and the Pembertons)[426] but it received Royal Assent on 19 June 1828. Lord Cawdor's support of the Bill had undoubtedly helped its passage through Parliament but, in return, he required covenants from the Llangennech Coal Company and Tunno, binding them to carry out specified measures.[427] An important clause also stipulated that permission to build would lapse if railway and dock were not completed within 5 years. Daniel Tower Shears, James Henry Shears, Margrave and Ellwand, proprietors of the Llangennech Coal Company, formed the first committee of management of The Llanelly Railway and Dock Company.[428]

The development plans for the two companies depended solely on the

* Daniel Tower Shears, James Henry Shears, Thomas Margrave and William Ellwand, junior were joined by Edward Landemann, Christopher Tilemann, William May Simons, Anthony De Horne junior, Francis Wadbrook, James Southgate Stevens and William Fry.

St David's pit reaching the Swansea Four Feet Vein but, after three years, only thin, uneconomic seams had been met. The capital to build the railway and dock had been subscribed before the Act was passed but, because the Vein had not been met, the start was delayed.

A legal case, drafted against Tunno, over loss of access to the River Loughor caused by the railway from Llangennech Quay to Spitty Bank[429] led to its abandonment in 1829,[430] the Company reverting to the earlier, costly barge-towing of coal. As 1829 ended the Company's situation was not good because, apart from the money expended with little return, there were signs of an industrial recession. The proprietors must have been encouraged by the growing demand for Llangennech coal, as its excellent steam-raising properties became more widely known,* but all depended on the St David's pit. They faced two bad years, which forced the abandonment of the Spitty Bank Copperworks by the Shears in 1831,[431] but, in June 1832, the Swansea Four Feet Vein was finally reached.[432] The railway and floating dock (called the "*New Dock*") was started and, for some 20 years, "*Llangennech*" steam coal was used extensively in home and overseas markets.

(10) *John Gryll and Partners*

In March 1829, John Gryll and others agreed to pay General Warde a nominal 5 shillings "*for permission to clear out a Water Level now in progress in the Great Vein under the lands of Maesyrdafen called Caebach Wayn Wilcock, Cwm Cefn Caie, Cae Cefn Caeiau and Parkydai or Wayn Whifon*" and a royalty of 5 shillings per wey of 108 heaped bushels for all coal worked.[433] Gryll and partners evidently intended working coal from an existing level in the Swansea Four Feet Vein, which Warde had already worked and was abandoning.** Its location is not definitely known, the "*Old Cefncaeau Slant*" (123) being the only entry located near the landshares described in the agreement.[434] (Fig. 18). No further reference to Gryll has been seen.

---

* In 1830 the Superintendent of the Indian Navy recommended the exclusive use of Llangennech coal for his steam ships and, in 1834, the Senior Assistant Examiner for the East India Company, responsible for its steamships, confirmed that Llangennech coal was used by them, having been recommended by the commanders of the Mediterranean Packets and the London breweries. (Report from the Select Committee on Steam Navigation to India, 14 Jul 1834, evidence of Captain Francis Chesney and T. L. Peacock).

** This agreement is significant. It is the first documentary proof seen of the existence of a small concern winning coal by removing or reducing coal pillars, left in old workings to support the roof. This late 18th/early 19th century practice of leaving pillars, when working a virgin coal seam, resulted in many safe old workings from which gleaners, working in small groups, could get a reasonable living. These pillar-robbers were a feature of the Coalfield after 1830 and still, occasionally, operate under licence from the National Coal Board.

CHAPTER 4 — REFERENCES

1 Reliable statistics are not available to estimate the actual growth in population within the area studied, between 1800 and 1829. J. L. Bowen says that the population of the Borough hamlet increased from an estimated 600 in 1795 to 4,137 in 1831 but his unreferenced statement must be treated with caution, as it is likely that he was considering dissimilar areas. ("History of Llanelly" by J. L. Bowen, 1886).
2 Calendar of the Diary of Lewis Weston Dillwyn, Vol 1 (4 Apr 1819) (NLW). The legal documents relating to the sale have not been seen.
3 Nevill 809, 810 (15 & 17 Dec 1802) and un-numbered Carmarthenshire lease (2 Oct 1802) (CCL). Also copy leases Stepney Estate 1112 and Nevill MS LVI of the same date.
4 The contents of Stepney's will have been abstracted from Stepney Estate 1104 — "An Act to enable William Chambers, Esquire and others, to grant Mining, Building and other leases of certain Estates in the counties of Carmarthen and Glamorgan devised by the will of Sir John Stepney, Baronet, deceased" (1840) and from "The Stepneys of Prendergast", Hist. Soc. of West Wales Transactions, Vol VII (1917-18).
5 The Estate did fortuitously return to the Stepney family in 1855. Unexpectedly, all six first named inheritors died without male issue and Eliza Gulston succeeded to the Stepney Estate in her 87th year.
6 *The Cambrian* (26 Oct 1805).
7 "Some Notices of the Stepney Family" by R. Harrison, private impression (1870) (LPL).
8 The evidence of Nevill MS IV, copy letter (21 Dec 1804) and *The Cambrian* (26 Oct 1805).
9 "Llanelly Parish Church" by A. Mee, p.lvii. Temesar, Hungary is probably the present Timisoara, Rumania.
10 Stepney Estate 1104 (1840).
11 Stepney Estate Office plan 46 (1814) (CRO).
12 Stepney Estate 1104 (1840). George James had become the Marquis of Cholmondeley prior to his death.
13 A. Mee, op.cit., p. 74.
14 "History of Carmarthenshire", edited by J. Lloyd, Vol II, p. 462.
15 Stepney Estate 1103 (1872) quoting a compromise agreement dated 7 Nov 1828.
16 "Cadets of Golden Grove" by Francis Jones, Trans. Hon. Soc. Cymmrodorion 1971, part II.
17 A. Mee, op.cit., p.1x.
18 Conflicting evidence is available regarding the date of Dame Mary Mansell's death. It was given as 18 Dec 1800 in Thomas Mainwaring's Common Place Book (LPL); as 1799 in Mansel Lewis London Collection 34 (1805); as 1800 in Mansel Lewis London Collection 32 (1825) and as 1801 in the Trans. of the Carmarthenshire Antiquarian Soc. and Field Club Part LIX (1934).
19 *The Cambrian* (8 Nov 1806).
20 Mansel Lewis 191 (1802) and Mansel Lewis 10 (17 Sep 1802).
21 Mansel Lewis 118 (30 Apr 1805), 170-171 (29 and 30 Apr 1805), 880 (Nov 1825).
22 A. Mee, op.cit., p.lix and *The Cambrian* (8 Nov 1806).
23 "The Old House of Stradey" by Francis Jones, Archaeologia Cambrensis Vol cxxii (1973).
24 Mansel Lewis 839-840 (5 Feb 1807).
25 "Some notes on the Mansel Lewis papers" by R. Craig, Carmarthenshire Antiquary Vol III Part 2 (1960).
26 "St Illtyd's Church, Pembrey" by E. Roberts and H. A. Pertwee (1898).
27 Mansel Lewis 839-840 (5 Feb 1807).
28 "The Old House of Stradey" by Francis Jones, op.cit. Descendants of the Shewens maintained, right up to the end of the 19th century, that Thomas Lewis had fraudulently acquired the Stradey Estate (see Accessions 4585 at CRO). Thomas Mainwaring wrote that Lewis was a "*man of will making notoriety*" and stated that Mary Mansell's will had been made under his instructions by "*Thomas of Laugharne*" (Thomas Mainwaring's Commonplace Book at LPL).
29 Francis Jones, op.cit.
30 Mansel Lewis London Collection 32 — "Case of Leases of Coal Mines under the Stradey Estate — Mr Pepys's opinion, 1825".
31 Francis Jones, op.cit.
32 "The Story of Carmarthenshire" by A. G. Prys-Jones (1972), Vol II.
33 Cawdor (Vaughan) 43/5847 (undated but c1803); Cawdor (Vaughan) 14/450 (26 Dec 1803); Cawdor 2/133 — Brief of a Bill in Chancery (1814).
34 Un-numbered Carmarthenshire lease (2 Oct 1802) (CCL).
35 Prys-Jones, op.cit.
36 Stepney Estate 1103 (1872) and Prys-Jones, op.cit.
37 Stepney Estate 1088, letter (30 Mar 1793).
38 Cawdor 2/262 (31 May 1804).
39 "An Act for inclosing Lands in Llanelly in the County of Carmarthen; and for leasing part of the said Lands and Applying the Rents thereof in improving the Town and Port of Llanelly in the said County" (47 Geo III Sess. 2 c107, 8 Aug 1807).

40 "History of Carmarthenshire" edited by J. Lloyd, Vol II.
41 Cawdor 2/64 memorandum (24 Jun 1813). Also "Lead Mining in Wales" by W. J. Lewis (1967).
42 "An Act for Inclosing Lands in the several Parishes of Llanelly, Llangennech and Llanedy, within the commote of Carnawllon, in the Lordship of Kidwelly in the County of Carmarthen" (52 Geo III c57, 1812).
43 Stepney Estate 1103 (1872).
44 Ibid.
45 Nevill 467 (20 Jun 1827) and 468 (24 Jun 1827).
46 Cawdor 2/70 — document entitled "Grant of a right of making and using a Rail Road for a term of 21 years" (1828) and Mansel Lewis 29 (7 Jan 1828).
47 Cawdor 2/70 — covenants between Earl Cawdor, E. R. Tunno and the Llangennech Coal Company dated 21 Apr and 24 Apr 1828.
48 No bill of sale relating to Symmons's purchase of Stepney's lands has been seen, but landshares at Llangennech, detailed in Stepney Estate rentals in the late-18th century (Stepney Estate 1099), were included in a rental of Symmons's Llangennech Estate in 1819 (Penller'gaer A944, Michaelmas 1819). Details of Symmons's purchase of Stradey Estate lands were given in Mansel Lewis 118 (30 Apr 1805), 170-171 (29 and 30 Apr 1805) and 880 (Nov 1825).
49 Llanelli Public Library plan 14 (undated but c1825); plan book and schedule of the Llangennech Estate (undated but c1825) (CCL); Penller'gaer A944 (Mich 1819).
50 Llanelli Public Library plan 14 (c1825); Mansel Lewis 880 (Nov 1825).
51 Penller'gaer A974 (16 Apr 1819); "Calendar of the Diary of Lewis Weston Dillwyn", Vol I, entry (4 Apr 1819).
52 (42 Geo III c80, 3 Jun 1802).
53 Cawdor (Vaughan) 43/5847 (undated but late 1803/early 1804) and 14/450, letter (26 Dec 1803). Symmons referred to Sir John Stepney as "*my friend*" in the letter of 26 Dec 1803.
54 "Reports of the Llangennech Collieries" by Edward Martin dated 21 Sep 1803, contained in Local Collection 37 (11 May 1824).
55 *The Cambrian* (28 Jan 1804).
56 Ibid, (29 Sep 1804).
57 Penller'gaer A.972 (8 May 1819).
58 *The Cambrian* (13 Oct 1804).
59 Carmarthenshire lease 81 (28 Jan 1805) (CCL).
60 Nevill MS IV, copy letter (3 Apr 1805).
61 Ibid., copy letter (24 May 1806).
62 Ibid., copy letter (13 Nov 1806).
63 Penller'gaer A972 (8 May 1819).
64 Nevill MS IV, copy letter (13 Nov 1806).
65 *The Cambrian* (18 Oct 1806, 28 Mar 1807).
66 *The Cambrian* (25 Apr 1807).
67 "History of Carmarthenshire" ed. by J. Lloyd, Vol II, p. 362; "The Industrial Development of South Wales" by A. H. John, p. 37; *The Cambrian* (12 Feb 1814).
68 *The Cambrian* (5 Nov 1808).
69 A. H. John, op.cit., quoting Public Record Office C/114/136.
70 *The Cambrian* (1 Oct 1808) referred to the existence of the Spitty Copperworks. In view of the fact that Davenport, Morris and Rees did not take over theLlangennech Estate's industrial interests until after April 1807, it is tempting to speculate that John Vancouver planned, and perhaps commenced, the Spitty Copperworks prior to surrendering the Estate back to Symmons. No confirmatory evidence of this possibility has been seen.
71 Plan of part of the first field survey for sheet 37 of the Ordnance Survey 1 inch to 1 mile plan, surveyed 1813/14 at a scale of 2 inches to 1 mile (NLW).
72 *The Cambrian* (12 Feb 1814).
73 Ibid,(23 Apr 1814).
74 Ibid,(4 Mar 1815).
75 Ibid,(26 Apr 1817).
76 Penller'gaer A947, letters (2 Mar, 10 Mar 1819); "Calendar of the Diary of Lewis Weston Dillwyn", Vol I, entries (4 Apr, 11 Apr 1819) (NLW).
77 Penller'gaer A954, letter (16 Jun 1819).
78 "Calendar of the Diary of Lewis Weston Dillwyn", Vol I, entry (27 Jun 1819) (NLW).
79 Ibid., entry (1 Jul 1819).
80 Penller'gaer A981, letter (undated but cJul 1819).
81 *The Cambrian* (19 Dec 1820).
82 Mansel Lewis 29 (7 Jan 1828). Local Collection 37 (11 May 1824) referred to Symmons as the former owner of the Estate.
83 Local Collection 37 entitled — "A letter from Richard Cort to John Taylor on the Smelting of Copper at Spitty Bank" dated 11 May 1824. Richard Cort had acted as manager of the Llangennech Estate. He was the son of Henry Cort the discoverer of the "*puddling*" process for pro-

ducing malleable iron with coal instead of charcoal. John Taylor was an influential industrialist and the Treasurer of the Geological Society.

84 Mansel Lewis 29 (7 Jan 1828); Nevill 466 (1 Jan 1827).
85 *The Cambrian* (16 Jun 1832).
86 Nevill 467 (20 Jun 1827), 468 (24 Jun 1827).
87 Mansel Lewis 29 (7 Jan 1828).
88 Cawdor 2/70, bonds and covenants between the Llangennech Coal Company, E. R. Tunno and Earl Cawdor dated 21 and 24 Apr 1828.
89 Cawdor (Vaughan) 103/8046 (Mich. 1802), 103/8048 (Lady Day 1803) and 103/8047 (Mich. 1803).
90 Cawdor 2/133 — Brief of a Bill in Chancery, R. A. Daniell and others against Lord Cawdor and others (1814).
91 Cawdor (Vaughan) 43/5847, letter (undated but considered to be late 1803/early 1804 in view of the events described in it).
92 Cawdor (Vaughan) 14/450, letter (26 Dec 1803).
93 Cawdor (Vaughan) 43/5847, letter (late 1803/early 1804).
94 Ibid.
95 Nevill 411 (29 Apr 1825) quoting an unseen agreement dated 6 Jan 1804 and Nevill MS IV, copy letter (16 Mar 1805).
96 *The Cambrian* (16 Jun 1804).
97 Cawdor 2/133 — Brief of a Bill in Chancery, R. A. Daniell and others against Lord Cawdor and others (1814).
98 Nevill MS IV, copy letter (10 Nov 1804).
99 Ibid., copy letter (12 Nov 1804).
100 Nevill 650 (26 Jan 1805).
101 Nevill MS IV, copy letter (1 Dec 1804).
102 Ibid., copy letter (12 Dec 1804).
103 Nevill 650 (26 Jan 1805).
104 Nevill MS IV, copy letter (16 Mar 1805).
105 Ibid., copy letter (3 Apr 1805).
106 Ibid., copy letter (23 Sep 1805).
107 Ibid., copy letter (24 Oct 1805).
108 Ibid., copy letter (14 Jan 1806).
109 Ibid., copy letter (15 May 1806).
110 *The Cambrian* (9 Aug 1806).
111 Nevill MS IV, copy letters (11 and 23 Aug 1806).
112 Ibid., copy letters (4, 15 Sep 1806).
113 Ibid., copy letter (24 Sep 1806).
114 Ibid., copy leter (28 Nov 1806).
115 Ibid., copy letter (12 Jan 1807).
116 Nevill 411 (29 Apr 1825) quoting an unseen agreement dated 16 Feb 1807.
117 Nevill MS IV, copy letters (10, 23 Feb 1807).
118 Cawdor 2/133 — "Instruction for Assignment of Leases of Collieries etc" between Messrs Roderick and Co. and Messrs. Daniell and Co. (1807).
119 Ibid., and Nevill MS IV, copy letter (17 Jun 1807).
120 Nevill 411 (29 Apr 1825). Roderick was described as a "*dealer and chapman*" in a notice calling for a meeting of his creditors in 1812 (*The Cambrian,* 1 Aug 1812).
121 "Llanelly Parish Church" by A. Mee, p. 107 gave Roderick's age at death as 85 years. *The Cambrian* (17 May 1823) gave his age as 87 years in his obituary notice.
122 *The Cambrian* (14 May 1809).
123 Mansel Lewis London Collection 98, letter (1 Mar 1799) and "Hanes Llanelli" by D. Bowen (1856).
124 Mansel Lewis London Collection 13 — Stradey Collieries, accounts of coal raised 1802 to 1828.
125 Ibid.
126 42 Geo III c80 (3 Jun 1802).
127 Museum Collection 387 (1802 to 1807) — Report of the Carmarthenshire Railway Committee dated 9 May 1807.
128 Cawdor (Vaughan) 43/5847, letter (undated but late 1803/early 1804).
129 Cawdor 2/262 (31 May 1804).
130 *The Cambrian* (2 Nov 1805); Mansel Lewis 56 (1 Sep 1806); Mansel Lewis 57 (15 Nov 1814).
131 *The Cambrian* (5 Nov 1808) announced the opening of the Box Colliery and work must have commenced on the sinking of the pits and the erection of the steam engines shortly after the granting of the lease, in September 1806.
132 "History of Carmarthenshire" ed. by J. Lloyd, Vol II, p. 334 and Thomas Mainwaring, op.cit.
133 Mansel Lewis London Collection 13 — Stradey Collieries, accounts of coal raised 1802-1828.
134 "Some Notes on the Mansel Lewis Papers Part II" by R. Craig, Carmarthen Antiquary 1962 and Museum Collection 387 (1802 to 1807).

135 Museum Collection 387 (1802-1807) — Report of the Carmarthenshire Railway Committee dated 9 May 1807.
136 Mansel Lewis London Collection 108, letters (1806 to 1808).
137 Mansel Lewis London Collection 3, copy of a draft release, assignment and deed of composition (10 Jun 1807).
138 Mansel Lewis London Collection 108, letter (5 Sep 1807).
139 Mansel Lewis London Collection 112, letter (18 Dec 1807).
140 *The Cambrian* (5 Nov 1808).
141 Ibid, (16 Sep, 7 Oct 1809).
142 Mansel Lewis 57 (15 Nov 1814) quoting an unseen agreement dated 3 Apr 1810 and Thomas Mainwaring, op.cit.
143 Nevill 64 (20 Oct 1810).
144 *The Cambrian* (22 Dec 1810, 16 Mar 1811).
145 Ibid,(4 May 1811).
146 In May 1815 the Committee of the Llanelly Copper Smelting Company directed R. J. Nevill to reduce Raby's debts to them as quickly as possible, thus confirming that they had lent him money (Nevill MS XXI, abstract from the 10th Annual Meeting of the Llanelly Copper Smelting Company held on 20 May 1815).
147 *The Cambrian* (6 Jul 1811).
148 Ibid,(28 Mar 1812).
149 Ibid,(4 Apr 1812).
150 King's Bench, Hereford Spring Assizes, action heard on 10 Jul 1812 — David Lewis and others v Alexander and Arthur Raby.
151 "The Canals of the Gwendraeth Valley (Part 2)" by W. H. Morris and G. R. Jones, The Carmarthenshire Antiquary, Vol. 8, 1972.
152 *The Cambrian* (9 Jan 1813).
153 Nevill 685 (25 May 1814); 686 (21 Jun 1814); 682 (30 Mar 1815).
154 Mansel Lewis 57 (15 Nov 1814).
155 *The Cambrian* (29 Jul 1815).
156 Mansel Lewis 1637, letter (28 Dec 1808).
157 "Old Llanelly" by J. Innes, quoting an unidentified source.
158 Mansel Lewis 203, Stradey Estate coal accounts 1811 to 1823.
159 Nevill MS XXI — Minutes of the 10th Annual Meeting of the Llanelly Copper Smelting Company (20 May 1815).
160 Nevill 1739 (27 Jan 1816).
161 Nevill 1736 (26 Feb 1816).
162 Nevill 1740 (26 Feb 1816).
163 *The Cambrian* (9 Jan 1819).
164 Nevill 410 (18 May 1819); Mansel Lewis London Collection 91 — "Plan by W. W. Bailey of Veins of Coal under the Sea Shore belonging to the Crown" (November 1818).
165 Nevill 689 (8 May 1820).
166 Mansel Lewis 204, account dated 28 Feb 1821.
167 Nevill 2790 (5 Feb 1849).
168 Nevill 1745 (5 Mar 1821).
169 *The Cambrian* (10 and 24 Mar 1821).
170 Mansel Lewis 206, coal accounts 1821 to 1823.
171 Ibid.
172 Mansel Lewis 204, letter (24 Sep 1822).
173 "Industrial South Wales", ed. by W. E. Minchinton, p.136.
174 Mansel Lewis 206, letter (30 Aug 1823).
175 Ibid. and Nevill 1746 (25 Apr 1823).
176 Nevill 1747 (1 Dec 1823).
177 Nevill 1743 (1 Dec 1823).
178 Nevill 1747 (1 Dec 1823).
179 Nevill 1744 (1 Dec 1823).
180 "Llanelly Parish Church" by A. Mee, p.xxxiii, stated that Alexander Raby died on 3 Mar 1835, aged 88. Raby's obituary notice in *The Cambrian* (7 Mar 1835) gave a date of 24 Feb 1835 and an age of 89.
181 Mansel Lewis 206, letter (18 May 1824).
182 "Old Llanelly" by J. Innes, p. 105.
183 Nevill 1732 (26 Feb 1825).
184 Nevill 72 (12 Mar 1825).
185 Nevill 74 (19 May 1826).
186 *The Cambrian* (17 Jun 1826).
187 Nevill 1735 (8 Jul 1826).
188 Nevill 158 (11 Jul 1826).
189 Nevill 1731 (14 Jul 1826).

190 *The Cambrian* (23 Sep 1826).
191 Mansel Lewis London Collection 108, letter (11 Oct 1826).
192 Nevill MS VII, loose letter 165 referring to letters written by R. J. Nevill to Arthur Raby on 22 Mar and 20 May 1827.
193 "Hanes Llanelli" by David Bowen (1856).
194 Nevill 2790, letter (5 Feb 1849).
195 "The History of Llanelly" by J. L. Bowen (1886) stated that Raby lost about £175,000 in his commercial transactions in the Llanelli area although the authority for this statement was not given. Mansel Lewis 34, document entitled — "Raby's reply to Thomas Lewis's complaint — In the Exchequer 1813" — confirmed that he had expended more than £100,000 on his works and buildings at Llanelli up to that time.
196 "Hanes Llanelli" by D. Bowen (1856); "The History of Llanelly" by J. L. Bowen (1886); "The South Wales Coal Trade" by C. Wilkins (1888); "Llanelly Parish Church" by A. Mee (1888); "Old Llanelly" by J. Innes (1904); "History of Carmarthenshire" ed. J. Lloyd (1939). It is likely that each subsequent author merely perpetuated the statement in "Hanes Llanelli' without researching the facts. David Bowen had obtained his information on the Rabys' industrial ventures by corresponding with Arthur Raby in Germany.
197 *The Cambrian* (7 Oct 1809) and Thomas Mainwaring, op.cit.
198 Un-numbered Carmarthenshire lease (2 Oct 1802) (CCL).
199 Nevill MS LVI, copy lease (2 Oct 1802).
200 Nevill 809 (15 Dec 1802); 810 (17 Dec 1802).
201 Nevill 811 (17 Dec 1802).
202 Mansel Lewis 378 (17 Dec 1802).
203 Nevill MS X, accounts for 1803/04.
204 *The Cambrian* (22 Jun 1804, 6 Jul 1811).
205 Nevill 809, 810, endorsements dated 2 Nov 1804.
206 Nevill MS IV, copy letters (15 Nov, 21 Dec 1804).
207 *The Cambrian* (23 Mar 1805).
208 Nevill 85 (12 Dec 1805); Penller'gaer B.18.64 (1817/18).
209 Nevill MS X, entries 1805/06.
210 Nevill 170 (30 Sep 1809).
211 Llanelli Public Library plan 8 (1 Jul 1812).
212 Nevill MS X, entry for 25 Jan 1808.
213 *The Cambrian* (14 Nov 1807).
214 Nevill MS X, entries in 1808.
215 Charles Nevill had remarked, in November 1806, that Warde's steam engine at Daven was so small that it had to be kept constantly at work to keep the water back (Nevill MS IV, copy letter 13 Nov 1806).
216 Nevill MS X, entries between 1803 and 1809.
217 Ibid, entries in 1808.
218 *The Cambrian* (17 Sep 1808); Nevill MS X, which was the accounts book for Warde's Collieries, was entitled — "Journal - Carmarthenshire Collieries".
219 *The Cambrian* (13 Aug 1808).
220 Nevill MS X, entry dated 24 Aug 1808.
221 The Act was passed in 1787 and regulated the length of a bowsprit to less than two-thirds of the length of the vessel.
222 *The Cambrian* (17 Sep, 1, 22, 28 Oct 1808).
223 Nevill MS X, entry dated 5 Dec 1808.
224 *The Cambrian* (7 Jan 1809).
225 Nevill MS X, entries in 1809.
226 All reference ceased to the Daven pit in Warde's accounts (Nevill MS X) after 1810.
227 Nevill MS X, entries late 1809.
228 The evidence of Nevill MS X, Llanelli Public Library plan 9 (1814), Stepney Estate Office plan 46 (1814) (CRO).
229 *The Cambrian* (20 Apr 1811). The Old Castle at Llanelli has not been located but late 18th and 19th century plans designated the area between the present John Street and Murray Street, where the Midland Bank stands, "*Little Castle*" (Old Llanelly by J. Innes), "*Pen-y-Castell*" (first edition of Ordnance Survey plans, surveyed 1878) and "*Old Roman Camp*" (Stepney Estate Office plan 47, 1841). The first Old Castle Colliery was situated within 300 yards of this site. For a full discussion on the possible siting of the Old Castle at Llanelli refer to an article by C. H. Glascodine in the *Llanelly Guardian* for 16 Sep 1920.
230 "The Industrial Development of South Wales 1750-1850" by A. H. John pp. 32 and 75, quoting Public Records Office C/114/136, C/114/139, letter (15 Nov 1814).
231 Nevill MS X, entries up to 1812.
232 Penll'rgaer B.18.64 (1817, 1818).
233 Llanelli Public Library plan 8 (1 Jul 1812).

234 Nevill MS X, entries in 1812 show that the Old Castle Colliery was still undergoing development.
235 Llanelli Public Library plan 8 (1 Jul 1812) records a scheme to deepen the Genwen pumping pit from 240 feet to 336 feet (73 to 102m). A report on the Genwen Colliery dated January 1836 (Nevill MS XVIII) stated that the pit was still only 240 feet (73m) deep at that time.
236 Mansel Lewis 1654 (1810-1822), document entitled "Account of trespasses by General Warde".
237 Mansel Lewis 57 (15 Nov 1814). The Box Colliery purchase was apparently transacted in the name of an Edward Hawkins who was acting as Warde's agent.
238 *The Cambrian* (15 May 1813).
239 Nevill 815, 816 (23 May 1814); 812 (21 Oct 1814); 813 (22 Oct 1814). A John Warde of Squerries Lodge, Kent was involved in these transactions and there is unsubstantiated evidence that he was George Warde's father. (Thomas Mainwaring, op.cit.).
240 Mansel Lewis 57(15 Nov 1814).
241 *The Cambrian* (29 Jul 1815).
242 Ibid. (12 Aug 1815).
243 Nevill 1765 (26 Aug 1815).
244 Nevill 1766 (6 Nov 1815).
245 Mansel Lewis 1654 (1810-1822).
246 Nevill 1767 (24 Mar 1816).
247 Nevill 1771 (13 Feb 1817).
248 Penll'rgaer B.18.64 (1817/18). Also the Catalogue of Plans of Abandoned Mines (1930) listed a plan, no longer extant, showing workings in the Fiery and Great Veins (Swansea Four and Five Feet Veins) at the "*Llwynhendy and Genwen*" colliery up to 1815. This could have been an abandonment plan.
249 Penll'rgaer B.18.64 (1817/18).
250 Nevill 817 (1 Jan 1818); also endorsement dated 20 Apr 1818 on Nevill 813 (22 Oct 1814).
251 Mansel Lewis 1654 (1810-1822).
252 Penll'rgaer B.18.66 (25 Nov 1817).
253 Nevill 67, 68 (14 Nov 1818).
254 *The Cambrian* (3 May, 14 Jun 1823).
255 Nevill 1769 (1 Jun 1820).
256 "Calendar of the Diary of Lewis Weston Dillwyn", entry dated 3 Jun 1820 (NLW).
257 Llanelli Public Library plan 11 (1821).
258 Nevill 818 (12 Jun 1822).
259 *The Cambrian* (14 Jun 1823).
260 Nevill 1870-1918 (Jan-Dec 1826); 40 (1 May 1827); 286 (3 Jul 1827).
261 Nevill 1064 (1 Jan 1824).
262 Nevill 1040 (29 Sep 1824); 1041 (16 Oct 1824).
263 Nevill 1705 (12 Dec 1824).
264 Nevill 72 (12 Mar 1825); 1714-1716 (22 Mar 1825); 1585, 1053, 1054 (all dated 23 Apr 1825).
265 Nevill 75 (4 Nov 1826).
266 Nevill 1625 (9 Nov 1826).
267 Nevill 327 (6 Jun 1828); 328 (7 Jun 1828).
268 Nevill 1764 (18 Oct 1828); 76 (20 Oct 1828).
269 *The Cambrian* (24 Jan 1829).
270 Nevill 821 (25 Oct 1831).
271 *The Cambrian* (6 Jun 1829).
272 Nevill 748 (24 Dec 1829).
273 *The Cambrian* (13 Mar 1830).
274 Nevill 747, undated but filed with 748 (24 Dec 1829); Thomas Mainwaring op.cit.
275 Llanelly Guardian (14 Jun 1888). This reference stated that he came to Swansea in 1795 but Nevill MS III suggests that he did not arrive until 1797.
276 "Non-ferrous smelting" by R. Roberts, contained in "Industrial South Wales 1750-1914" edited by W. E. Minchinton (1969).
277 Nevill MS IV, copy letter (1 Nov 1804).
278 Ibid, copy letters (Oct to Dec 1804).
279 Ibid, copy letter (12 Dec 1804).
280 Nevill 650 (26 Jan 1805).
281 *The Cambrian* (31 Aug 1805).
282 Nevill 157 (13 Aug 1805); 108 (1805).
283 Nevill MS XX (1805-06).
284 Nevill 526 (5 Mar 1805).
285 Nevill 1667 (22 Jan 1807).
286 *The Cambrian* (9 Feb 1805).
287 Nevill MS IV, copy letter (23 Sep 1805).
288 Ibid., copy letter (4 Feb 1806).
289 Ibid., copy letters throughout 1806.

290 Cawdor 2/133 (1807) referring to two unseen agreements dated 11 May, 21 May 1807.
291 *The Cambrian* (11 Jul 1807).
292 No documentary evidence of Furnaess's appointment has been seen but he was R. J. Nevill's colliery agent in 1816 (Mansel Lewis 1654, 1810 to 1822).
293 Nevill MS XI (1807-1817).
294 Nevill MS XLIX (1807-1813).
295 Nevill MS XX (1805/1806). It was stated in this document that 20 weys (approx 200 tons) of coal were required to smelt the intended 120 tons of copper ore per week.
296 *The Cambrian* (6 Aug 1808).
297 Llanelli Public Library plans 4 to 7 (July 1806).
298 Mansel Lewis 57 (15 Nov 1814) quoting an unseen agreement dated 3 Apr 1810.
299 *The Cambrian* (8 Sep 1810).
300 See previous discussion on Alexander and Arthur Raby.
301 Mansel Lewis 57 (15 Nov 1814).
302 Nevill MS XLVII, accounts for 1813. Thomas Mainwaring, op.cit., stated that excessive water inflow defeated this early attempt to exploit the coal resources at Machynis.
303 "History of Carmarthenshire", J. Lloyd (editor), Vol II, p. 369.
304 Nevill 240 (21 Jul 1812).
305 Cawdor (Vaughan) Box 228, bundle headed "Papers on the subject of a proposed Canal in Carmarthenshire".
306 *The Cambrian* (11 Sep 1813).
307 *The Cambrian* (27 Nov 1813) gave Charles Nevill's date of death as 20 Nov 1813 whereas A. Mee, op.cit. gave a date of 10 Nov 1813.
308 *The Cambrian* (18 Jan 1856), reporting R. J. Nevill's death, gave his age as 70 years.
309 Nevill 685 (25 May 1814), 686 (21 Jun 1814), 682 (30 Mar 1815).
310 Cawdor 2/133 — brief of a Bill in Chancery (1814) and Cawdor 2/110 (undated but c1814).
311 Nevill MS XXI, loose statement (19 May 1815).
312 Nevill 1765 (26 Aug 1815), 1766 (6 Nov 1815); Mansel Lewis 1654 (1810-22).
313 Nevill 1767 (24 Mar 1816).
314 Nevill 1736, 1740 (26 Feb 1816).
315 Nevill MS XI (1807-1817). No details relating to the abandonment of the colliery have been seen but the accounts relating to it ceased in 1817.
316 Nevill 688 (3 Jun 1816).
317 Nevill 675 (14 Dec 1816).
318 Nevill 8 (26 Mar 1817); 42 (29 Mar 1817).
319 Nevill 677 (2 Jun 1817).
320 Nevill 817 (1 Jan 1818).
321 Mansel Lewis 1654 (1810-22).
322 Nevill 67, 68 (14 Nov 1818), 70 (Aug 1819).
323 *The Cambrian* (14 Jun 1823).
324 Llanelli Public Library plan 11 (1821).
325 Nevill 450 (20 Feb 1819), 1525-1557 (1819-1822).
326 Nevill 684 (Apr 1819).
327 Nevill 1745 (5 Mar 1821).
328 "The Industrial Development of South Wales" by A. H. John pp. 43-44.
329 Nevill 1116 (1 Aug 1821).
330 Thomas Mainwaring, op.cit.
331 Nevill 243 (24 Jun 1823).
332 A. H. John op. cit., p. 44.
333 Thomas Mainwaring, op.cit.
334 Nevill 43 (6 Apr 1824), 74 (19 May 1826).
335 *The Cambrian* (14 Jun 1823).
336 Nevill 1747 (1 Dec 1823).
337 Nevill 74 (19 May 1826), 1735 (8 Jul 1826), 158 (11 Jul 1826).
338 Nevill 1032 (1 Jan 1827), 1034 (7 Apr 1827).
339 A. H. John, op.cit., p. 44.
340 Nevill 748 (24 Dec 1829), 820 and 821 (25 Oct 1831).
341 Mansel Lewis London Collection 97, letter (Jan 1798). The "*Mr Pemberton*" was probably John Pemberton.
342 Cawdor 2/110, Complaint in Chancery (undated but c1814).
343 Cawdor (Vaughan) 14/450 (26 Dec 1803).
344 Nevill 411 (29 Apr 1825) quoting an unseen agreement dated 6 Jan 1804.
345 "Old Llanelly" by J. Innes and Llanelli Public Library plan 6 (1806).
346 Cawdor 2/133 — Brief of a Bill in Chancery (1814).
347 Nevill MS IV, copy letter (27 Apr 1805).
348 Ibid., copy letter (12 Nov 1804).

349 Ibid., copy letter (16 Mar 1805).
350 Llanelli Public Library plan 7 (Jul 1806) showed the Bres pit to be in an abandoned condition at that time.
351 Nevill 650 (26 Jan 1805).
352 Nevill MS IV, copy letter (19 Aug 1806).
353 Ibid., copy letter (12 Jan 1807).
354 Nevill 411 (29 Apr 1825), quoting an unseen agreement dated 16 Feb 1807.
355 Nevill MS IV, copy letter (23 Feb 1807).
356 Cawdor 2/133 — "Instructions for assignment of Leases of Collieries etc" between Messrs. Roderick and Co. and Messrs. Daniell and Co. (1807).
357 *The Cambrian* (16 Jan 1808).
358 Ibid., (21 May 1808).
359 Thomas Mainwaring, op.cit.
360 Stepney Estate Office plan 46 (1814) (CRO).
361 Cawdor 2/133 (1814), 2/110 (c1814).
362 Nevill MS XXI, loose statement (19 May 1815).
363 Nevill MS XI (1807-1817).
364 Thomas Mainwaring, op.cit.; Stepney Estate 1117 (c1878) — copy abstract from an unseen Deed of Covenant dated 1 June 1812; Stepney Estate 1103 (1872).
365 *The Cambrian* (1 Nov 1817); Thomas Mainwaring, op.cit.
366 *The Cambrian* (6 Sep 1817).
367 Local Collection 252 (1828).
368 NCB Abandonment plan 5244 (1908).
369 Thomas Mainwaring, op.cit.
370 Stepney Estate 1103 (1872).
371 Nevill 89 (25 Mar 1825), 187 (26 Mar 1825).
372 Nevill 411 (29 Apr 1825).
373 Ibid.
374 Thomas Mainwaring, op.cit.
375 Ibid.
376 Local Collection 252 (1828).
377 Calendar of the Diary of Lewis Weston Dillwyn, Vol II, entry (30 Apr 1828).
378 9 Geo IV c91 (19 Jun 1828).
379 Stepney Estate 1103 (1872) and Nevill 610 (4 Sep 1837), both quoting an unseen agreement dated 7 Nov 1828.
380 *The Cambrian* (28 Feb 1829).
381 Thomas Mainwaring, op.cit.; Nevill 1/6 (1 Jan 1830) (CRO).
382 *The Cambrian* (7 Apr 1838).
383 Stepney Estate 1103 (1872).
384 Mansel Lewis 731 (1 Apr 1800).
385 Thomas Mainwaring, op.cit.
386 Mansel Lewis 1928 (c1804/05).
387 Mansel Lewis 2082, account (8 Nov 1805).
388 Mansel Lewis 198 (27 May 1805).
389 Mansel Lewis 2082, account (8 Nov 1805).
390 Mansel Lewis 2538 — "Map of Ystradeu Demesne" (1805). This pit was still in use in 1908 as an air pit to the Cwmmawr Colliery (NCB Abandonment plan 12827 — 1910).
391 Thomas Mainwaring, op.cit.
392 Brodie Collection 68 (1808).
393 Thomas Mainwaring, op.cit.
394 Mansel Lewis London Collection 97, letter (4 Nov 1797).
395 42 Geo III c80 (3 Jun 1802).
396 Thomas Mainwaring, op.cit.
397 47 Geo III, c107 (8 Aug 1807).
398 Mansel Lewis 1654 (1810 to 1822).
399 Mansel Lewis 732 (19 Oct 1813).
400 Museum Collection 177 (1818-1832), 176 (1819-1829).
401 Nevill 1784 (30 Nov 1836), quoting an unseen lease.
402 The evidence of: NCB Abandonment plan 3564 (1896); Penller'gaer B.29.21 (c1828); Llanelli Public Library plan 10 (c1822).
403 Local Collection 37 (11 May 1824).
404 It has been stated that Roberts and Nevill were relations ("Canals of South Wales and the Border" by C. Hadfield p. 38) but no confirmation of this has been seen. Nevill was at least a close friend of the Roberts Family though, because John Roberts, senior, appointed him the trustee to administer Martyn John Roberts's marriage settlement in 1829 (Nevill 1/18, 2/22 both dated 20 Dec 1853 — CRO).

405 "The Canals of the Gwendraeth Valley (Part 2)" by W. H. Morris and G. R. Jones, The Carmarthen Antiquary Vol. 8, 1972 and Charles Hadfield, op.cit.
406 Nevill 1732 (26 Feb 1825).
407 Thomas Mainwaring, op.cit.
408 Nevill 1009 (3 Dec 1836).
409 Nevill 1731 (14 Jul 1826).
410 Nevill 2790 (5 Feb 1849).
411 Charles Hadfield, op.cit.
412 Museum Collection 181 (1827-34).
413 Nevill 1784 (30 Nov 1836), 51 (21 Mar 1839).
414 Nevill 2774 (2 Jul 1849), 2692 (27 Jul 1849).
415 Local Collection 37 (11 May 1824).
416 Mansel Lewis 29 (7 Jan 1828). This document made reference to the fact that "*Messrs. Shears and Jones*" took over the Spitty Bank copperworks, but no other reference to Jones has been seen and it is not known who he was.
417 Penlle'rgaer A. 974 (16 Apr 1819).
418 "Report of the Llangennech Collieries" by Edward Martin dated 21 Sep 1803, contained in Local Collection 37 (11 May 1824).
419 *The Cambrian* (16 Jun 1832).
420 British Almanack 1835, article entitled "Railway and Dock of the Llangennech (St David's) Company".
421 Mansel Lewis 29 (7 Jan 1828).
422 Nevill 73 (1 Feb 1826).
423 Mansel Lewis 29 (7 Jan 1828).
424 Nevill 466 (1 Jan 1827).
425 Nevill 467 (20 Jun 1827), 468 (24 Jun 1827).
426 "Calendar of the Diary of Lewis Weston Dillwyn", Vol II (30 Apr 1828) and House of Commons Journals, LXXXIII (1828), 202-3, quoted in "Some Notes on the Mansel Lewis Papers", Part III by R. Craig, Carmarthen Antiquary, Vol IV (1963).
427 Cawdor 2/70, covenants dated 21 and 24 Apr 1828.
428 "An Act for making and maintaining a Railway or Tramroad from Gelly Gille Farm, in the Parish of Llanelly in the County of Carmarthen, to Machynis Pool in the same Parish and County; and for making and maintaining a Wet Dock at the Termination of the said Railway or Tramroad at Machynis Pool aforesaid" (9 Geo IV c91 dated 19 Jun 1828).
429 Mansel Lewis 29 (7 Jan 1828).
430 Letter dated 16 May 1829 enclosed in Nevill 73 (1 Feb 1826).
431 Y Diwygiwr, Vol XII, Nov 1847 stated that the Spitty Copperworks had been idle for 16 years. Additionally, the copper ore purchase statements in *The Cambrian* in 1831 showed that Messrs. Shears had stopped buying ore by June of that year. (I am obliged to Mr R. O. Roberts for bringing this to my notice).
432 *The Cambrian* (16 Jan 1832).
433 Nevill 77 (25 Mar 1829).
434 NCB Abandonment plan 9265 (1927) showed the "*Old Cefncaeau Slant*" into the Swansea Four Feet Vein but nothing is known of the origins of this colliery. Hunts Mineral Statistics also listed a "*Cefn Caen*" colliery between 1858 and 1870 and this may have been the level originally worked by Warde, taken over by Gryll and his partners and later worked by others.

CHAPTER 5

CANALS AND RAILWAYS

INTRODUCTION

The local demand for Llanelli's coal did not begin to equate to its production until the later 19th century, when the coal industry was in decline and iron and tinplate works consumed large quantities of coal. Before this, coal output was directly controlled by export trade demand, and, as the trade was conducted almost solely by shipping up to 1830, it meant bulk transportation of coal from pit-heads to shipping places. For this to be easily achieved, transport systems were necessary but, despite the early rise of Llanelli's coal industry, they were not provided until after 1750, when canals and railways replaced horses, mules and oxen, laden with panniers or drawing carts, and, perhaps, men with wheelbarrows where pits were near shipping places.[1]

This chapter examines canals and railways, constructed and proposed, in the Llanelli area before 1830.

Canals are interpreted as all artificially constructed waterways (navigable rivers, creeks or pills will be dealt with under Shipping Places). Railways are interpreted as all specially constructed waggon-roads, waggon-ways, tramroads, railroads and railways, along which horse-drawn coal waggons could operate (steam locomotive haulage was not introduced to the Llanelli area until 1840).[2] The known routes of all the canals and railways used to move coal in the area before 1830 are shown on Figure 20 and Appendix I.

(I) CANALS ACTUALLY CONSTRUCTED

Llanelli's period of canal construction probably lasted from 1766 to 1798, although it might have commenced in the 1750s. Only five small canals appear to have been built, although many were proposed, some of them major enterprises. As far as can be ascertained, they were simply constructed, without locks, and situated mainly in tidal areas. Parliamentary approval was not needed for their construction. They had probably

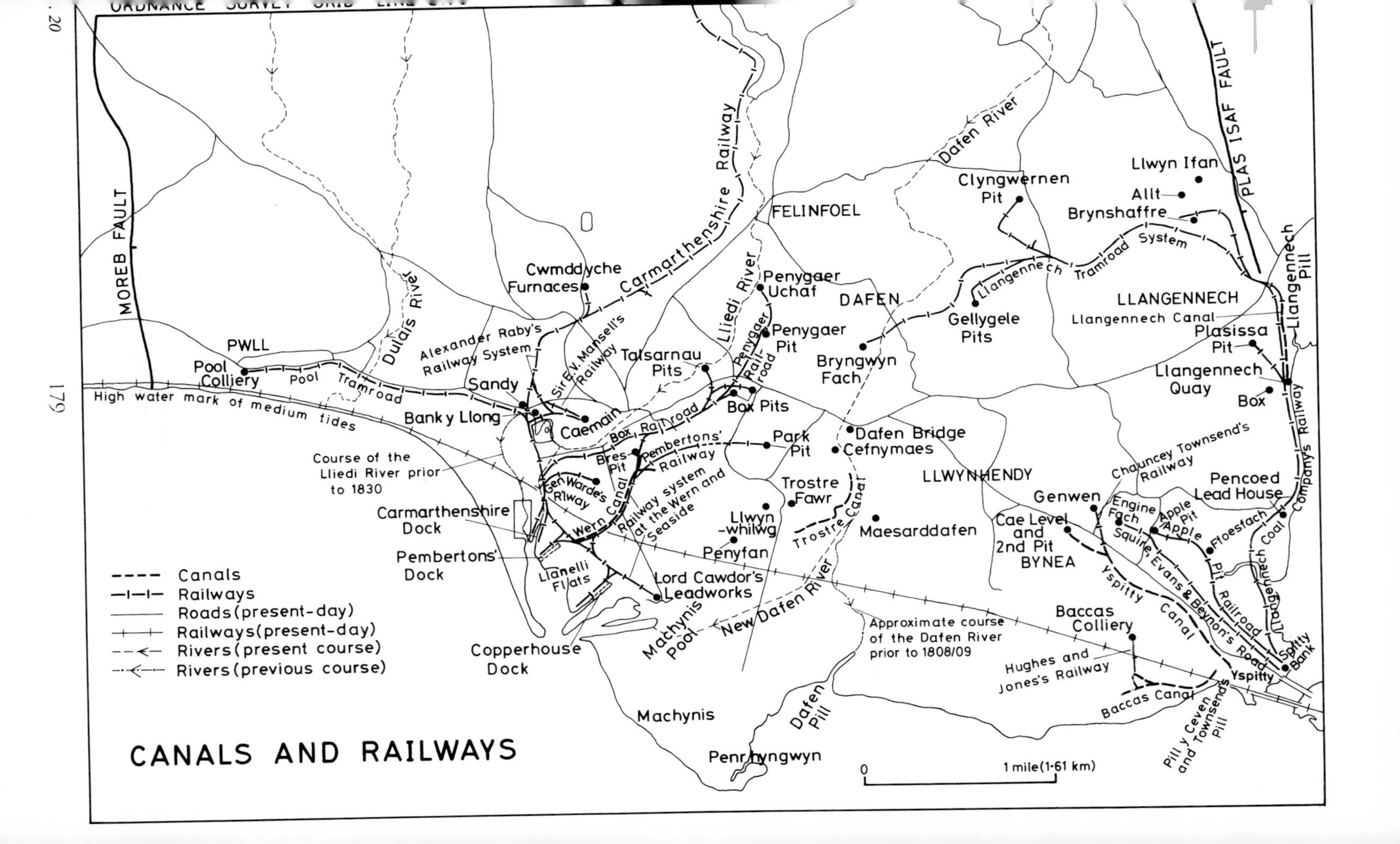

CANALS AND RAILWAYS
MOREB FAULT
PLAS ISAF FAULT
PWLL
Pool Colliery
Pool Tramroad
High water mark of medium tides
Dulais River
Alexander Raby's Railway System
Sandy
Bank y Llong
Cwmddyche Furnaces
Carmarthenshire Railway
Sir E. V. Mansell's Railway
Caemain
Talsarnau Pits
Box Railroad
Box Pits
Penygaer Railroad
Penygaer Uchaf
Penygaer Pit
Lliedi River
FELINFOEL
DAFEN
Bryngwyn Fach
Dafen River
Clyngwernen Pit
Gellygele Pits
Llangennech Tramroad System
Llwyn Ifan
Allt
Brynshaffre
LLANGENNECH
Llangennech Canal
Plasissa Pit
Llangennech Quay
Box
Llangennech Pill
Company's Railway
Chauncey Townsend's Railway
Pencoed Lead House
Llangennech Coal
Ffoesfach
Apple Pit
Apple
Engine Fach
Squire, Evans & Beynon's Road
Pit Railroad
Spitty Bank
Yspitty
Yspitty Canal
Genwen
Cae Level and 2nd Pit
BYNEA
LLWYNHENDY
Dafen Bridge
Cefnymaes
Park Pit
Pembertons' Railway
Bres Pit
Gen Warde's R'lway
Wern Canal
Railway system at the Wern and Seaside
Trostre Fawr
Trostre Canal
Maesarddafen
Llwyn-whilwg
Penyfan
Lord Cawdor's Leadworks
New Dafen River
Machynis Pool
Approximate course of the Dafen River prior to 1808/09
Baccas Colliery
Hughes and Jones's Railway
Baccas Canal
Pill y Ceven and Townsend's Pill
Course of the Lliedi River prior to 1830
Carmarthenshire Dock
Pembertons' Dock
Llanelli Flats
Copperhouse Dock
Machynis
Dafen Pill
Penrhyngwyn
0
1 mile (1·61 km)
Canals
Railways
Roads (present-day)
Railways (present-day)
Rivers (present course)
Rivers (previous course)

. 20

all ceased to be used regularly by 1810, although the re-use and extension of two were considered in the 1820s.[3]

Unlike coal mining, well documented because of royalty income, little reference was made to canals in contemporary documents and it is difficult to draw positive conclusions in all cases.

The five canals known to have been constructed were:-

1. The Yspitty Canal, built by Chauncey Townsend between 1766 and 1770, from his colliery at Genwen to his shipping place at Yspitty.
2. The Trostre Canal, built by Thomas Bowen c1770 (or perhaps earlier), connecting with the Dafen River in the Trostre region.
3. The Llangennech Canal, probably constructed by Thomas Jones c1775 and later extended to the site of Llangennech Quay by William Hopkin, c1790.
4. The Baccas Canal, constructed by the Rev. David Hughes and Joseph Jones c1795, connecting the Baccas or Carnarvon Colliery to Pill y Ceven.
5. The Wern Canal, constructed by William Roderick, Thomas Bowen and Margaret Griffiths c1795, from their collieries in the Bres/Wern region to their shipping place at Seaside.

(1) *Yspitty Canal*[4]

The full route of the Yspitty Canal is given in Figure 21. Between 1752 and 1762 Chauncey Townsend was granted most of the unleased coal in the Llanelli area by the Mansells and by the various inheritors of the Llanelli Vaughans' original estate. Townsend invested little capital in the Llanelli Coalfield until he embarked upon the development of a major colliery at Genwen, in 1766.

To convey his coal to ships, Townsend intended constructing a canal from the colliery to a pill, at Yspitty, which led into the River Loughor. In 1765, he entered into a draft agreement with Richard Vaughan of Golden Grove for a lease of part of the common land in the Hamlet of Berwick, to construct this canal.[5] The final agreement, dated July 1766, leased the land for 21 years at an annual rent of £3.3.0. The canal was not to exceed 8 yards (7.3m) breadth and at least twelve men were to be employed each day, for at least a hundred days in each year, until it was completed. It was also agreed that work would begin immediately and that, because of the great expense involved, Vaughan would renew the lease, as he could not grant leases for more than 21 years.[6] Townsend probably commenced his canal in July/August 1766 and there is evidence that it was completed before 1770, Townsend's grandson, Henry Smith, stating in 1810 "*I also know that, previous to the year 1770, my grandfather had sunk pits, erec-*

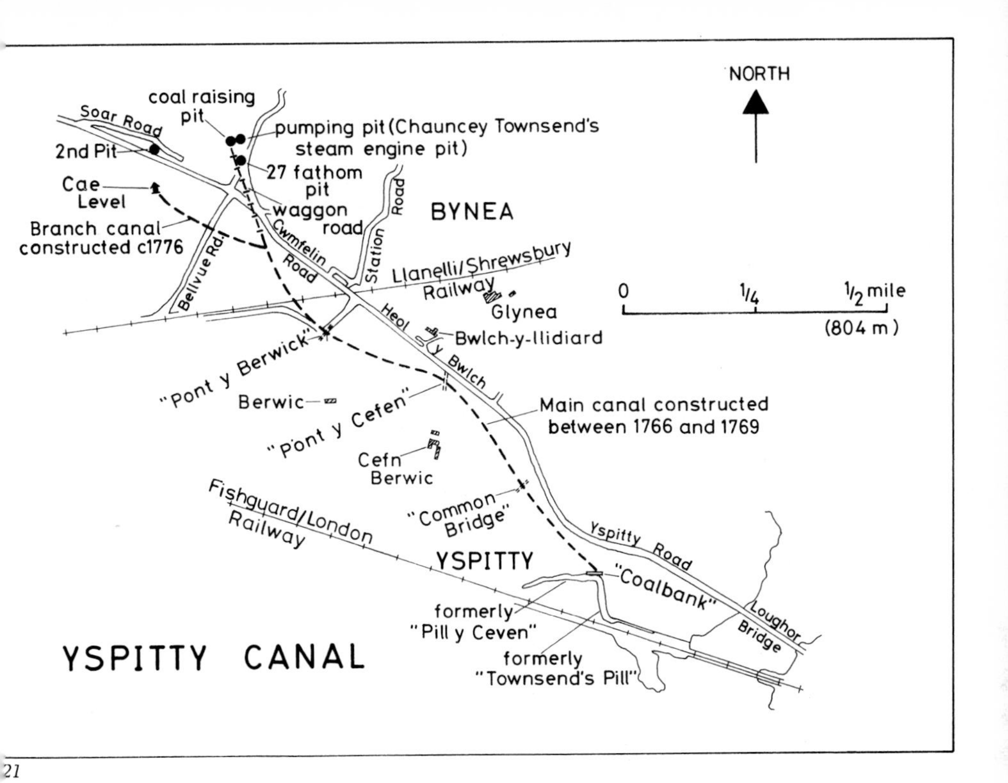

21

ted a fire engine, and had made a Canal for leading Coals down to the water-side, near Llanelly in Carmarthenshire".[7] Townsend died in 1770, when the management of his industrial interests passed to his son-in-law John Smith. The extent of Townsend's canal at this time was shown on plans of 1772.[8] The canal was linked by a waggon-road to two coal raising pits (181, 183) and ran 1700 yards (1553m), ending at Yspitty, short of a creek called Townsend's Pill which led into the River Loughor. The canal width scaled 30 feet (9.1m) on these plans, although 24 feet (7.3m) was specified in the 1766 lease from Vaughan. The canal was crossed by three bridges, "*Pont y Berwick*", leading to the Berwick Farms, "*Pont y Cefen*", leading to Cefen Farm and "*Common Bridge*", probably leading on to the common lands of Bynea Marsh. The coal bank at Townsend's Pill suggests that the coal was taken by barge down the canal, unloaded at the coal bank, and then loaded into small boats, or into keels which took the coal to larger ships lying at the deeper anchorages in the River Loughor, off Penclawdd and Whitford Point. (Plate 21)

John Smith hardly worked the colliery at Genwen, but he probably opened the "*2nd pit*" (177)[9] and the "*Cae Level*" (178)[10] into the Swansea Five Feet Vein and constructed a westerly extension, some 400 yards (365m) long, from Townsend's canal to the entry to the Cae Level.[11]

Plate 21 — Part of John Thorton's 1772 plan showing the Yspitty Canal. The canal was referred to as *"Townsend's Canal"* on Thornton's plan.

This was constructed before 1785, a plan, made that year,[12] showing part of a canal in a position corresponding to the westerly extension of Townsend's canal. As Smith had intended working to the west of the colliery at Genwen in 1776,[13] it is possible that the extension was built about this time. Information, superimposed at a later date on a 1751 plan,[14] termed the extension *"Canal for 2nd Pit"* and it must have been constructed solely to convey coal from the new winnings in the Swansea Five Feet Vein to Townsend's Pill. In 1786, Smith was leased lands in the Hamlet of Berwick, to extend the present canal westwards to the Dafen Bridge area, where he intended sinking new pits,[15] and it has been interpreted that the western extension to the Cae Level was built as a result of this lease.[16] The direction of the extension, on a northward curve towards the Cae Level and the 2nd Pit, and away from a true westward route to Dafen Bridge, and the evidence of the 1785 plan, tend to discredit this view, however, and it is considered that the extension to Dafen Bridge was never started.

Smith probably worked little coal after the 1770s[17] and the Yspitty Canal must have been seldom used. Nevertheless, he continued to rent it, probably up to his death in 1797,[18] when it passed to his two sons Charles and Henry. They assigned their interests to George Warde, in 1802, and ownership of the canal passed from Townsend's family for the first time since its construction, 36 years earlier.

Warde initially worked coal from inherited pits, while forming his plans to exploit the Bynea, Llandafen and Old Castle coal areas and he must have used the Yspitty Canal and Townsend's Pill, which were named "*General Warde's Canal*" and "*General Warde's Shipping Place*" in later plans.[19] The canal was certainly in use in October 1807, his accounts for that month containing the entry "*Paid Morgan Davies for bringing a piece of Oak timber down canal, 6/4 ½ d*".[20] It probably fell into disuse after Warde's new Apple Pit Tramroad and his new shipping places at Spitty Bite became operative, in August 1808, as a plan of 1813 termed it "*Old Canal*".[21]

In 1821, it was proposed to extend the canal westwards for almost a mile, to just south of Maesarddafen,[22] but this proposal was not acted upon. In 1824, a report on the smelting of copper ore at Spitty Bank suggested that coal from the Baccas (Carnarvon) Colliery could easily be delivered to the Spitty Copperworks, if the two canals "*already cut*" in the vicinity were linked up.[23] These canals must have been the Yspitty Canal and the Baccas Canal, constructed by Hughes and Jones c1794, but again the proposal was not acted upon. No further reference to the Yspitty Canal has been seen and all available evidence supports the view that it fell into disuse shortly after August 1808.

Most of the canal has now been filled in or built over but a short length of the original route still exists in 1979. (Plate 22)

(2) *Trostre Canal*[24]

A plan of Trostre Fawr Farm, dated 1785,[25] showed a canal, approximately 16 feet (4.9m) wide and half a mile (804m) long, connecting with the River Dafen. The canal was of unusual shape (Fig. 22), consisting of two straight arms joined to a central curved portion, which may have been

Plate 22 — Aerial view of the remains of the Yspitty Canal today. A short length of the canal route shows up clearly to the right of the main road, opposite Bowden's Factory.

an old meander course of the River Dafen, and which linked with the river in two places. The arms must have led to convenient loading places for nearby pits, where coal was put into barges for transport to the River Dafen, and then down Dafen Pill to a coal bank, or directly to ships anchored in the Estuary. (Plate 23)

No lease has been seen relating to this canal and its construction date has not been positively determined. We know, however, that Thomas Bowen was taking coal, in keels, to ships anchored in the Estuary off Penclawdd in 1752,[26] that he was leased the "*Llwynwilog Vein*" by John Vaughan in 1754[27] (Llwynwhilwg was close to the southern end of the canal), and that he had a "*Colework*" (a coal shipping place) at Dafen Pill in 1757.[28] Additionally, tradition in Llanelli has perpetuated the opinion that Thomas Bowen built a canal, in 1750, to convey his coal from Penyfan (near Llwynwhilwg) to Penrhyngwyn Point (on Dafen Pill).[29] All this suggests that Bowen built the Trostre Canal (or its southern arm?) in the 1750s, but the evidence is circumstantial and Golden Grove Estate rentals for this period[30] make no reference to what would have been Wales's earliest known canal.

Documentary evidence exists, however, to prove that Bowen used the

Plate 23 — Plan of "*Trawstre in Llanelly, 1785*" showing the Trostre Canal.

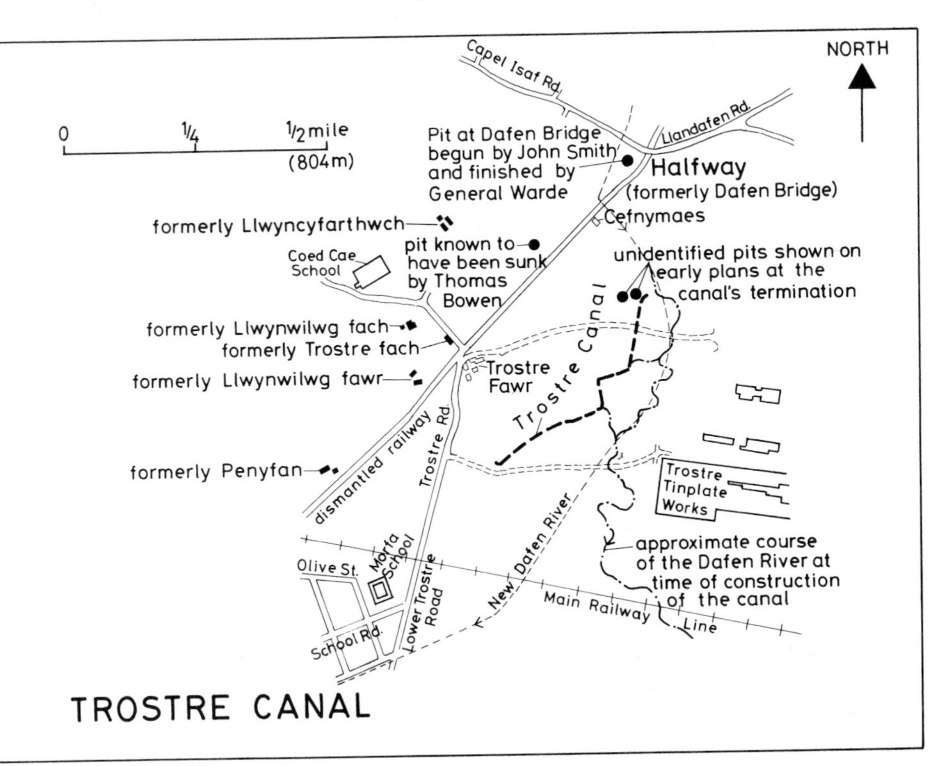

Trostre Canal in the 1770s. In 1768 he was leased the coal under Trostre, Llwyncyfarthwch and Cae Cefn y Maes,[31] a letter of April 1769 saying "*Mr Bowen I hear has got a fresh pit into coal on the same estate as before but cannot make any head thereof without making a Water Course*[32] *over part of Sir Edward's Estate which is Drawsdre or Llwynwhilwg Vach*".[33] A further letter, of November 1771, said that Bowen was allowed to work coal under Stradey Estate lands, the correspondent also stating "*If you will put me into a method of vending more coal than we do, we certainly shall work more. and if we had Trade to take of Mr Edward's Coals near Davon Bridge it should not be long before the Canal was carried up there but unfortunately at present we want Ships to take of even the Coals we now raise*".[34] As "*Davon Bridge*" was the area now called Halfway, the Trostre Canal (or its northern arm?) must have been built before November 1771. Bowen paid rent to the Golden Grove Estate for a canal and a watercourse, between 1772 and 1781,[35] and to the Stepney Estate for Penrhyngwyn Point, between 1772 and 1780.[36] We also know that Bowen worked a pit near to Cefn y Maes (92), close to the northern branch of the canal,[37] and two pits (121, 122), shown on old plans at the end of the northern branch,[38] were situated in an area included in Bowen's coal leases. Golden Grove Estate rentals for 1781/82, 1782/83, 1787/88 and 1790 listed Bowen's canal and watercourse as "*not used*".[39] Bowen must have ceased operations in the Trostre region in 1781/82, when his canal fell into disuse.

Both Smith and Warde considered incorporating the Trostre Canal into their schemes[40] but it was never used after Bowen abandoned it. The entire route of the canal was still plainly visible on the first 1 : 10560 scale Ordnance Survey plan (surveyed in 1878 and published in 1891) but only parts of the southern branch remain today and are marked as "*drains*" on the present editions of the Ordnance plans.

(3) *Llangennech Canal*

A plan of c1790[41] showed an "*Old Canal*" about 360 yards (329m) long, running from just south of Llangennech to Llangennech Pill (Fig. 23). The canal ended at the pill, where barges must have off-loaded coal for further loading into barges, keels or small ships able to sail up the pill. No record of the canal's construction has been seen but available evidence suggests that it was formed by, or for, Thomas Jones of Llwyn Ifan.

In 1773 Thomas Jones applied to William Clayton (owner of the Alltycadno Estate) and to Sir John Stepney, 8th Bart., to take over the coal interests at Brynsheffre and the Allt which Sir Thomas Stepney, 7th Bart.,

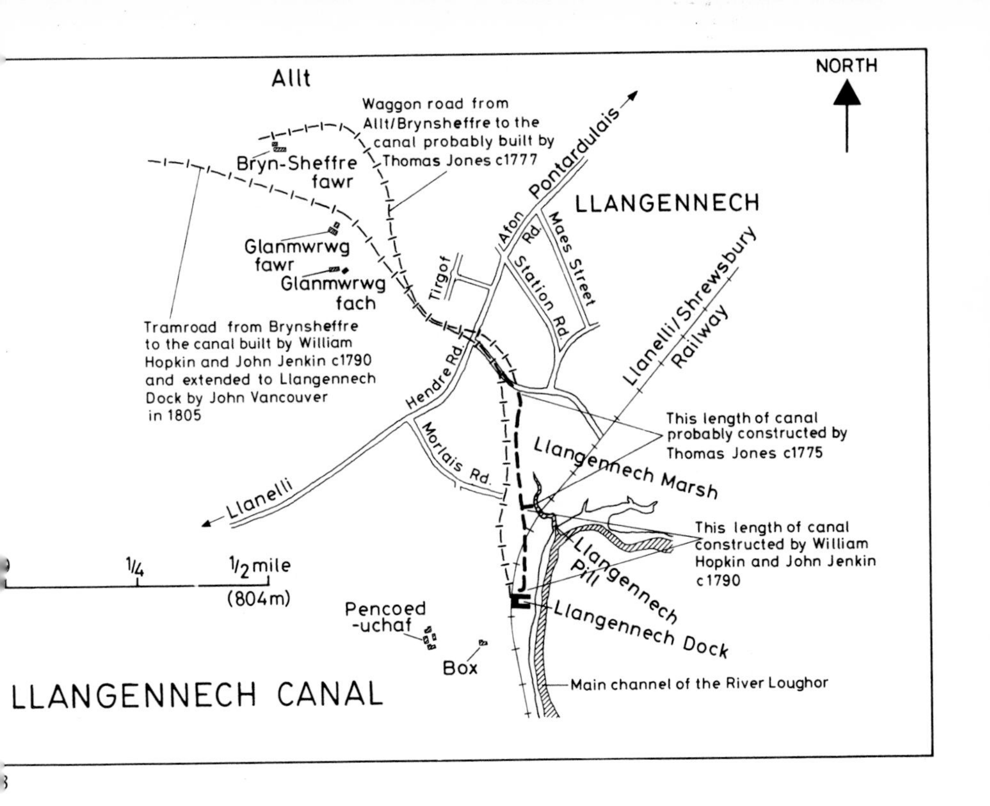

had started, shortly before his death in October 1772. Jones was leased the Brynsheffre coal by Clayton, in 1774, when Stepney probably also sub-leased the Allt coal. Jones intended to construct a waggon road from his colliery to a coal bank alongside the River Loughor, on Clayton's land near the tenement of Box, where coal could be loaded directly into ships. Jones mined at Brynsheffre between 1775 and 1783 and was involved, in 1777, in the construction of waggon roads from Brynsheffre to a new shipping bank, situated on "*Llangennech Marsh*"[42] and not opposite Box tenement, as originally intended. Jones ceased working about 1783 and assigned his lease of Brynsheffre coal to William Hopkin before 1793.[43] Hopkin, therefore, took over Thomas Jones's interests and the plan already referred to showed that, before c1790, a waggon road or tram-road had been formed from the Allt/Brynsheffre region, to connect with a short canal which ended at Llangennech Pill on Llangennech Marsh. Thomas Jones must have used this canal to barge coal from Llangennech to Llangennech Pill between 1775 and 1783, and it was probably cut by specialist canal contractors. When Clayton leased Brynsheffre to Jones, he also leased his coal at Talyclyn to a John Browne, who later lived at Llanedy Forest.[44] Browne informed Clayton, in April 1775, "*I have hired an Engineer and two other Persons in different departments, from*

England, to direct the making of a Canal over the Marsh",[45] the actual site being at Talyclyn Uchaf and Talyclyn Isaf, on the evidence of a dispute between Browne and Clayton's tenants, whose lands were cut to form the canal.[46] It is highly probable that the builders of the Talyclyn Canal also built Jones's canal, as both were made at the same time and both carried coal from Clayton's estate to the upper reaches of the navigable River Loughor.

When William Hopkin (and John Jenkin) took over Jones's operations at the Allt and Brynsheffre, between 1789 and 1793, they proposed constructing new tramroads and extending the canal to the north and south. The plan of c1790 showing these proposals, was entitled "*Plan of Mr Hopkin's intended Canal*"[47] and there is no doubt that the existing canal was meant to play a major part in Hopkin's enterprise. No contemporary evidence of the implementation of Hopkin's schemes has survived, but subsequent information shows that he extended and used the Llangennech Canal. A plan of 1824 showed the canal with a 300 yard (274m) long southerly extension leading to "*Llangennech Dock*" (Quay), a shipping place on the River Loughor opposite the tenement of Box, and the publication which contained the plan made reference to "*the little canal which Mr Hopkins formerly cut*".[48] Hopkin (and Jenkin) were therefore responsible for finally implementing Sir Thomas Stepney's 1772 scheme. The canal was not used for long, because William Hopkin surrendered his lease of the Allt coal in 1794, and of Brynsheffre coal before 1805.[49] The Llangennech Canal fell into complete disuse in 1805, when the Earl of Warwick and John Vancouver completed their new railroad, to transport Llangennech Estate coal directly to Llangennech Quay.

The entire 700 yards (640m) of the Llangennech Canal are still visible on aerial photographs and traces can be seen at ground level, although sedimentation has largely concealed its profile. (Plate 24)

(4) *Baccas Canal*[50]

In September 1794, John Vaughan leased coal under "*Morfa Baccas*" to the Rev. David Hughes of Ffosfach and Joseph Jones of Bryn Carnarvon, and permission was given to make a "*navigable cut or Canal from the said Collieries through and over the Common or Waste called Baccas to and as far as the River called Loughor adjoining the said waste for the purpose of conveying the coal from the said Collieries in as large and ample a manner only as the said John Vaughan can or may lawfully Grant or Lease.*" Liberty to use the canal, at the rate of ½ pence per ton, was reserved to John Vaughan and Sir John Stepney.[51]

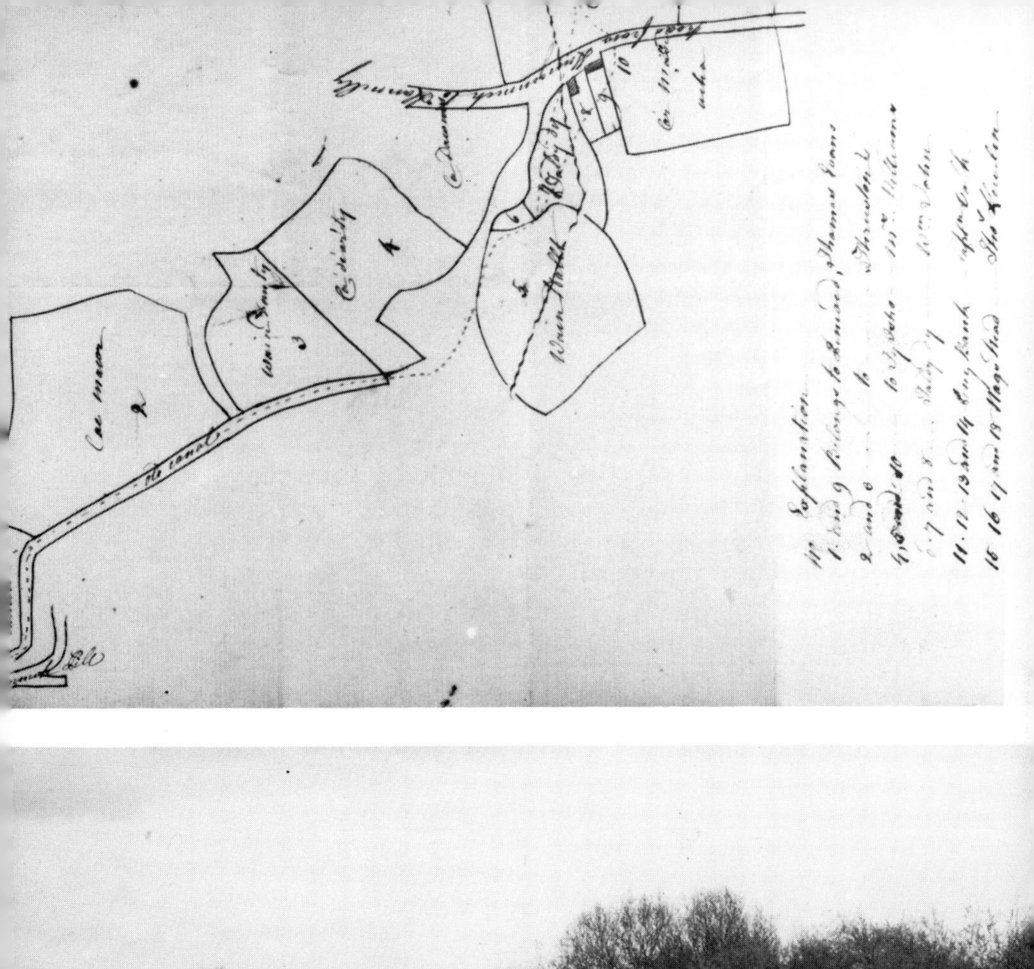

Plate 24 — The Llangennech Canal. (Top) Plan of c1790 showing an "*Old Canal*" finishing at Llangennech Pill. (Bottom) Present-day remains of the canal — part of the straight portion running past "*Caemawr*" in the top plate.

Hughes and Jones sank the Baccas (later called the Carnarvon) Colliery and constructed a tidal canal, to transport coal to Pill y Ceven, which joined Townsend's Pill. Although no contemporary plan of the canal appears to have survived, a number of plans, of later date,[52] showed the canal's route in detail. An existing water course, in the form of a loop, was probably incorporated at the western end of the canal and a railway built from the colliery to this water course. The canal was just over half a mile (804m) long and curved northwards at its eastern end, to meet Pill y Ceven (Fig. 24).

The period over which the canal operated is not known. Rental was paid by "*Messrs. Hughes*", in 1797, for the trespass of the canal over Stepney Estate lands[53] but a schedule of the Estate, dated 1802,[54] made no reference to it. A report on the "*Baccas Colliery*", drawn up before 1824,[55] said that the managing partner of the concern (Joseph Jones?) had died suddenly and that the "*London proprietors*" had discontinued working there. Plans of 1824 and c1825[56] certainly referred to the canal as "*Old Canal*", so it must have been abandoned several years before that time.

Although sedimentation has concealed its profile, the canal's route is

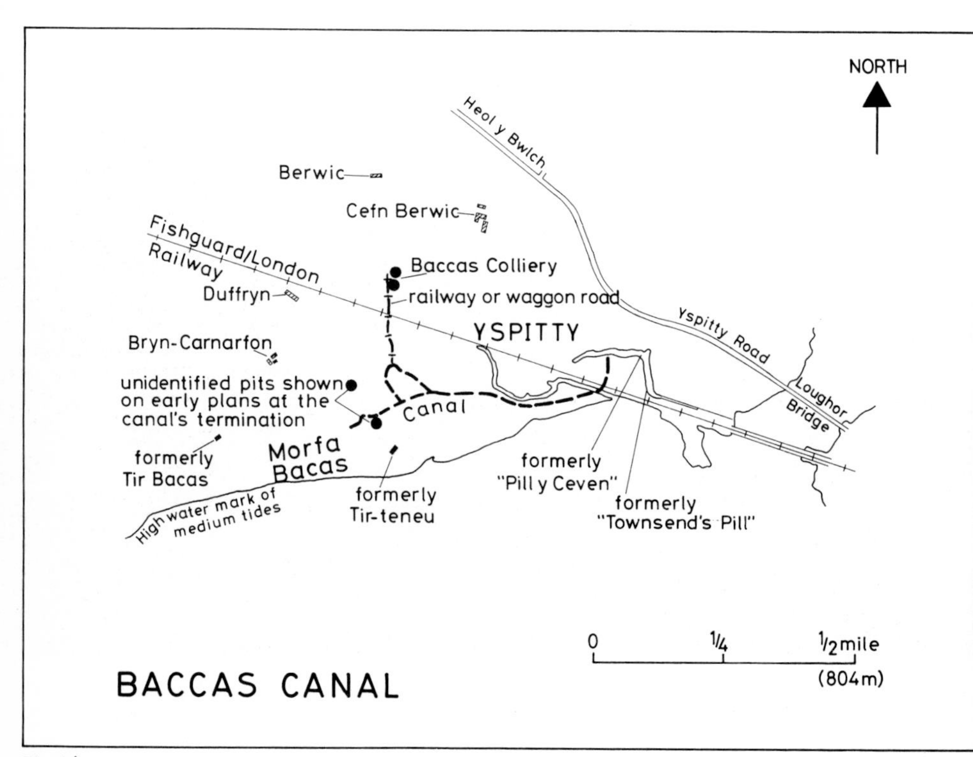

Fig. 24

still visible at ground level, shows up clearly on aerial photographs but, surprisingly, has not been shown on any edition of the Ordnance Survey plans. (Plate 25)

(5) *The Wern Canal*

William Roderick was leased coal under Stepney Estate lands in the Llanelli Town region by Sir John Stepney, and coal under the common of the Wern by John Vaughan, on 1 October 1794. Both leases granted permission for "*trenches, canal or canals*" to be constructed.[57] Roderick and his partners commenced sinking the Bres pit (49), on Stepney's lands at Bres Fawr, in late 1794 or early 1795, when a canal to take coal to a shipping place at Seaside was probably started. There is evidence that a water course had previously been formed, from the River Lliedi to the Wern, to impel water engines[58] and it is probable that this supply, with water raised from the pit being sunk, was used to fill the canal, although tidal water could have been used in the lower reaches.[59] When finished, the canal was approximately three-quarters of a mile (1206m) long, linking the Bres pit to a shipping place on Llanelli Flats, with a short eastern branch, at the Wern, connected to the water course from the Lliedi[60] (Fig. 25). It ran from just east of Bres Road, into the yard of Waddle Engineering and Fan Co. Ltd., with the branch to the water course running just south of Columbia Street. It then went southwards, ran immediately west of Ann Street, crossed over Station Road, along Heol Fawr, across the main railway just west of the gates, and ended in Bryn Terrace some 50 yards (46m) north west of St John's Church.(Plate 26)

The Bres pit sinking stopped about 1797, as the engine could not cope with the water, further sinking being abandoned until the Pembertons installed a steam engine, in 1807. The existence of a branch of the canal to the Bres pit, therefore, implies that the canal was constructed between the granting of the leases, in October 1794, and the abandonment of the Bres pit, in 1797, although the first contemporary evidence of its existence was a plan dated 1800.[61] It was being used to carry coal in April 1801,[62] and also in 1804, when the Burgesses of Llanelli required "*Bowen and Roderick to repair the road leading from the town to the Flats between Penishayrwern and Heolfawr which is out of repair due to the water breaking out of the Canal and running over the road*".[63] Although this suggests that the canal was poorly maintained, it was not abandoned. Charles Nevill stipulated, in January 1805, in an agreement for Roderick to supply coal to the copperworks "*the coal must be delivered from the Canal by a Railway*";[64] the Burgesses of Llanelli levied a rent of £2 in

Plate 25 — The Baccas Canal. (Top) Plan of c1822 showing an "*Old Canal*" over Morfa Baccas common. (Bottom) The route of the canal shows up clearly on present-day aerial photographs.

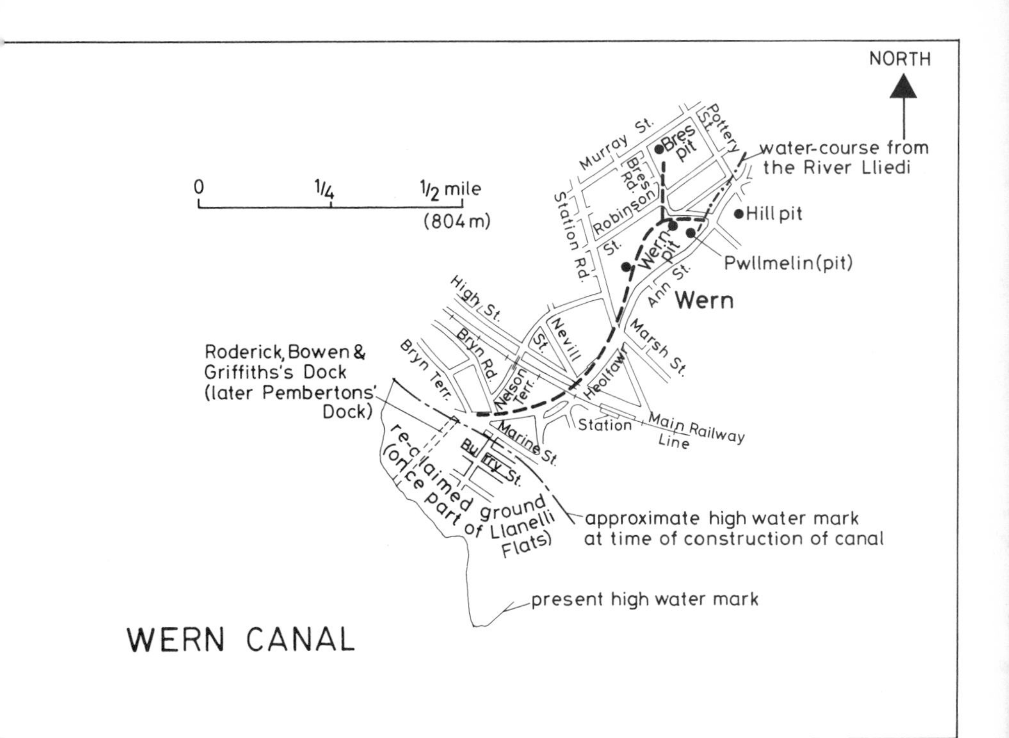

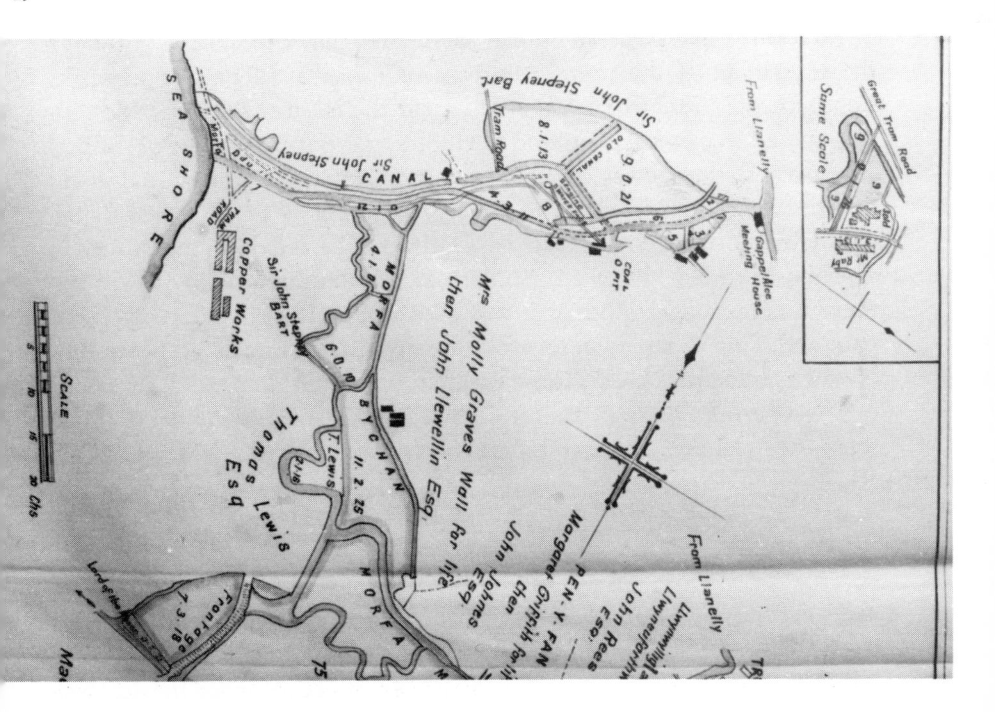

Plate 26 — The Wern Canal in 1810.

October 1805 on the "*present Canal and Railways*" used by Bowen and Roderick;[65] and Thomas Bowen said, when advertising his and his daughter's half share of the colliery for sale, in August 1806, "*the Coals are raised within half a mile of the shipping place to which there is an easy communication by canal and railroad*".[66]

Roderick and partners sold out to the Pembertons and the Copperworks Company in 1807, after which the canal probably fell into increasing disuse. The Wern Colliery, which had previously used the canal to transport its coal, was re-named the "*Wern Railway Colliery*" by the Copperworks Company in September 1807,[67] implying that the railway along the Wern to the Copperworks, constructed when the Copperworks Company lent money to Roderick in 1805,[68] had superseded the canal as their main transportation route. The Pembertons, who re-developed the Bres pit from 1807, still regarded the canal as a transportation route though, because a sale of lands at the Wern to the Copperworks Company, in September 1809, stated that the ground occupied by the Canal was being sold "*subject and without prejudice to the right of John Pemberton, Esq., his heirs, executors, administrators or assigns to the use of the said Canal*".[69]

The canal was probably not used, however, because a plan, of October 1810, termed the northern branch to the Bres pit, "*Old Canal*", (although the remainder of the route was termed "*Canal*").[70] The Copperworks Company paid rent of £4.4.0, in 1811, for the lease of the canal and railway and a piece of waste ground at Heolfawr,[71] but the linking of the Box Tramroad to the railway at the Wern ensured that the Wern Canal would not be redeveloped, plans of 1814[72] showing a partly filled canal, paralleled along its entire route by a railway. The canal was mentioned in a proposal to sell the Wern Colliery to the Pembertons in 1815,[73] but there was no suggestion that it was then in use, the Copperworks Company referring to it only in terms of the provision of roads over it to R. J. Nevill's residence (Field House or Glanmor House).

The Wern Canal was, therefore, used between c1795 and 1807, but, thereafter, fell into disuse and was probably abandoned between 1810 and 1814.

(II) PROPOSED CANALS

In addition to the five known canals in the Llanelli area before 1830, a further thirteen canals were proposed but not constructed. These proposed canals fell into two distinct categories: those intended for the transportation of local coal to local shipping places and works; and those intended to transport coal and minerals from the Carmarthenshire hinterland to the works and shipping places of the Llanelli area. Many parts of South Wales were developed because canals were built during the late 18th and early 19th centuries, and men of influence in Llanelli would have seen the canal as the means to industrialise the entire area. Consideration of the proposed canals is, therefore, important, because it provides a valuable insight into the way local landowners and industrialists wanted the Llanelli area to develop between 1770 and 1830.

(1) *Chauncey Townsend's proposed canal at Maesarddafen*[74]

A few weeks after he had been leased the coal under extensive areas in and around Llanelli, Chauncey Townsend was granted the right to construct "*one or more Water Course or Water Courses, Waggon Way or Ways in, over or through the said Tenement called Maesyrddaven*" by Owen Howell for ten shillings per year.[75] In 1757, Townsend was shipping coal at Dafen Pill (called "*Esq. Townsend's Barge way*")[76] and we know that he was mining at Llwynhendy (168 and probably 170 to 172).[77] It is tempting to speculate that Townsend constructed a canal over Maesarddafen lands, to link his collieries to Dafen Pill, but no confirmatory evidence has been seen. Nevertheless, the straight water course, shown on the first 1 : 10560 edition of the O.S. plan, running east from the top of Dafen Pill, and then north towards the Llwynhendy/Hendre area, is puzzling because, prior to the building of the Machynis embankments and the formation of the New Dafen River Cut in 1809, Dafen Pill would have connected with the old course of the River Dafen, to the north-west. Unfortunately, no contemporary plan of this area has survived, and subsequent embanking and railway construction over the head of Dafen Pill has completely obscured the old profiles. It has been claimed that Townsend constructed this canal[78] but the evidence is purely circumstantial, and the Estate rentals and correspondence of the period make no reference to its existence.

(2) *William Fenton's proposed canal from Pencoed Lead House to Pantyffynon Mill*

In 1770, William Fenton petitioned the House of Commons for permission to construct a canal, at his own expense, from "*Pantyfunnon Mill*" (south of Ammanford) to Pencoed Lead House. Fenton worked anthracite in the Llandybie and Llanedy regions and wished to provide an economical transport route to ships in the River Loughor. Fenton's petition stated that the river was navigable for large ships only as far as

Pencoed.[79] The canal was not constructed, and its proposed route is not known in detail, although it was up the Loughor Valley.[80] Similar schemes were proposed in later years.

(3) *John Smith's proposed canal from Ffosfach to Spitty Bank*

John Smith became part-owner and manager of Chauncey Townsend's coal interests in the Llanelli area, when Townsend died in 1770. His main colliery, at Genwen, was connected by canal to a shipping place (Townsend's Pill), which Smith considered to be "*very inconvenient*". Instead, he wished to use Spitty Bank as his shipping place and approached Sir E. V. Mansell on this point in 1776.[81] Spitty Bank, then used as a shipping place by David Evans, was not Mansell's legal property, however, although the Bynea Farm lands adjacent to it were. These were also leased to Evans, the lease expiring in 1781. Smith informed Mansell, in 1780, that he wished to lease Bynea Farm and requested permission to construct canals, waggonways and shipping places on the lands.[82] Shortly after Evans' lease expired, in May 1781, Mansell agreed to lease Bynea Farm to Smith, stipulating that coal could be worked as soon as Smith had constructed a canal to the lands adjoining Spitty Bank, and built a wharf or shipping place there. The canal's route was not set out in detail but it was to pass over the common between Ffosfach and Bynea Farms.[83] Smith did not construct the canal or shipping place and the agreement for the lease was never finalised.

(4) *John Smith's proposed canal from Dafen Pill to Penrhyngwyn*

In the 1780s, John Smith surrendered the lease, granted by Sir Thomas Stepney and others to Townsend in 1752, and negotiated a re-lease, to include coal under new lands near Dafen Bridge (Halfway). Smith agreed to sink pits at Dafen Bridge, within two years of the signing of the new lease in 1785.[84] This coal had previously been leased to Thomas Bowen, who had constructed a canal which ended some 300 yards (274m) from the proposed pits and connected with the navigable reaches of the River Dafen. Smith intended to set up his shipping place at Penrhyngwyn and to construct a canal from the top of Dafen Pill to Penrhyngwyn.[85] This would allow coal to be taken in barges from the new pits, down Bowen's Trostre Canal into the River Dafen, and then down the new stretch of canal to the coal bank at Penrhyngwyn. Smith commenced at least one pit at Dafen Bridge (119) but stopped sinking before coal was reached.[86] The canal was not constructed but a similar scheme was proposed by George Warde some 20 years later.

(5) *John Smith's proposed canal from Cwmfelin to Dafen Bridge*

Smith must have intended providing two shipping outlets for the coal from his Dafen Bridge colliery because, in November 1786, he was leased portions of "*Wain Techon, Morva yr unis and Dole Vawr y Bynie*", to extend the existing Yspitty Canal westwards to Dafen Bridge. The lease, granted for 75 years by John Vaughan, Sir John Stepney, Sir E. V. Mansell and others, stipulated that the width of the canal should not exceed forty feet (12.2m) plus the width required for banks.[87] This proposed canal was to start from the existing branch of the Yspitty Canal, run west over the low-lying lands and then swing north-west towards the colliery at Dafen Bridge.[88] The combination of lessors, the length of the canal (1¼ miles - 2010m) and its substantial width, all suggest that this scheme was very important to Smith's plans to develop his coalfield. As already seen, Smith stopped sinking at Dafen Bridge before coal was reached and the canal was not constructed. George Warde proposed a similar scheme some 35 years later.

(6) *Proposed canal from Spitty Bank or Pencoed Lead House to Llandilo or Llandovery*

In 1793, a group of Carmarthenshire landowners considered the construction of a canal from the River Loughor to Llandilo or Llandovery, with branches up the Gwilly River to Mynydd Mawr and down the Gwendraeth Valley to Kidwelly, which would take minerals from the Carmarthenshire hinterland to ships for export. A meeting was held at Carmarthen, in March 1793, when 500 guineas were subscribed to engage surveyors to investigate the intended canal route.[89] By July 1793, the intention was to construct the canal from the Pencoed Lead House to Llandovery, William Hopkin, agent to the Stepney Estate, being involved in the survey, carried out by a Mr Cockshott.[90] Cockshott's report was completed in 1793, and he stated that he had surveyed as far down as Spitty and that the canal could "*conveniently communicate*" with the River Loughor at either Pencoed or Spitty.[91] A further meeting was held at Llandilo, in November 1793, but dispute arose as to whether a short canal from Spitty to Llandilo ("*proposed by Mr Jones of Duffryn and Mr Phillips the Attorney*") or a long canal from Spitty to Llandovery ("*proposed by Mr Cambell, Mr Vaughan, Lord Dinevor and the generality of the Gentlemen in the County*") should be adopted.[92] The Stepneys were not convinced that the canal would be of advantage to them and were also afraid that it would create a flooding hazard to collieries on their lands near Derwydd.[93] It is not known whether it was the Stepneys' opposition, or a lack of capital, which caused the scheme to fail but, by February 1794, all activity had ceased and the proposal was completely abandoned in 1798.[94]

This proposed canal was a more ambitious version of William Fenton's scheme of 1770. Fourteen years went by before the next canal up the Loughor Valley was projected.

(7) *Alexander Raby's proposed canal from Sandy to Llanelli Flats*

Raby was granted the Bushy Vein coal, under Stradey Estate lands west of the River Lliedi, by Dame Mary Mansell in March 1797.[95] Raby agreed to construct a waggon-way or railway, from the iron furnace at Cwm-ddyche to a canal, which he would build over the marsh lands and sea-shore, to connect with the shipping place he intended forming on Llanelli Flats. Raby was required "*to make and finish a Canal from the said Waggon Way or Railway over the Lands of the said Dame Mary Mansell to communicate with the Sea*".[96]

Dame Mary Mansell's business correspondence, written immediately before the lease was granted, referred to the proposed canal, which was to begin by the coal steward's house at "*Cae Pump Mawr*", then to Morfa Bach and end at the shipping place.[97] The canal would, therefore, have run two-thirds of a mile (1073m) from Sandy, over the Old Castle Pond area, across the old foreshore west of Albert Street to the Carmarthen-shire Dock. Raby was obliged "*to make the Railway and Canal within a limited time, at furthest within 12 months from the date of the Lease.*"

Raby did not construct this canal but built a railway over the same route, although no reference has been seen to any agreement with Dame Mary to substitute the railway for the canal.

(8) *General Warde's proposed canal from Dafen Bridge (Halfway) to Penrhyngwyn*

General Warde, Francis Morgan and James Morgan were re-leased coal, under lands in the Llanelli area, on 2 October 1802.[98] It was specified that "*for the promoting of the business of the said colliery*" a canal was to be commenced, within two years, for "*the Navigation of Boats and Barges up to the said Bank called Penrunwen Bank*". The width of the canal, its banks and towing paths was not to exceed 30 feet (9.1m). This scheme was virtually identical to John Smith's of 1785. Warde later confirmed his intention to construct the canal when he said, in 1814, "*I finished the Dafen Pit for the purpose of shipping at it by a Canal to Penrhunwen, a line of Canal heretofore used by Bowen".*[99] The Dafen Pit (119) was the pit at Dafen Bridge (Halfway) which John Smith had started but abandoned before reaching coal. Warde re-commenced the sinking, installed a steam engine[100] and built, or started to build, a shipping place at Dafen Pill.[101] A plan of 1808[102] showed "*Gen*[l] *Warde's Davin Engine*" linked, by means of "*Gen*[l] *Warde's Canal*", to "*Gen*[l] *Warde's Shipping Place*" (which was shown extending from the head of Dafen Pill to Penrhyngwyn Point) but there is little doubt that the canal shown represented Warde's intentions only, although part of it already existed through his inheritance of Thomas Bowen's canal. The steam engine at the Daven pit was too small to cope with the water, however,[103] and Warde abandoned the pit c1810,[104] when he must also have abandoned his plan to construct the canal to Penrhyngwyn.

(9) *The Kidwelly and Llanelly Canal Company's proposed branch canal from Kymer's Canal to Old Castle House*

The difficulties of access into the Gwendraeth Estuary, and its effect upon the trade of Kidwelly Harbour, caused great concern in the early years of the 19th century and meetings were held, in 1809 and 1811, to consider methods of improving the situation. On 24 June 1811, a scheme, devised by Edward Martin and David Davies, was placed before a meeting held at Kidwelly. The scheme proposed improvements to Kidwelly Harbour, and included plans to develop the mineral resources of the area by constructing a canal system up the Gwendraeth Valley, with an extension across Pinged Marsh and through Pembrey to Llanelli.[105] The same scheme was considered at Llanelli, in the following week, and was apparently accepted by the meeting, (including Lord Cawdor, John Rees of Cilymaenllwyd, Alexander Raby and Charles Nevill).[106] An Act of Parliament to implement the scheme was obtained, in June 1812, and the Kidwelly and Llanelly Canal Company came into being.[107] The Company was granted powers to extend the existing Kymer's Canal up the Gwendraeth Valley, to just beyond Cwmmawr, and to construct a canal, from Kymer's Canal, through Pembrey, to link with the Carmarthenshire Railway near Old Castle House (between the Carmarthenshire Dock and Sandy, at the end of the present Old Castle Road). Short canals or tramroads, to link the main canal with pits and works, and the provision of wharfs at the west side of the Carmarthenshire Dock, were also authorised.

Kymer's Canal was extended up the Gwendraeth Valley and a number of the canal branches were formed, but the canal to Llanelli was finally constructed only as far as Pembrey New Harbour (Burry Port), with a railway provided over the remainder of the proposed route.[108]

(10) *Proposed canal from Machynis via Spitty Bank to the Carmarthenshire and Glamorganshire hinterland*

A notice in *The Cambrian,* in September 1812, said that application was to be made to the next session of Parliament for an Act to make a canal or railway or tramroad, or partly a canal and partly a railway or tramroad from Machynis to Llandovery, with the provision of shipping places at Machynis Pool and Spitty Bank. It was also stated that "*collateral branches*" were intended to "*be made into, or pass through the several parishes of Llanelly, Llangennech, Llanedy, Llanon, Bettws, Llangadock, Llandebye, Llanarthney, Llanfihangel Aberbythick, Llandefysant, Llanthoisant, Llandilo, Llansadwrn, Llanwrda and Llandingat in the County of Carmarthen; and also through the parish and township of Lougher, and the parishes of Llandilo Talybont, Llangevelach and Llanguke, in the county of Glamorgan*".[109]

This was another scheme for a canal up the Loughor Valley but the initiator of the notice is not known with any certainty. The terminal location of Machynis, and the intention of forming a shipping place at Spitty Bank, suggests that it could have been General Warde, but the Copperworks Company and John Symmons could have been involved. It was probably proposed in opposition to the Kidwelly and Llanelly Canal Company, because of fears that their developments would divert coal exports from Llanelli to Pembrey. There is no evidence that application was made to Parliament and the scheme probably lapsed shortly after its announcement.

(11) *Proposed canal from the Copperhouse Dock via Spitty Bank to Gwilly Bridge, Llanedy*

The Cambrian, in September 1813, announced that application was to be made to the next session of Parliament for an Act to make and maintain a canal or a railway and tramroad, or partly a canal and partly a railway and tramroad from Copper-House (Copperworks) Dock to Gwilly Bridge in the parish of Llanedy. Collateral branches and the provision of shipping places and wharfs at Spitty Bank were also proposed.[110]

Because of the reference to the Copper-House Dock, the Llanelly Copperworks Company undoubtedly initiated this notice, but it is not known if the proposal was a less ambitious version of that proposed a year earlier (from Machynis to the Carmarthenshire and Glamorganshire hinterlands), or an entirely separate scheme. The Copperworks Company certainly hoped to make their dock the premier shipping place in the area,[111] and rivalry with the Kidwelly and Llanelly Canal Company must have been an added stimulus. Charles Nevill, who probably initiated the scheme, died within two months of the publishing of the notice[112] and the application to Parliament was not made.

The scheme was re-introduced four years later, in September 1817, coinciding with another proposed expansion of the Copperworks Dock.[113] There is no evidence that this scheme was ever taken beyond the declaration of intent stage and the proposed canal was not built.

(12) *General Warde's proposed canal from Cwmfelin to Maesarddafen*

A plan of 1821[114] showed an "*intended canal*", commencing at the existing extension to the Yspitty Canal and running west for just under one mile (1.6km), before ending some 300 yards (274m) south of Maesarddafen Farm (now occupied by the Trostre Tinplate Works). The purpose of this proposed canal is not known, especially as the Yspitty Canal was termed "*Old Canal*" at this time, and it is doubtful if Warde would have considered shipping again from Townsend's Pill, which had fallen into disuse c1808. It must, therefore, be assumed that the canal was intended for the transportation of coal from the Cae Level (178), and perhaps from other proposed pits along its line, for shipping at Dafen Pill.

The proposal was similar to John Smith's of 1786, except that the canal did not turn north towards Dafen Bridge. Like Smith's scheme, it was not implemented.

(13) *Richard Cort's proposed canal from Llangennech Marsh to Mynydd Mawr*

Richard Cort submitted a report (probably commissioned by the owner of the Llangennech Estate), in May 1824, advertising the merits of smelting copper ore on commission at the Spitty Bank Copperworks, using Llangennech coal.[115] The report also contained a proposal to construct a canal from a dock to be built on Llangennech Marsh at Viney, to the meeting point of the Rivers Loughor and Gwilly, and then up the Gwilly Valley to Mynydd Mawr. It was proposed to divert the channel of the River Loughor to a straight course between the Spitty Copperworks and the dock at Viney.

This scheme, proposed 54 years after Fenton's original plan, was the final attempt to raise support for a canal up the Loughor Valley and, like all its predecessors, it was not implemented.

Tramroads or railways were eventually built instead of some of these 13 projected canals, their routes closely following the proposed canal routes.

(III) RAILWAYS

Many of the roads and pathways, shown on old plans of the area, were probably built to haul coal, by horse and pannier or horse and cart, from early pits and levels but the full extent of this construction will never be known, because of lack of documentary evidence. As mining activity increased, and the quantities of coal to be transported grew, it became necessary to form special routes which made coal haulage easier. This was achieved by constructing prepared tracks, from the collieries to the shipping places or works, which provided an even and relatively frictionless surface and which normally followed a downhill gradient or level route for the loaded waggons. The available evidence shows that these tracks were provided in the Llanelli Coalfield from 1750 onwards. Very little evidence has survived about the mode of construction adopted so the generic term Railways has been used in the text, in the absence of any specific description of the track.[116]

Only railways whose existence has been proved by documentary and/or plan evidence will be considered, although it is thought that others were constructed.

(1) *Squire, Evans and Beynon's waggon road from Engine Fach to Spitty Bank*

Henry Squire, David Evans and John Beynon were leased coal in the Bynea/Pencoed region in December 1749 and February 1750.[117] They sank pits and constructed a waggon road, almost 1¼ miles (2010m) long, to take coal to their shipping place at Spitty Bank. The waggon road was probably built in 1750, because a portion of it, termed "*Mr Squire's Waggon Road*", was shown on a plan of 1751.[118] (See Plate 38)

The full route of the waggon road was shown on a plan of 1772,[119] which termed it "*Dr Evans' Waggon Road*" (Squire and Beynon were no longer partners). In present-day locations, it ran from Engine Fach, across Station Road, Bynea, between Bwlch-y-Llidiard and Glynea Farms and then straight to its end, at the side of Yspitty Road, near Loughor Bridge (Fig. 26). (See Plate 21)

Golden Grove Estate rentals listed the "*Waggon Way*" in 1757/58,[120] and as late as 1781/82,[121] but termed it "*not used*" in 1782/83[122] and 1787/88,[123] showing that it was abandoned in 1781, the year the coal leases expired.

This waggon road was probably the first railway track laid in the Llanelli area, and it possibly set the pattern for the others constructed, in and around around Llanelli, in the second half of the 18th century. It scaled between 14 and 18 feet (4.3 and 5.5m) wide on the 1772 plan[124] and there is unsubstantiated evidence that the track was formed from wooden rails, cased with iron and set into stone sleepers.[125] The size of coal waggon or cart is also known, because, when Chauncey Townsend was leased coal in 1752, he agreed to pay a royalty of "*3 shillings per weigh of 24 carts, each cart containing the same measure now used at Swansea and by Henry Squire and Co., at Spitty Bank and which said cart is in the clear Long 4 Feet, Breadth 2 feet 1 Inch, Deep 1 foot 3 inches filled top full*" [126] A cart of these dimensions would carry just under half a ton of coal and, as

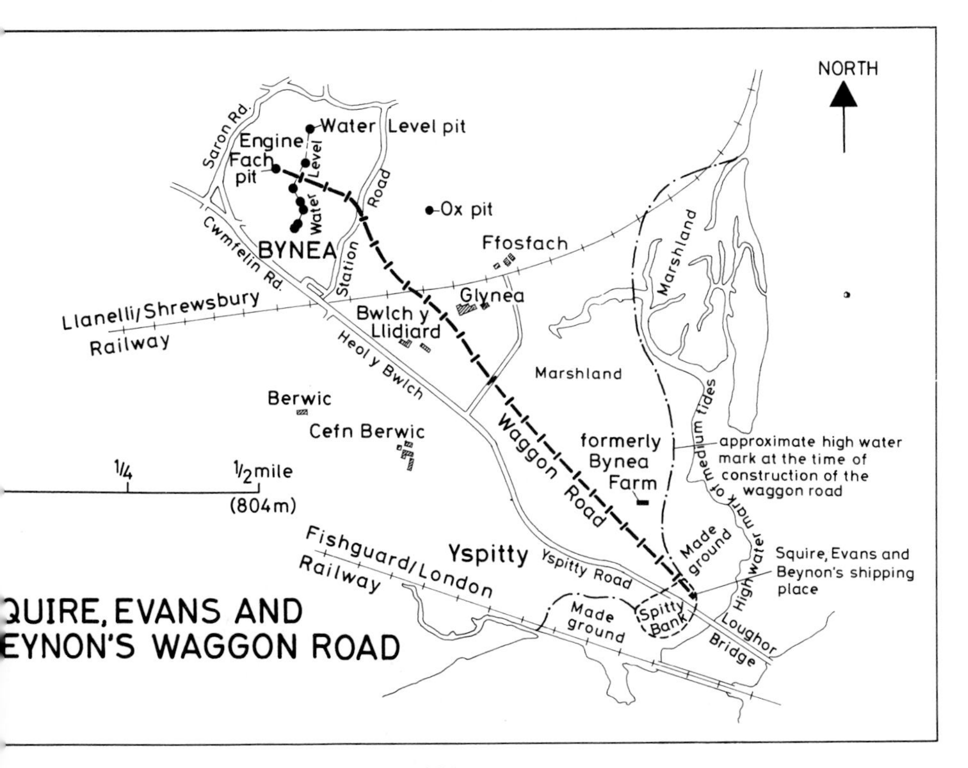

the Swansea wey was 10 to 11 tons, 24 such carts would be needed to each "*weigh*".

(2) *Sir E. V. Mansell's railway from Caemain to Bank y Llong*

A plan of 1787[127] showed a pit (46), worked by a water-engine at Caemain. A straight track, some 260 yards (238m) long, led from the pit to the main Llanelli to Pembrey turnpike road, approximately 100 yards (91m) from "*Bank y Llong*" on the River Lliedi (Fig. 20). The track was not described on the plan but it must have been some form of railway, as it led directly from a "*Landing Pit*" (coal raising pit) to a shipping place. Dame Mary Mansell was probably referring to it when she leased the Rosy, Fiery and Golden Veins to Alexander Raby (in 1797) and required that "*The Water Engine, Rail Road, Coal Barge etc., to be valued by 2 indifferent persons and taken at that valuation*".[128] Further confirmation of this was provided, in 1804, when Raby was leased Morfa Bach Common to erect his forge, the land leased being described as "*that which lies between a certain Old Rail Road called Lady Mansell's Railroad and a new Rail Road made lately over the said Common under an Act of Parliament called the Carmarthenshire Rail Road Act*".[129] (See Plate 10)

As it is considered that Sir E. V. Mansell took over the water-engine pit from Charles Gwyn, the railway may have been built in the early 1760s and used until 1794/95.

(3) *Chauncey Townsend's railway from Genwen to the Yspitty Canal*

Chauncey Townsend erected a steam engine at Genwen between 1766 and 1769, sank an engine or pumping pit (182) and other pits to raise coal. His major coal-raising pits appear to have been the 27 Fathom pit (183) and a pit west of the engine pit (181). A plan of 1772[130] shows a 250 yards (228m) long track from these pits to the top of the Yspitty Canal. This track is not described but its representation is identical to that of "*Dr Evans' Waggon Road*" on the same plan, and this undoubtedly was also a waggon road, constructed to carry Townsend's coal to his canal (Fig. 21).

(4) *The Llangennech Tramroad System*

The railway system in the Llangennech/Bryn region was started in the 1770s and was expanded and improved during the next 50 to 60 years. Because the location of the Llangennech Estate records is not known, one cannot be certain about facts relating to this railway, but sufficient information exists to give a general account of its history.

Sir Thomas Stepney, 7th Bart., was the first to consider building a railway in the Llangennech area. In December 1771, he asked permission from William Clayton to construct a waggonway over part of his lands. William Roderick, Clayton's agent, wrote "*Sir Thos. would be glad to have leave granted him by agreement, for to make a Level on Bryn Jeffrey Lands, to drive under a Colliery which he is now going to carry on by Llangennech, and to have the same grant of making a Waggonway, over one of your fields, to Land and carry away the said Cole to the river*". By August 1772, Stepney's plans were more advanced, Roderick reporting that he wished to construct a waggonway from the pithead to the water side "*which place where they mean to ship the Cole will be a spot of land part of the Tenement of Box*".[131] Stepney died two months later, in October 1772,[132] and his plans were never implemented.

By August 1773, Thomas Jones of Llwyn Ifan was negotiating with Roderick for a lease of the Brynsheffre coal,[133] Roderick informing Clayton, in October, that Jones was also negotiating with Sir John Stepney for the Allt coal and for "*a way over Sir John's land to carry all the Alt Cole and others adjoining to the road, which road will go over a little slang of yours and so will take all the Cole landed on your grounds as if it came from the Allt*".[134] Brynsheffre coal was leased to Jones in October 1774[135] and, in November, Roderick reaffirmed Jones's intention of adopting Stepney's plan to construct a shipping place, near the hamlet of Box.[136] It is considered that Thomas Jones constructed railways from Brynsheffre and the Allt to join the Llangennech Canal, which led to Llangennech Pill. This is deduced from a plan of c1790,[137] which showed "*Old Levels*" (interpreted as being the level ways formed for a railway track) leading towards the Allt and Brynsheffre and an account book, which itemised coal mining activity and the construction of "*new roads*" from the Brynsheffre region to a shipping bank on Llangennech Marsh, in 1777.[138] The length of the "*new roads*" was given as one mile and seventeen chains (1950m) in this account book, which corresponds very closely with the length of the route indicated by the "*Old Levels*" on the c1790 plan. The probable route of this initial railway system in the Llangennech area is shown in Fig. 27.

The working life of this railway is not known, but it had certainly been abandoned by c1790, when William Hopkin proposed constructing "*Roads or Canals*" in the Llangennech region,[139] to transport the Allt coal, leased to him and John Jenkin in 1789,[140] and the Brynsheffre coal, which was leased to them in 1793.[141] Hopkin and Jenkin proposed building their "*Roads or Canals*", from Brynsheffre and the Allt to the top of the existing old canal, which they intended extending southwards.[142]

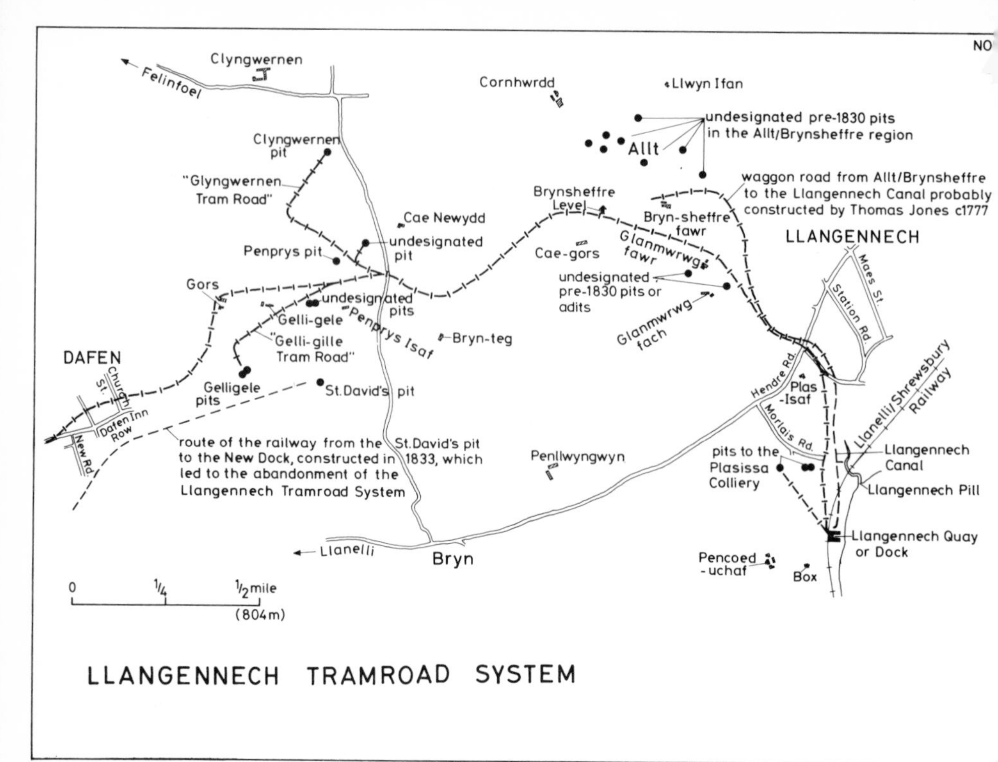

Fig. 27

Although they surrendered their sub-lease of the Allt coal in 1794,[143] the available evidence shows that the railways were constructed and the canal extended.

The Stepney Estate lands in the Llangennech area were purchased by John Symmons, in the late 18th/early 19th century, when Hopkin and Jenkin probably ceased working. Symmons inherited their railway system because he paid £3,000 for the existing "*railroads*" over the purchased lands[144] but there is evidence that he shipped at Llangennech Pill, because land at the Pill was described as "*This piece sold to John Symmons, Esq., for a coal bank and shipping place*".[145] In 1803, Symmons commissioned Edward Martin of Morriston to report on the Llangennech Estate. His report, of September 1803,[146] proposed that a "*tram road laid down on the improved plan*" be constructed from Clyngwernen, where a colliery was being developed, to Spitty. Symmons advertised his Estate for sale or letting, in January 1804,[147] and the Earl of Warwick and John Vancouver purchased it for £70,000 shortly afterwards.[148] They immediately commenced large scale development of the Estate's coal, probably concentrating their efforts into the collieries at Glanmwrwg, Clyngwernen, Brynsheffre and perhaps Gelli-gele. By April 1805, it was reported that

"*Vancouver is now proceeding in making preparations for shipping coal*" and "*Mr Vancouver has begun a Railway from on his colliery to the Sea which is expected to be compleat in six weeks.*"[149] He built either a new railway system or branch railways, from Clyngwernen (and Gelli-gele) to the shipping place near the tenement of Box, where a dock (Llangennech Quay) was being built. Warwick's creditors foreclosed upon him, however, and he had to surrender the Estate to Symmons before the close of 1806, because of his failure to complete the purchase.[150]

When Symmons repossessed the Estate, he must have completed Vancouver's development plans, as he advertised, in April 1807, that the "*collieries at Llangennech Park are now in full work and Large Quantities of Coal and Culm ready for shipping at their New Dock on the Loughor or Burry River*".[151] Richard Davenport, George Morris and William Rees took over the Llangennech Collieries and erected the Spitty Copperworks, Symmons being financially involved with them. It is probable that Davenport, Morris and Rees extended (or constructed) the branch of the railway, which ran from Gelli-gele to Bryngwyn Fach Farm, to take their coal, perhaps via the Box Railroad, to the docks at Llanelli. A plan of 1813/14[152] certainly showed the representation of such a railway branch. Davenport, Morris and Rees were bankrupt by 1814, and Symmons, again controlling his Estate's collieries, used its railways until 1823/24, when he sold the Estate to Edward Rose Tunno. Some four miles of "*Tram Road*" then existed on the Estate[153] but it is not certain when it was all constructed. Plans of the period[154] show that the railway system ran from Llangennech Quay along its previous route past the Glanmwrwg and Brynsheffre Collieries (144, 142), then branched into two lines north of Penprys Isaf, the north-western branch (termed the "*Clyngwernen Tram Road*")[155] running to the Clyngwernen pit (133),[156] with a small branch off it to an un-named pit near Cae-Newydd (148). The south-western branch (termed the "*Gelli-gille Tram Road*")[157] also branched into two lines, one arm following the previous route past Gors Farm[158] and the other leading to the Gelligele pits (152, 153).[159] It is not known whether both these south-western branches were used at the same time. A short branch line also ran directly from Llangennech Quay to the Plasissa Colliery (165).[160] Tunno leased the Estate coal to the Llangennech Coal Company, between 1825 and 1827, and they initially worked the inherited collieries and used the Llangennech Tramroad. The railways were regularly used until the Llanelly Railway and Dock Company built their railway in 1833.

Surprisingly little is known about the construction of the railway track used on the Llangennech Tramroad. It probably started as a wooden

Plate 27 — Remains of the Llangennech Tramroad System. Much of the route is used today as country paths.

waggon-way and then became an iron track railway, but it is not known when this occurred, nor when the branches were built. Nevertheless, for many years, the Llangennech Tramroad was the longest single railway system used to carry coal in the Llanelli area. In 1979 much of its route is still used as country paths. (Plate 27)

(5) *Hughes and Jones's railway from the Baccas Colliery to the Baccas Canal*

In 1794 the Rev. David Hughes of Ffosfach and Joseph Jones of Bryn Carnarvon were given permission to construct a canal over Morfa Baccas, to carry coal from their intended pits on Bryn Carnarvon Farm lands.[161] They constructed the Baccas Canal, from Pill y Ceven to a point about 300 yards (274m) south of their colliery (228, 229). No contemporary plan of the area has been seen but a plan of c1822[162] shows a straight track, 300 yards long, connecting the colliery to the top of the canal. This is probably some form of railway, which carried coal to the canal (Fig. 24).

(6) *Alexander Raby's railway system west of the River Lliedi*

When Raby obtained leases of the Rosy, Fiery, Golden and Bushy Vein coal under Stradey Estate lands west of the River Lliedi, in 1797 and 1798, he inherited working pits and probably a railway, from the water-engine pit at Caemain (46) to Bank y Llong. Raby initially worked some of the inherited pits and must have used Bank y Llong as his shipping place but his intention was to build an industrial complex, west of the River Lliedi, consisting of iron furnaces at Cwmddyche, a forge at Caemain, new collieries at Caemain, Caebad and Caerelms, and a dock at Llanelli Flats, all interconnected by a transportation network.

Work on his network probably commenced as early as 1797 because, in November, Raby requested Dame Mary Mansell to "*give Mr. Walker leave to begin to take up the Old Railroad as I want to use them to lay the new Iron Rail Way upon*".[163] Progress was not all that rapid because Raby was still constructing branch lines in April 1801, when the Llanelli Burgesses granted "*free liberty to Alex. Raby through Morva Bach for the purpose of making a Waggon Way and also to continue the present rail-road*".[164] By June 1802, when the Carmarthenshire Railroad Act was passed, the network consisted of "*a Railroad now made from Stradey Furnace to the said place called the Flats with Branches therefrom to Sundry Collieries and Iron Works belonging to Alexander Raby Esq*".[165] Raby was a prime mover in the formation of the Carmarthenshire Rail-road Company, and the Act stipulated that the proprietors should con-tract with him for his main railroad line, from Furnace to Llanelli Flats. He was paid £3,117 for this part of his railway system,[166] which was adopted as the first stretch of the Carmarthenshire Railroad.

Raby continued to develop his network west of the River Lliedi and, by 1806, branches had been constructed to pits and to the Old and New Slip Collieries.[167] Raby, who virtually controlled the activities of the Carmarthenshire Railroad Company in Llanelli, misused the Company's funds to complete his railway network, however,[168] and also avoided pay-ment of tolls to the Company for the coal and iron he transported along their main line.[169] Demands for the return of the misused monies, and the depression after the Napoleonic Wars, caused continued crises in Raby's affairs and his entire interests in Llanelli were advertised for sale, in October 1809. (His pits and adits west of the Lliedi were termed the "*Llanelly Colliery*" in the sale notice which stated "*to each pit is laid an iron railway of the best construction*").[170] The Copperworks Company helped Raby out of his immediate difficulties and he retained control of his interests. By 1814, an underground tramroad, almost three-quarters of a mile (1206m) long, had been constructed in the Fiery Vein to join the Caemain Colliery to the Pembertons' new pit at Talsarnau.[171] This was probably the final railway branch built by Raby, and, although underground (and not shown in Fig. 28), is mentioned because of its unusual length.

Raby's railway and the present Cynheidre line traversed approximately the same route from behind Albert Street to a point near the bottom of Constitution Hill. The line of Raby's railway then followed the lane between Furnace Rugby Club House and the Temple Works of Isaac Jones, across Furnace Square and along Pentrepoeth Road. Raby Street and Park View Terrace mark the routes of branch lines to the Caemain and Caebad Collieries and to the Forge. (Plate 28)

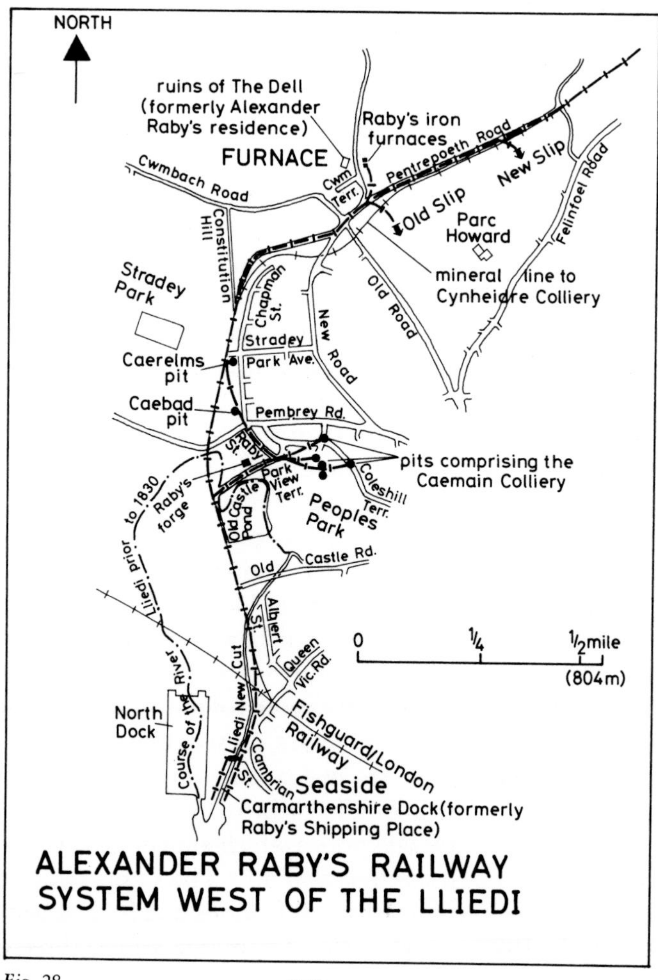

Fig. 28

Plate 28 — Alexander Raby's railway system. (Top) Park View Terrace. (Bottom) Raby Street. Raby's branch railways linking the Caemain, Caebad and Caerelms pits to the Carmarthenshire Railway ran along these streets.

(7) *The Carmarthenshire Railway*

Proposals had been made, in 1790 and 1793, to build a transport system which would bring minerals from the Carmarthenshire hinterland to ships in the Burry Estuary. They were not implemented but, in 1801, a new scheme was forwarded to construct a railway or tramroad, approximately 16 miles (25.7km) long, from Llanelli Flats to Castell-y-Garreg (2 miles north of Cross Hands), where limestone was abundant. As Alexander Raby needed limestone and iron ore to operate his furnaces — the proposed railway would also pass through an area where ironstone existed — he was doubtless the prime mover of the scheme. Although no correspondence has been seen, he must have been advocating it in 1800, as matters were well advanced in 1801, when plans were prepared.[172] In April 1801, the Llanelli Burgesses gave permission for "*the New Rail or Tram Road Company*" to make rail or tramroads with branches over Morfa Bach common[173] and, in November, subscribers met at Llannon,[174] E. W. R. Shewen, husband of Mary Martha Ann Mansell, acting as Chairman. The meeting resolved to petition the House of Commons that the railroad and dock would be of "*great Public Utility and Benefit to the landed proprietors on the line of the said intended Rail Road*" and passed a vote of thanks to Raby, who said he would use only Castell-y-Garreg limestone in his ironworks.

The Carmarthenshire Railway or Tramroad Company came into being when the Act was passed, in June 1802.[175] Raby's influence was clearly reflected in its wording, which said that it "*will be of very great Advantage to several Collieries, Iron Mines and Iron Works in the Neighbourhood by opening a cheap and easy Communication for the conveyance of Coal, Iron, and other Goods*" and that the proprietors were to contract and agree with Raby for the purchase of his existing railroad and dock. It named 57 proprietors and authorised share capital of £25,000 in £100 shares. A further £10,000 could be raised by mortgage, if necessary. Raby probably held 42 shares when the Act was passed.[176] He was paid £3,117 for his railway and dock, which the Company took over,[177] but it is possible that not all of his main line was included, as a plan of 1803[178] showed a proposed new line, west of Raby's existing railroad, near Old Castle House. Raby was also given the contract to supply tramplates.[179]

The railway, which would be 16 miles (25.7km) long, was started immediately. In July 1803 James Barnes, the Company's engineer, reported that 5 miles (8km) of the line, plus "*the Branches from the sea*" (probably Raby's railways), were in use, that a further 6½ miles (10.5km), with branches and passing places, had been formed and were awaiting only the final laying of the rails, leaving only 1¼ miles (2km) of the proposed line untouched.[180] Barnes forecast a return of 20% per annum on each share once the line was completed and the anticipated quantities of coal, stone coal and culm (anthracite), ironstone, limestone and manufactured iron were being carried. This optimistic forecast seemed to be capable of realisation as the line was laid further northwards and colliery development undertaken. By November 1803, the line had reached the Brondini/Cynheidre region[181] and, in February 1804, it was reported that Brondini Estate coal had been let to "*a respectable company of gentlemen* (Raby's brother-in-law Thomas Hill Cox and others), *who will be enabled to bring to market, by virtue of the Carmarthenshire Rail Road, an immense quantity of that article, which otherwise would have remained of little or no use to the community*".[182] In the same month, it was reported that "*four crops of valuable stone coal*" had been discovered during the railroad's construction and the "*Cynheidre Colliery and Farm*" were advertised for letting.[183] In March 1804 it was reported that Raby's Trevithick steam engine had been operated successfully and economically using anthracite[184] and Brondini Colliery anthracite was advertised for sale, in June 1804, when it was stated that large quantities of it were obtainable "*near the Dock on Llanelly Flats*".[185] The improvements to Raby's original shipping place were also progressing, being completed in 1805,[186] when the railway was also completed as far as Gorlas,[187] some 1¼ miles (2km) from its specified end at Castell-y-Garreg (Fig. 29).

In four years the Company had undoubtedly made significant progress, but serious financial problems were developing due to Raby's misuse of the Company's affairs and funds. The Act had specified that a General Assembly of the Company could be convened if proprietors in possession of, or holding by proxy, at least 126 of the 250 shares were in attendance. Raby had 42 shares and held proxies for a further 96, giving him control of 138. He could, therefore, convene a General Assembly anytime he wished and decisions could be taken without the knowledge of proprietors who would be likely to question them. He had misused this power, probably since 1802/03, constructing branch lines to his own concerns lying along the railway, with the result that all share and loan capital was spent before the line reached Castell-y-Garreg.[188] There is also evidence that Raby avoided payment of tolls for goods carried over these lines,[189] significantly reducing the Company's revenue, as Raby was probably the Railway's major user.[190] Affairs came to a head, in August 1806, when dissatisfied shareholders, who had probably not received any dividends, convened a General Assembly and formed a sub-committee to

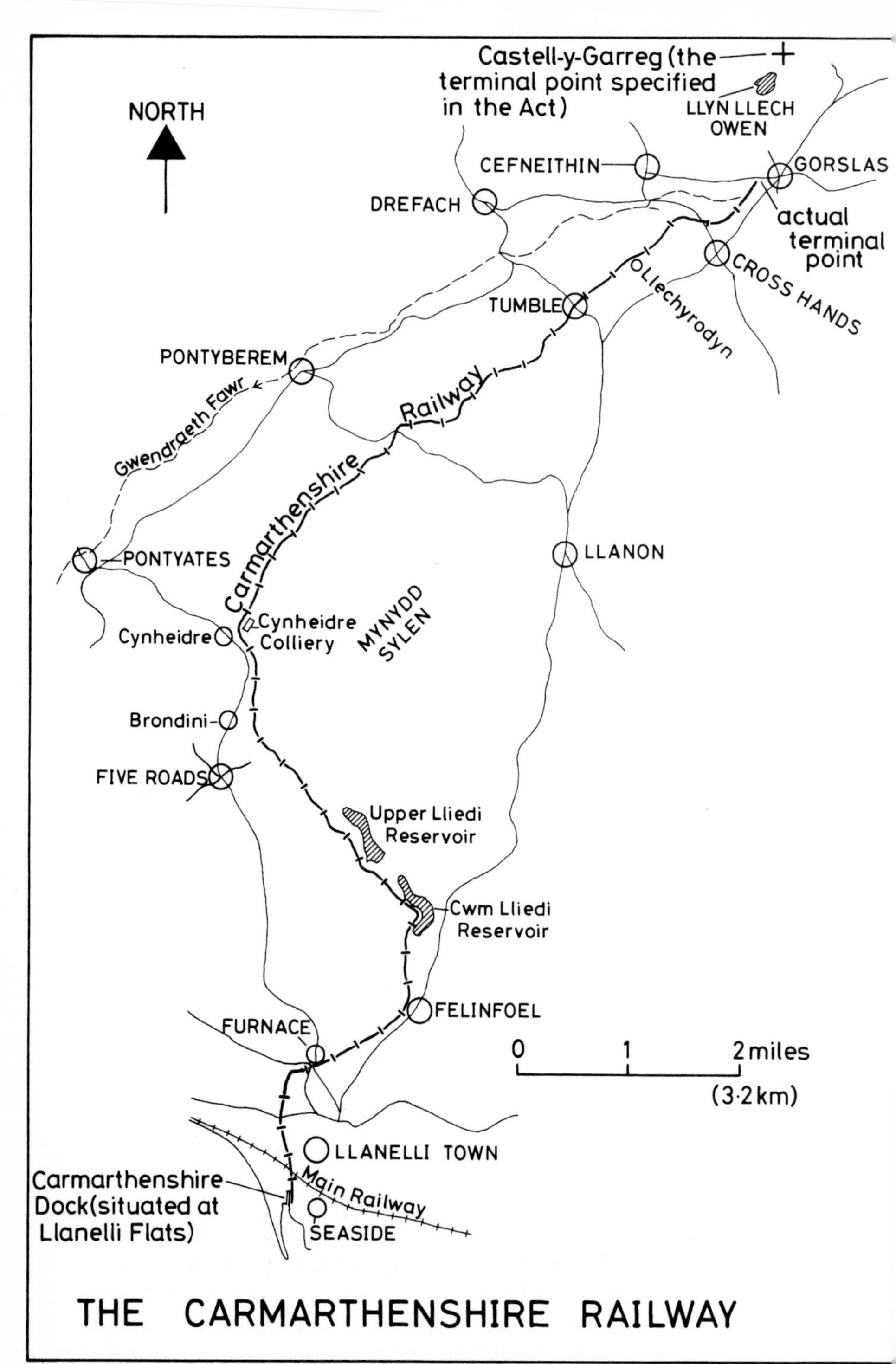
Castell-y-Garreg (the terminal point specified in the Act)
LLYN LLECH OWEN
NORTH
CEFNEITHIN
GORSLAS
DREFACH
actual terminal point
CROSS HANDS
TUMBLE
Llechyrodyn
PONTYBEREM
Gwendraeth Fawr
Carmarthenshire Railway
PONTYATES
LLANON
Cynheidre Colliery
Cynheidre
MYNYDD SYLEN
Brondini
FIVE ROADS
Upper Lliedi Reservoir
Cwm Lliedi Reservoir
FELINFOEL
FURNACE
0
1
2 miles
(3·2 km)
LLANELLI TOWN
Carmarthenshire Dock (situated at Llanelli Flats)
Main Railway
SEASIDE
THE CARMARTHENSHIRE RAILWAY

Fig. 29

report on the extent of Raby's branch lines. The irregularities revealed were described as "*unsufferable*" in their report[191] and a further report, dated 9 May 1807,[192] said that expenditure had exceeded all estimates and there was no surplus to divide among the proprietors. They said that this was because Raby had built some 8,700 yards (7.9km) of branches to his own works, which were unauthorised by the Act. They recommended that, in future, all committees should be chosen from London proprietors who had no ancillary interests at Llanelli. Raby was asked to submit his accounts to arbitration so that the cost of all unauthorised branches could be deducted, the gravity of the situation being summarised by the statement "*Among other measures recommended by the Committee in London to the Committee at Llanelli was the investigating the accompt of Mr. Alexander Raby, an accompt of such magnitude and importance, as is to be feared involves in it the ruin or recovery of the Company's affairs, as far as any profits to the Proprietors are to be expected.*"

The proprietors were in an extremely difficult situation. Raby had spent all his capital on his ironworks and collieries but they had not prospered and he was deeply in debt. The Committee's report must have brought matters to a head, because Raby convened a meeting of his creditors, in June 1807. He assigned his personal estates to his principal creditors, so that they could sell them to reduce his debts, and also transferred his leasehold coal and metal interests to them, to sell if necessary.[193] Raby's action left the Company without capital, and public confidence in its future prospects must have been shaken. Forty shares were advertised for sale, by auction, in July 1807.[194] It is quite possible that these were Raby's shares, put up by his creditors, but their sale could also have indicated a loss of confidence by other shareholders. The Company tried to satisfy its urgent need for money, when it advertised the dock dues, the receivable tolls and the right invested in the proprietors to construct the railway to Castell-y-Garreg, for sale or letting, by auction in July 1807.[195] The sale was postponed but was advertised again in September 1807,[196] but no evidence has been seen that any purchase resulted.

The Company and Raby survived this crisis and remained in business, although Raby must have lost control of the Company's affairs at Llanelli. He had the effrontery to criticise the Company over the safety of their dock in March 1808 and stated that he, his son and his brother had "*many Cargos of Culm and Bricks, Clay etc., ready, and bid fair to be great Customers to the Road and Dock if the same is properly executed, if not, every Idea either of the Road or Dock answering will be done away*".[197] The dock dues and the rail-road tolls were again advertised for letting, in July 1809,[198] but it is not known if a lessee was obtained.

Although Raby was no longer involved in running the Company's affairs he was still probably the railway's major user. His continual financial crises, and periods of standstill or little activity at his ironworks, collieries, and ironstone mines, must have affected the Company's income significantly, although the lower part of the railroad and the dock were used by other industrialists at Llanelli. The upper reaches of the railway probably carried smaller tonnages each year, but it is not known when this part of the line actually became disused. It was used in 1810, when the Company advertised anthracite and ironstone collieries at Llechrodyn and Hirwen-uchaf (near the northern end of the railway)[199] and in 1811, when *The Cambrian* referred to a proposed expansion of the dock and the anthracite collieries "*on the line of the Carmarthenshire Rail-Road*".[200] The upper reaches were also mentioned, in *The Cambrian* in 1822, when an advertisement for letting the Brondini Colliery stated that the works lay five miles (8km) from Llanelli "*having a direct communication therewith by means of the Carmarthenshire Rail-Road, the Tolls of which are low*".[201] An 1822 plan of the area north of Felinfoel also showed the "*Carmarthenshire Railway or Tramroad*",[202] but it is doubtful if much traffic used the upper reaches of the line at this time. An application to Parliament, in 1834, for a new Act to convert the tramroad into a railway for locomotive engines, stated that the tramroad was "*greatly out of repair*"[203] and a plan of 1835 also described the railway, north of Felinfoel, as "*broken up*".[204] The lower part of the railway, from Pentrepoeth, remained in constant use because the Rabys' collieries on the Stradey Estate, the Box Colliery, the original Old Castle Colliery and the Pool Colliery all connected with it, between 1802 and 1829.

Some details of the construction and operation of the Carmarthenshire Railway are known. The track was formed of flanged tramplates laid on longitudinal stone slabs.[205] From late 1804, the tramplates were probably fastened to the slabs using a method, invented by Charles le Caan, which did not require the use of wooden or wrought iron plugs, and facilitated the laying of a straight track.[206] The loaded waggons weighed 3 tons[207] and were drawn by two horses, one return journey, between Llanelli and Gorslas, being made each day.[208] From 1810/11, the waggons probably used an improved wheel design, manufactured by Raby,[209] and a patented braking system, devised by Charles Le Caan.[210]

It is possible that the venture was too ambitious for its time, and the demand for ironstone, limestone, and particularly anthracite, from the Carmarthenshire hinterland, greatly overestimated.[211] But any chance of success was greatly diminished by the fact that, in its early years, Raby misused its funds and held the Company's affairs to be of secondary

Plate 29 — The Carmarthenshire Railway. (Top) Part of the route from Sandy Bridge to Furnace. The present mineral line to Cynheidre Colliery (formerly the Llanelly and Mynydd Mawr Railway) is probably slightly to the right of the original route. (Bottom) The straight part of Pentrepoeth Road was the original route of the Carmarthenshire Railway. The present mineral line on the left was not built until the Llanelly and Mynydd Mawr Railway was formed in 1880/81.

importance to his own. The railway and dock were sold to the Llanelly Harbour Commissioners in 1844 and the line was rebuilt as the Llanelly and Mynydd Mawr Railway in the 1880s. (Plate 29)

(8) *Railway system at the Wern and the Seaside constructed by Roderick, Bowen and Griffiths and The Llanelly Copperworks Company*

Roderick and partners commenced mining at the Bres in 1794/95 and constructed a canal from their Bres pit (49) to Seaside.[212] Their attempts to sink the Bres pit to coal were defeated by water and, c1797, they moved to the Wern, where the seams were less deep.[213] They took coal down the Wern Canal and, apparently, did not construct a railway until after 1801, a meeting of the Llanelli Burgesses, in 1801, referring only to the trespass of their canal.[214] However, they constructed a railway from a pit at the Wern to the canal, between 1801 and 1803, as the Burgesses of Llanelli allowed them, in November 1803, to "*make use of the Rail Road already made by them over part of Gwern Caeswddy from a certain Coal Pit called Pwllycornel to the Canal*".[215] (The location of Pwllycornel is not known with any certainty but a plan of 1814[216] showed a straight track, some 200 yards (183m) long, from the canal to a point at which an un-named pit (71) was shown on a plan of c1810).[217] This may well have been their first railway on the Wern (Fig. 30).

Charles Nevill and his son Richard Janion came to Llanelli in 1804, to establish a copperworks. By December 1804, they had decided to build it at Seaside and had provisionally agreed, with Roderick and partners, to supply coal to the works.[218] They had also arranged to use a third of the partners' shipping place and were considering constructing their own dock.[219] The construction of the copperworks required large quantities of locally quarried stone and Charles Nevill approached the Stepney Estate, in December 1804, for permission to construct a tramway over Stepney lands, to convey stone from the Carmarthenshire Railway. Nevill wrote "*I flatter myself you will name a low sum as a compensation for our making a Tram Way from the Carmarthenshire Rail Road to Roderick & Co's Dock ... our Erections will want several Thousand Tons of Stone and it would be a considerable saving to us by laying down a Railway as above*".[220] Nevill also wished to have the coal from the colliery at the Wern delivered to the Copperworks by railway and wrote, in January 1805, "*the coal must be delivered from the Canal by a Railway — Roderick says he will make one to the Colliery by the side of the Canal ... I am much afraid R & B too poor to do their part*".[221] When the partners

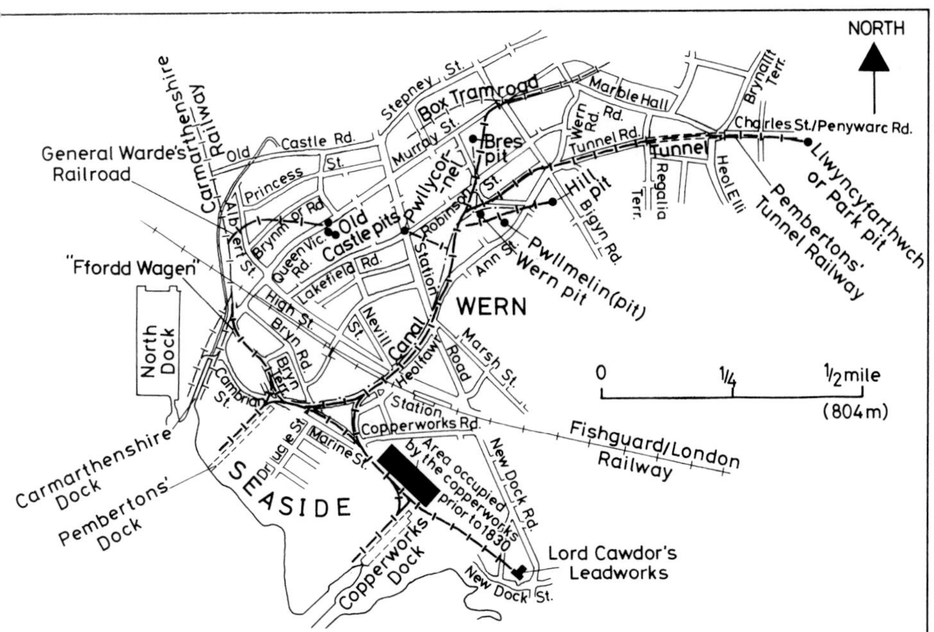

RAILWAY SYSTEM AT THE WERN AND SEASIDE AND THE PEMBERTONS' TUNNEL RAILWAY (ALSO WARDE'S RAILROAD)

made a formal contract to supply coal to the copperworks, they were advanced £500 to improve their colliery and make a railway to the copperworks.[222] Construction of the copperworks commenced in February 1805[223] and the dock was nearing completion by August.[224] The railway from the Carmarthenshire Railway to the Copperworks Dock, with a branch up to the Wern, was also undoubtedly constructed in 1805, as Nevill reported that the line to the dock was nearing completion, in April 1805,[225] and a plan, dated August 1805, showed a railway running over Heolfawr.[226]

In 1806/07, Roderick and partners sank a new pit, and constructed a branch railway from it to the main railway and canal, Charles Nevill writing in February 1807 "*They have compleated a Railway from the New Pit in the Field and have sent us some coal from thence... This Pit certainly gives them a facility of working of coal*".[227] A plan of 1808[228] showed a "*Rail Road*", some 250-300 yards (228 to 274m) long, from the canal/main railway on the Wern to an un-named "*Coal Pit*" east of the Wern. A plan of 1806[229] called it the Hill pit (82), probably the new pit referred to by Nevill.

The Copperworks Company and the Pembertons took over Roderick and partners' collieries in 1807, the Company renaming the South Crop

Wern Colliery the Wern Railway Colliery, which implies that the railway had then superseded the canal to carry coal.[230] The extent to which the Pembertons used the railway over the Wern is not known, but the Copperworks Company were undoubtedly the major transporters of coal along the line. The Wern railway was joined with the Box Tramroad, between 1810 and 1814, and coal from the Box and Talsarnau pits was carried over the Wern to the docks at Llanelli.

In 1811, Lord Cawdor consulted Charles Nevill about restarting the Carmarthen Leadworks and, in 1812, it was decided to dismantle and re-erect it at Llanelli,[231] on a site some 400 yards (365m) south east of the copperworks. A branch railway was made from the Copperworks Dock to the lead works,[232] probably in 1812, as the works was operating by January 1813.[233]

No further main branches were added before 1830 to this quite extensive railway system over the Wern and Seaside,[234] although branches must have been built to works and to new jetty heads at the docks. Some 2 miles (3.2km) of railway were built, between 1804 and 1814, from the Bres/Wern region to shipping places and works at Seaside, and it is only in recent years that the original routes disappeared, when locomotive engine rails, which superseded the original tramplates, were taken up. One tangible reminder of this railway system has survived in a street name — "*Ffordd y Wagen*", at Seaside, marks the route of the railway built by Charles Nevill in 1805 to carry stone for the construction of the copperworks. (Plate 30)

(9) *General Warde's Apple Pit Railroad*

When George Warde gained coal leases from Townsend's grandsons in 1802 he decided, initially, to concentrate his efforts at Llandafen and Genwen. He installed a new Boulton and Watt pumping engine at Genwen, dewatered the old workings through Townsend's original pumping pit, and extracted coal from pits sunk by Townsend and Smith. The coal was taken along the inherited Yspitty Canal and shipped at Townsend's Pill. This was only a short-term measure for Warde, his long-term plan for this area being to sink a new coal winning pit (208), east of the pumping engine, to the Swansea Five Feet Vein, with a mile (1.6km) long railroad from the pit to a new shipping place at Spitty Bite, north of Spitty Bank.

Sinking probably started in late 1806/early 1807,[235] and the construction of the railroad and shipping place about the same time. The railroad's route passed near at least two old pits, which were to be used to raise coal, the Ox pit (209) and the 21 Fathom pit (210) (Fig. 31). When

Plate 30 — The Railway System at Seaside. (Top) Ffordd Y Wagen or Railway Place — the route of the branch railway, built 1805, linking the works and docks of the Seaside area to the Carmarthenshire Railway. (Bottom) The route of the railway to the Llanelly Copperworks past Bethel Chapel.

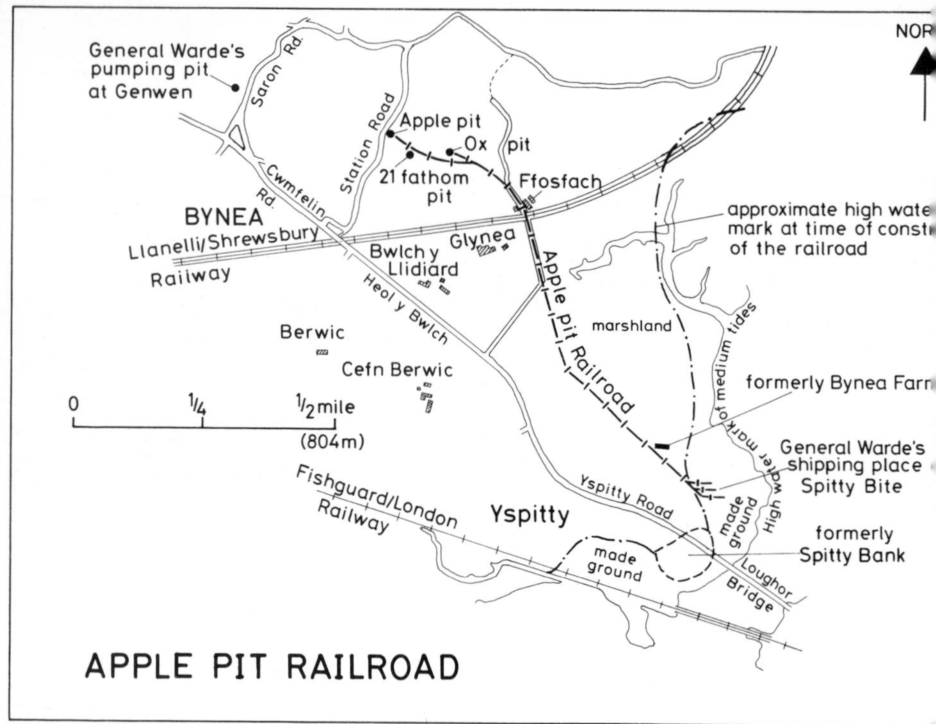

Fig. 31

the railroad and shipping place were finished, in August 1808, the new pit was still incomplete,[236] but Warde was anxious to announce the opening of his new enterprise, which was to be called "*The Carmarthenshire Collieries*", before the coal shipping season ended and was, therefore, prepared to rely on coal from the inherited pits. Accordingly, on 9 August 1808, the Carmarthenshire Collieries were formally opened and the first coal carried on the railroad to the new shipping place. The contemporary press reported the event, and stated "*We understand the railroad is of a new construction of plate under the General's own direction, whereby it is his intention (when a peace will allow him to reside in this country, and follow up experiments) to introduce a new mode of carrying coal. The present advantages, we learn, are the convenience of receiving any carriage, the width being that of a common axle-tree, and the plate opening such an angle as precludes the striking of a wheel against the flanch of the plate, by which, in the old mode, waggons get so frequently thrown out of the road*".[237] The railroad's route is shown in detail on plans of c1810 and c1822.[238] It ran east from the Apple pit (208), near the present Station Road in Bynea, turning south between Ffosfach and Glynea Farms. A short branch led to the Ox pit just north of the route. It was then carried on an embankment over the marshlands, ending at

Plate 31 — The Apple Pit Railroad. The present road over the marshlands to Ffos-fâch Farm near Bynea was once part of General Warde's railroad from the Apple pit to Spitty Bite.

Spitty Bite, where branches led to three shipping stages. It was built by Rosser Rees and John William, who were paid £370.17.0 by Warde in September 1808, for "*forming and laying 1745 yards of rail road and branches at 4/3 per yard*". By September 1809, it was referred to as the "*Apple Pit Rail Road*"[239] but a plan of c1822 called it "*General Warde's Tram Road*".[240] It was used until Warde ceased operations in the Bynea area, about 1820.

Today, some 300 yards (274m) of its original route survive as a road between Glynea and Ffosfach Farms. (Plate 31)

(10) *General Warde's railroad from the first Old Castle Colliery to the Carmarthenshire Railway*

In 1809, Warde started to exploit his coal west of Llanelli Town, where he began sinking at least one new pit and re-opened one or two old pits on Erwfawr lands, naming the colliery the "*Llanelly Pit*".[241] By February

1811 Warde, or his agent Rhys Jones, decided that the coal would be carried on a railroad linking the pits to the Carmarthenshire Railway and thence to the docks at Llanelli. On 19 February 1811, Jones came to an agreement with Hendry David and David Hugh "*for levelling, laying and completing a New Rail Road leading from a certain Coal-pit on the East Part of Erwvawr to the Llanelly Rail Road near the flats being about 300 yards at 2s.4d. per yard*". Jones also wrote to David and Hugh "*You will proceed without fail in making the Tramroad for General Warde over Hengastell Farm near Llanelly in the direction marked out*".[242] The tram-plates were purchased in the same month for £319.18.0, the railroad being completed by 16 April 1811, when David Hugh was paid. Payment was made for 416 yards (380m) of railroad plus "*85 yards of branches to the Pit*",[243] which was longer than originally specified. The opening of the colliery, re-named the Old Castle Colliery, was announced on 20 April 1811.[244]

The railroad's route is shown on plans of 1814.[245] In present day locations, it ran from the three pits constituting the colliery (63 to 65), which were between Queen Victoria Road and Lakefield Road, crossed over Queen Victoria Road and Brynmor Road and then curved south over the bottom part of Albert Street to the Carmarthenshire Railway (the River Lliedi had not then been diverted through the head of the Carmarthenshire Dock) (Fig. 30). The tenant of Old Castle Farm and Erwfawr lands, Thomas Rees, issued a writ, because of damage caused by the colliery and railroad, against Rhys Jones, John Hugh and Jenkin Hugh, in November 1811,[246] but no evidence of action has been seen. The life of the railroad is not definitely known. Warde sublet the Old Castle Colliery, c1813, to a George Walker, who worked it until c1817[247] and must have used the railroad, but it is not shown on a plan of c1822.[248]

(11) *The Box and Penygaer Railroads*

Raby leased coal under Box, Bryngwyn and Llanlliedi and commenced sinking the original Box Colliery pits (97, 98) in 1806, when he also probably started to build a railroad from the colliery to the Carmarthenshire Railway, as he intended shipping coal from the Carmarthenshire Dock. The railroad was probably completed by the Summer of 1808, because Raby paid rental to the tenant of Old Castle Farm from 1 August 1808,[249] the Box Colliery being opened in November 1808.[250] Raby's financial position grew worse and, eleven months after its opening, the colliery was offered for sale, the advertisement stating "*a newly-constructed iron railroad is laid from the Box Pit to the shipping dock near Llanelly*".[251]

The colliery and railroad were sold to Thomas Hill Cox and Stephen Jones by February 1810. The available evidence shows that Richard Janion Nevill was the real purchaser, with Cox and Jones acting for him. A letter written to Thomas Lewis of the Stradey Estate said "*The Trustees have sold the Leases of the Box Colliery and Railroad to Mr Richd Nevill who will, I presume, apply to you to have the latter transferred to him*".[252]

The railroad was just over a mile (1.6km) long, starting from the Box Colliery (97) and joining the Carmarthenshire Railway a quarter of a mile (402m) above the Carmarthenshire Dock.[253] In present day locations, it ran from west of Clifton Terrace, down Glenalla Road into Als Street, through the Market area, along the north side of John Street, along Lloyd Street and Old Castle Road backlanes, across the bottom of Albert Street, to the railway line to the Cynheidre Colliery (Fig. 32). Nevill quickly incorporated it into the Copperworks Company's railway system, joining it to the Wern Railway by a short link over the Bres by 1814,[254] when another branch was also constructed to the Pembertons' new pit at Talsarnau (95).

In 1813, Warde took over the Box Colliery, Nevill managing it by 1817. It was substantially developed, with the sinking of new pits to the

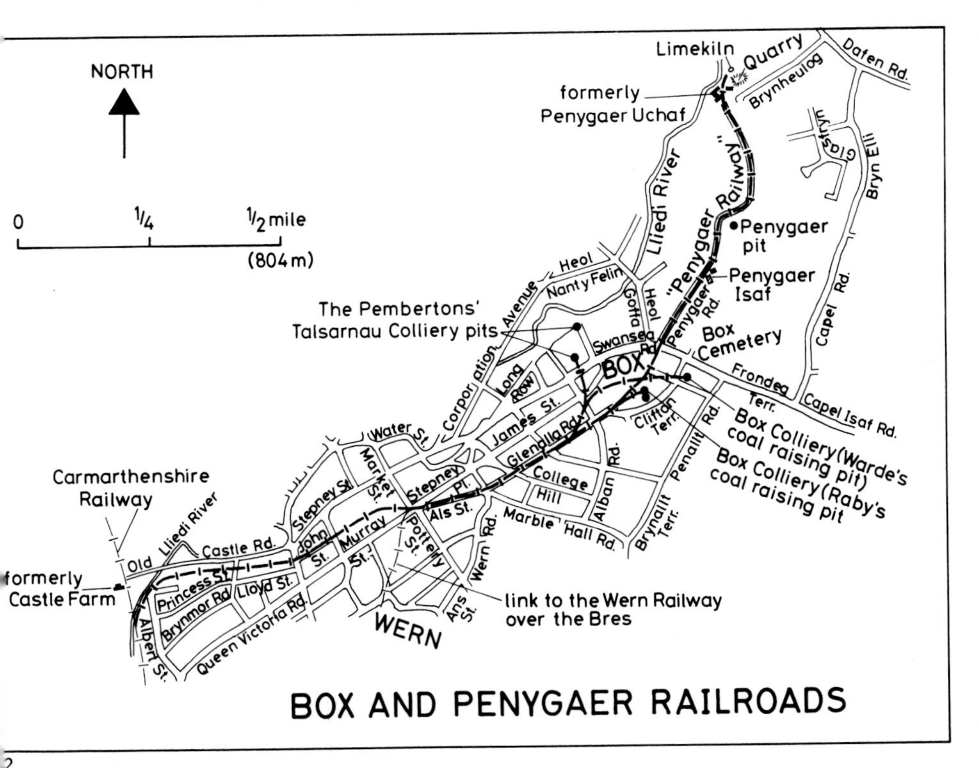

BOX AND PENYGAER RAILROADS

Plate 32 — The Penygaer Railroad. Penygaer Road was formerly the route of the railway linking General Warde's Penygare pit to the Box Railroad.

Swansea Five Feet Vein, between 1819 and 1823, a 300 yards (274m) long branch line being made to the new coal raising pit (100).[255] The Penygare pit (29) was sunk to complement the Box Colliery workings in the Swansea Five Feet Vein in 1824/25,[256] and a branch line was built, from the Box Railroad to the Penygare pit, and probably as far as a quarry and limekiln near Penygaer Uchaf, at this time. Reference was certainly made to this "*railway or tramroad recently built by Ge. Warde*", in July 1827,[257] and the entire two-thirds of a mile (1.1km) long branch is shown on a sketch plan of 1836, which termed it the "*Penygaer Railway*".[258]

By 1830, the bulk of the coal transported along the Railroad probably passed over the short link at the Bres and then down the Wern Railway, the original route from the Bres to the Carmarthenshire Railway probably being disused, as R. J. Nevill considered the legal implications of abandoning "*the road from Bresfawr to the Carmarthenshire Railway Dock*" in 1833.[259]

The original route of this two mile (3.2km) long railroad system through the now built-up areas of Glenalla Road, Als Street, John Street, and the backlanes behind Lloyd Street and Old Castle Road, remind us of the transport links, made in the early 19th century, to take coal to local works and docks. (Plates 32 & 33)

Plate 33 — The Box Railroad. (Top) Glenalla Road. (Bottom) Als Street. Both Glenalla Road and Als Street were formerly the route of the railway, built by Alexander Raby, from the Box Colliery to the docks at Seaside.

(12) *The Pembertons' Tunnel Railway from Llwyncyfarthwch to the Wern*

The Pembertons purchased lands at Llwyncyfarthwch from Lucinda and William Chute Hayton and from John Rees between 1812 and 1816 and started to sink a pit there, c1822.[260] To take the coal to their dock at Llanelli, the Pembertons were faced with the choice of a 1½ to 2 mile (2.4 to 3.2km) downhill route towards Trostre Farm and then around to Sea-side, or a straight and shorter route, to meet the railway at the Wern, in which they had a part interest. They chose the shorter route, which involved driving a tunnel through the high ground of the Marble Hall area between the pit and the Wern, and must have made it while their pit was being sunk.

The railway was about two-thirds of a mile (1.1km) long, and ran from the Llwyncyfarthwch pit (93), through a 300 yard (274m) long tunnel under Marble Hall, and then downhill to the Wern Railway, near the Wern pit.[261] (In present day locations, it started near the bottom of Penywarc Road, entered the tunnel east of the junction of Charles Street and Brynallt Terrace, ran immediately underneath Llanelli General Hospital, and emerged at the end of the tunnel at the junction of Regalia Terrace and Tunnel Road. It then ran down Tunnel Road, across the top of Pottery Street, down the side of Elizabeth Street and into Waddle Engineering and Fan Co. Ltd.) (Fig. 30).

The Pembertons apparently abandoned the sinking of the Llwyncyfarthwch pit when it had reached 300 feet (91m) but had not located a *"regular vein"*.[262]

In 1828, the Pembertons proposed that the Llangennech Company should link the St David's pit to their railway at Llwyncyfarthwch,[263] and unsuccessfully opposed the Company's Railway and Dock Bill.[264] The Pembertons left the Llanelli area in 1829 and it seems likely that the railway and tunnel were never used to carry coal. Only Tunnel Road now remains to delineate part of the route and to serve as a reminder of the Pembertons' ambitious but abortive venture. (Plate 34)

(13) *The Pool Tramroad*

Their 1812 Act authorised The Kidwelly and Llanelly Canal Company to construct a canal from Pembrey, to join the Carmarthenshire Railroad near Old Castle House. Although part of the Act had been implemented, the communication between Pembrey and Llanelli had not been built, but there was renewed interest in this link, about 1825, when the Company

Plate 34 — **The Pembertons' Tunnel Railway.** (Top) Tunnel Road — formerly the route of the railway linking the Pembertons' Llwyncyfarthwch pit to the railway at the Wern. (Bottom) Site of a recently demolished house at the top of Tunnel Road. The house was sited on the back-filled western entry to the Pembertons' tunnel under Marble Hall.

and the Carmarthenshire Railway Company decided to build a railway system between Llanelli and Pembrey.[265] There is evidence that R. J. Nevill had been active behind the scenes in promoting this renewal of interest[266] and, in 1825 or 1826, Martyn John Roberts, who was financially associated with Nevill, re-opened the Pool Colliery,[267] which was on the proposed route.

When work was due to start in November 1825, Thomas Lewis of Stradey objected to the railway passing over his lands, possibly because he feared that a transfer of trade to Pembrey would decrease his wayleave and royalty payments, or because he was holding out for high compensation. James Guthrie, acting for the Carmarthenshire Railway Company, wrote to him, on 26 November 1825, "*As we are now ready to commence forming the new Rail Road along the Beach from the River Lliedi to the Stradey Bridge I feel it my duty before taking any step in the work to beg that you will allow Mr. Pinkerton and myself to wait upon you as proprietor of the soil to explain the line to be pursued*".[268] Lewis continued to object to the railway and terms of compensation offered, but work proceeded, and it has been reported that a railway was constructed as far as the Pool Colliery, in 1826, at a cost of £2,770, paid by the Carmarthenshire Railway Company.[270] The continuation of the railway to New Lodge, to join with the Pembrey transport network, was not carried out at this time and, when the Pool Colliery was advertised for letting in October 1830, it was stated that "*A Tram Road leads immediately from the Pit to the different Docks in the Harbour of Llanelly*".[271]

The route of the Pool Tramroad was shown on the first edition of the 1 inch to 1 mile Ordnance Survey plan of 1830 and on James Green's plan of 1833.[272] It ran from the Carmarthenshire Railway (near Sandy Bridge), along the seashore just south of the main road from Llanelli to Pembrey, ending at the Pool Colliery, a distance of some one and a half miles (2.4km) (Fig. 20).

(14) *The Llangennech Coal Company's railway from Llangennech Quay to Spitty Bank*

The Llangennech Coal Company started mining coal at their inherited collieries on the Llangennech Estate c1825, and also commenced sinking the St David's pit. At least two of the directors of the Company were involved in the Spitty Copperworks. The Company inherited two shipping places, Llangennech Quay and Spitty Bank, and the extensive railway system which connected the Llangennech Estate collieries to Llangennech Quay. Spitty Bank could take fairly large ships but Llangennech Quay, a mile (1.6km) further up the River Loughor, only small ships. As their coal was delivered to Llangennech Quay, the Company had to transport it from the Quay to the Spitty Bank Copperworks, to ships at Spitty Bank and, more importantly, to large vessels using the docks at Llanelli.

At first, the Company loaded their coal into barges at Llangennech Quay and then towed them by steam tug to Spitty and to Llanelli,[273] but this double and even treble-handling was uneconomic. To alleviate the situation, the Company decided to construct a railway from the Llangennech Tramroad, at Llangennech Quay, to Spitty and approached the local landowners for permission to construct the railway over their lands. The landowners were reluctant to give their permission, because they were afraid that their access to the navigable River Loughor would be permanently lost, if the railway became the sole property of E. R. Tunno, the proprietor of the Llangennech Estate.[274] Tunno, on behalf of the Llangennech Coal Company, then approached Lord Cawdor, who held the commonlands and foreshore as Lord of the Manor, and obtained his permission to construct a railway between Llangennech Quay and Pencoed.[275] They also obtained Warde's permission to carry the railway over his lands between Ffosfach and Bynea Farms.[276] The exact date of construction of the railway is not known but it was built in 1826/7, a document, dated 7 January 1828, stating that the Llangennech Coal Company *"have made their Rail Road over the frontage of the lands lying between the Llangennech Quay and Spitty Bank"*.[277] At least one case was drafted against Tunno, alleging that the railway had deprived local landowners of access to the River Loughor. The Llangennech Coal Company probably stopped using it in May 1829, when they surrendered Warde's lease, which had allowed them to pass over his lands.[278] By this time, the Llanelly Railway and Dock Company had obtained its Act of Parliament[279] to construct a railway from the St David's pit to Machynis Pool, and the Llangennech Coal Company were probably content to revert to towing coal barges from Llangennech Quay, as a short-term measure. Three years passed, however, before the St David's pit reached coal and the Llanelly Railway and Dock Company went ahead with their railway. The Llangennech Coal Company, therefore, had cause to regret their decision to abandon the railway to Spitty as early as 1829, although they may have been legally forced to do so.

The probable route taken by this railway is shown in Figure 33. The 1828 lease, between Tunno and Cawdor, showed a "*Railroad*" running from Llangennech Quay to the region of Pencoed Isaf, along the high water mark, and the 1826 agreement with Warde showed the route,

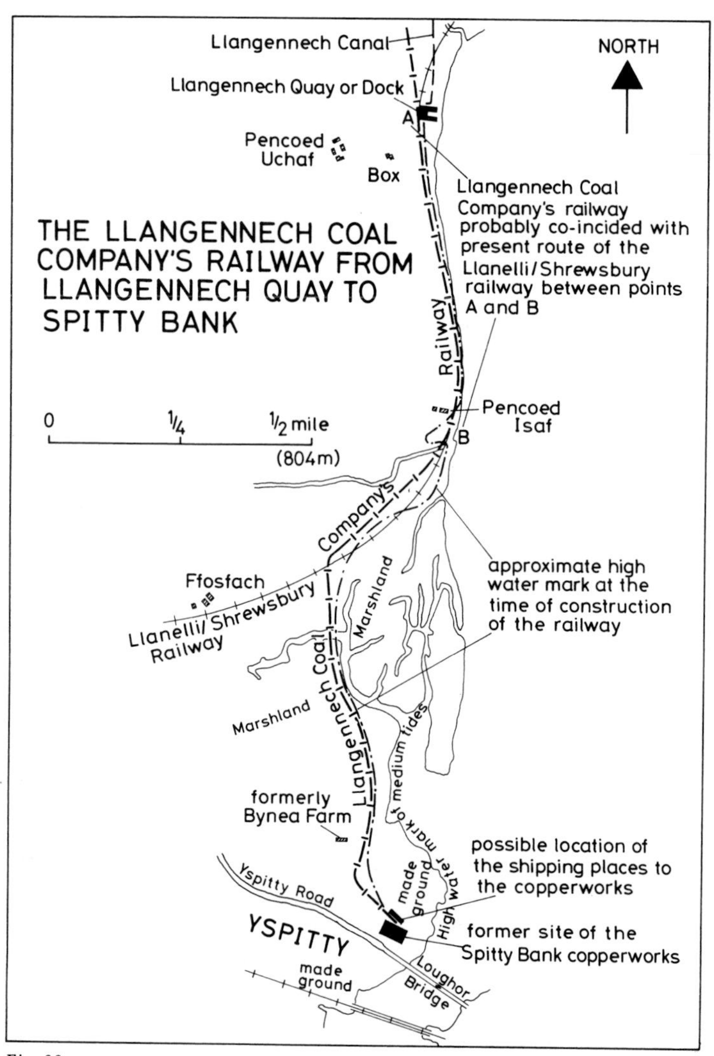

Fig. 33

running from near Ffosfach to near Bynea Farm. The railway must have followed the high water mark for most of its length, because the Lord of the Manor could only grant wayleave over the foreshore. It must, therefore, have closely followed the present line of the Llanelli/Shrewsbury railway from Llangennech Quay to some 400 yards (365m) east of Ffosfach Farm, where it branched south over the marshlands towards Spitty. A continuous embankment was shown over the marshlands, at this location, on the first 1:10560 scale Ordnance Survey plan (published 1891 but surveyed 1878), and its line corresponded closely to that of the

Plate 35 — The Llangennech Coal Company's railway from Llangennech Quay to Spitty Bank. The remains of the railway embankment running over the marshlands are clearly shown in the centre of the photograph, taken in 1923. The St. Davids Tin Plate works (also known locally as the Yspitty Works) occupied the site of the Spitty Bank Copperworks and might have incorporated some of its buildings. The present Loughor Bridge is shown under construction.

"High Water Mark" shown on a 1772 plan of the area.[280] The embankment still exists, although much of its southern length has disappeared, due to land reclamation. Examination of its structure shows that it is composed mainly of copper slag, which must have been obtained from the Spitty Copperworks. This railroad was between 1½ and 2 miles (2.4 to 3.2km) long, and its abandonment, after only two years, must have meant a serious financial loss to the Company. Although no confirmatory evidence has been seen, it is probable that the Llanelly Railway and Dock Company incorporated the route between Pencoed and Llangennech Quay into their Llanelli/Llandilo Railway in the late 1830s. (Plate 35)

CHAPTER 5 — REFERENCES

1 Bowen claims that John Allen, who mined in the Llanelli area between 1733 and c1761, carried coal to the "*seaside*" on the backs of mules. ("Hanes Llanelli" by D. Bowen, 1856, quoting an unidentified source).
2 *The Cambrian* (17 Apr 1840).
3 Llanelli Public Library plan 11 (1821) and Local Collection 37 (11 May 1824).
4 This canal was referred to as "*General Warde's Canal (Yspitty)*" in "Canals of South Wales and the Border" by Charles Hadfield.
5 Cawdor (Vaughan) 14/414 (Sep 1765).
6 Cawdor (Vaughan) 53/6140 (Jul 1766).
7 Report from Committee on Petition of the Owners of Collieries in South Wales 1810. Evidence of Henry Smith, Esq., M.P., given on 16 May 1810.
8 Llanelli Public Library plan 3 — "A Map of Colliery Lands to the East of Llwynhendy Fire Engine upon Lease to the Executors of the late Chauncey Townsend, Esq. Surveyed April 1772 by Jno Thornton"; Mansel Lewis 2565 (1772).
9 Llanelli Public Library 1 (1751). Information superimposed on this plan at a later date, probably by Rhys Jones of Loughor.
10 Llanelli Public Library plan 11 (1821) shows an un-named level at this location but oral tradition in the Llwynhendy area has retained the name of Cae Level for this entry — See "History of Llwynhendy and District" by the Rev. T. J. Euryn Hopkyns, published in the South Wales Press February to April 1932. (Written in Welsh. English translation available in LPL).
11 The westerly extension was shown in detail on Llanelli Public Library plan 11 (1821). The extension which was termed "*Old Canal*" at that time, was traversed by one bridge.
12 Cawdor (Vaughan) 8660 "Map of Three Llwynhendys" (1785).
13 Mansel Lewis London Collection 114, copy letter (10 Feb 1776).
14 Llanelli Public Library plan 1 (1751).
15 Nevill 409 (6 Nov 1786).
16 Charles Hadfield, op.cit.
17 The evidence of Mansel Lewis London Collection 34 (1805) and Mansel Lewis London Collection 10 — "Stradey Collieries and Binie-Queries for Mr Rees Jones dated February 1828".
18 Cawdor (Vaughan) 53/6139 (1782), 103/8038 (1783), 103/8039 (1788), 103/8040 (1795).
19 Llanelli Public Library plan 11 (1821) and 10 (c1822).
20 Nevill MS X, entry dated 28 Oct 1807.
21 Mansel Lewis 2511 (17 Apr 1813).
22 Llanelli Public Library plan 11 — "A Plan for the Continuation of Genl Warde's Canal 1821".
23 Local Collection 37 (11 May 1824).
24 This canal was termed the "*Penyfan Canal*" by Charles Hadfield, op. cit.
25 Cawdor (Vaughan) 5854 — Golden Grove Estate Plan Book, vol. IV - Plan of "Trawstre in Llanelly" (1785).
26 Cilymaenllwyd 135 (28 Feb 1753).
27 Cawdor (Vaughan) 21/623 (28 Jun 1754).
28 Brodie Collection 69 (1757).
29 "Old Llanelly" by J. Innes and the Hopkin Morgan MS (LPL).
30 Cawdor (Vaughan) 53/6124 (1751-54), 49/5960 (1758), 102/8028 (1759-61), 112/8394 (1763).
31 Stepney Estate 1099, quoting an unseen lease dated 21 Nov 1768.
32 A water course may have been a water supply for activating a water-wheel, as distinct from a canal.
33 Mansel Lewis London Collection 98, letter (14 Apr 1769).
34 Mansel Lewis 2082, letter from Thomas Lewis to Joseph Shewen dated 19 Nov 1771.
35 Cawdor (Vaughan) 63/6572 (1771/72) and 16/466 (1780/81).
36 Stepney Estate 1099, rent rolls for 1772, 1773, 1776 and 1780. Rentals for the years before 1772 and between 1780 and 1790 do not appear to have survived.
37 Llanelli Public Library plan 1 (1751). Information superimposed on this plan at a later date.
38 Llanelli Public Library plan 10 (c1822) and Cawdor (Vaughan) 8660 (1785).
39 Cawdor (Vaughan) 112/8396 (1781/82), 103/8038 (1782/83), 103/8039 (1787/88), 125/8649 (1789/90).
40 Nevill 27 (27 Jun 1785) and Thomas Mainwaring's Commonplace Book (LPL), copy letter from General Warde to Mr Goodeve dated 21 Mar 1814.
41 Stepney Estate 1110, undated but c1790.
42 Alltycadno and Gwylodymaes Estate rentals (1769-1785) (MS4.830) and Local Collection 465 (1773 to 1792).
43 Alltycadno and Gwylodymaes Estate rentals (1769-1785) (MS4.830) and Carmarthenshire leases 126, 127 (20 Feb 1793) (CCL).

44 Alltycadno and Gwylodymaes Estate, correspondence of William Roderick (1770-1786) (MS4.831), letters (10 Mar, 1 May, 26 Aug 1774).
45 Carmarthenshire deed 65 (13 Apr 1775) (CCL).
46 Ibid and Alltycadno and Gwylodymaes Estate rental for 1774/75 (MS4.830). The first 1:10560 scale Ordnance Survey plan, surveyed 1878 and published 1891, showed an unusually straight watercourse, approximately 450 yards (411 metres) long, leading from an old mine near Talyclyn Uchaf into the River Loughor. This may well have been the Talyclyn Canal.
47 Stepney Estate 1110, undated but c1790.
48 Local Collection 37 (11 May 1824).
49 Cawdor (Vaughan) 87/7350 (1 Oct 1794/95) and Carmarthenshire deed 81 (28 Jan 1805) (CCL).
50 Charles Hadfield, op.cit., termed this canal "*Hopkin's Canal*" on the assumption that it was the canal referred to in Local Collection 37.
51 Cawdor (Vaughan) 5/118 (1 Sep 1794).
52 Llanelli Public Library plans 10 (c1822) and 28 (c1840). Also Local Collection 37 (11 May 1824) — "Plan of Llangennech and Talyclyn Marshes and adjacent lands".
53 Stepney Estate 1099, "Sir John Stepney's Rental for the year 1797".
54 Ibid, "Draft schedule of Sir John Stepney's Estate", 1802.
55 Report by Rees Jones on the "Baccas Colliery situated in Berwick Marsh" contained within Local Collection 37 (11 May 1824).
56 "Plan of Llangennech and Talyclyn Marshes and adjacent lands" contained in Local Collection 37 (11 May 1824) and Llanelli Public Library plan 10 (c1822).
57 Cawdor (Vaughan) 8112 and 21/615 both dated 1 Oct 1794.
58 Stepney Estate 1099, "A particular of Several Estates in Carmarthenshire and Glamorganshire, part of the Estates of Sir John Stepney, Bart", undated, but c1791, made reference to "*Course for diverting the Water to the Water Engines*" over Bres Fawr. Llanelli Public Library plan 9 dated 1814 showed such a course from the Lliedi (near Mill Lane) to the Bres/Wern region. This water course may have been formed by John Allen or Thomas Bowen for the purposes of coal mining or could have been constructed as early as 1663, when reference was made to a "*new mill*" on the Bres (Cwrt Mawr 979 dated 20 Sep 1663).
59 An un-dated enclosure, contained within Cawdor (Vaughan) 43/5847 (c1803), made reference to the coal being carried down the canal in two 48 feet long boats (barges?).
60 Mansel Lewis 2510, "A rough plan of part of Morfa Mawr taken March 1808 by Joseph Jones"; Mansel Lewis 2538, "Map of Heolfawr" by M. Williams (1805) and Stepney Estate Office 74 — Copy of the Enclosure Award of 24 October 1810 (CRO).
61 The evidence of Cawdor (Vaughan) 43/5847 (c1803); Llanelli Public Library plan 7 (July 1806); *The Cambrian* (16 Jan 1808); and Mansel Lewis London Collection plan 57 — "Map of Heolfawr", surveyed by Wm. Maurice 1800.
62 Cawdor (Vaughan) 109/8292, Court Leet held at Llanelly on 30 Apr 1801.
63 Ibid, Court Leet held at Llanelly on 25 Oct 1804.
64 Nevill MS IV, copy letter from C. Nevill to R. Michell (7 Jan 1805).
65 Cawdor (Vaughan) 109/8292, Court Leet held at Llanelly on 24 Oct 1805.
66 *The Cambrian* (9 Aug 1806).
67 Nevill MS XI, Accounts of the Wern Railway Colliery (30 Sep 1807 to 22 Mar 1817).
68 Nevill 650 (26 Jan 1805).
69 Stepney Estate Office plan 74, Copy of Enclosure Award dated 24 Oct 1810 (CRO).
70 Ibid., plan at front of Enclosure Award.
71 Nevill 124 (15 Aug 1811).
72 Llanelli Public Library plan 9 (1814) and Stepney Estate Office plan 46 (1814) (CRO).
73 Nevill MS XXI, loose statement entitled "Proposal made to Mr R. S. Pemberton" (19 May 1815).
74 Charles Hadfield, op. cit., termed this proposed canal "*General Warde's Canal (Dafen)*".
75 Nevill 709 (20 Nov 1752).
76 Brodie Collection 69 (1757).
77 Llanelli Public Library plan 1 (1751). Information superimposed at a later date, probably by Rhys Jones of Loughor.
78 Charles Hadfield, op.cit.
79 Transactions of the Carmarthenshire Antiquarian Society and Field Club, Part 50, 1928, quoting a petition dated 10 Feb 1770.
80 A letter from William Hopkin to Lady Stepney in 1793 (Stepney Estate 1088, letter dated 12 Nov 1793), regarding a proposed canal from Spitty to Llandovery, stated that part of it would run from Llandilo to Rhosmaen and then down to Bearthy, "*where Mr Fenton intended his Canal and at which place the Cwmamman Collieries may have an excellent communication*".
81 Mansel Lewis London Collection 114, letters (25 Sep 1778 and 10 Feb 1776).
82 Ibid, letter (16 May 1780).
83 Mansel Lewis 1866 (24 Oct 1781).

84 Nevill 27 (27 Jun 1785).
85 Ibid.
86 Information superimposed at a later date on Llanelli Public Library plan 1 (1751).
87 Nevill 409 (6 Nov 1786).
88 Route superimposed at a later date, probably by Rhys Jones, on Llanelli Public Library plan 1 (1751).
89 Stepney Estate 1088, letter from William Hopkin to Lady Stepney dated 24 Jul 1793 and letter from William Hopkin to Sir John Stepney dated 31 Mar 1793. It was stated that "*Mr Cambell and Lord Dinevor subscribed 20 guineas each and the rest of the gentlemen 10 guineas each*" implying that at least 48 people in the county were interested in this canal project.
90 Ibid, letter from W. Hopkin to Lady Stepney dated 24 Jul 1793.
91 Cawdor (Vaughan) Box 228 — bundle headed "*papers on the subject of a proposed canal in Carmarthenshire — with correspondence of Mr Cockshott an engineer, thereon in 1793*".
92 Stepney Estate 1088, letter from W. Hopkin to Lady Stepney, 12 Nov 1793.
93 Ibid, and un-dated plan entitled "Plan of parts of Derwydd and Middle Glancennen" contained in the same bundle.
94 Charles Hadfield, op.cit.
95 Mansel Lewis 480 (25 Mar 1797).
96 Ibid.
97 Mansel Lewis London Collection 97 — "Business correspondence and papers of Lady Mansell relating to the Stradey Collieries", letter (16 Feb 1797).
98 Un-numbered lease (2 Oct 1802) (CCL) and Nevill MS LVI, copy lease of the same date.
99 Thomas Mainwaring's Commonplace Book (LPL). Copy letter from General Warde to Mr Goodeve dated 21 Mar 1814.
100 Nevill MS X, entries for 1805/06.
101 Nevill 170 (30 Sep 1809) made reference to a new jetty at the mouth of the Dafen River.
102 Brodie Collection 68 — "A chart of Burry Bar and Harbour Engraved for the use of Gen[l] Warde's Colliery and given gratis to Vessels loading at it, 1808".
103 Nevill MS IV, copy letter (13 Nov 1806).
104 Nevill MS X, entries for 1810/11.
105 Carmarthenshire Antiquary, Vol 8 (1972) — "The Canals of the Gwendraeth Valley (Part 2)" by W. H. Morris and G. R. Jones.
106 *The Cambrian* (6 Jul 1811).
107 52 Geo III c173 — "An Act for the Improvement of the Harbour of Kidwelly and for making and maintaining a Navigable Canal or Tramroads in Kidwelly and Llanelly and other Parishes therein mentioned in the County of Carmarthen, 20 June 1812".
108 Charles Hadfield, op.cit.
109 *The Cambrian* (12 Sep 1812).
110 *The Cambrian* (11 and 25 Sep 1813).
111 The evidence of Nevill 240 (21 Jul 1812).
112 "Llanelly Parish Church" by A. Mee, p. lxv.
113 *Carmarthen Journal* (5 Sep 1817), quoted in "Some Notes on the Mansel Lewis Papers — Part III," by R. Craig, Carmarthen Antiquary 1963 and Nevill 242 (8 Feb 1817).
114 Llanelli Public Library plan 11 — "A plan for the continuation of General Warde's Canal 1821".
115 Local Collection 37 (11 May 1824).
116 Contemporary plans and documents were not consistent in their description of an individual railway but generally "*road*", "*waggon road*" and "*waggon way*" were used synonymously. Similarly "*tramroad*", "*railroad*" and "*railway*" were synonymous terms. A further problem in description is that, once established, a route could remain in use for a very long period with the track itself being modernised at different times.
117 Stepney Estate 1099 quoting an unseen lease dated 13 Dec 1749, and Mansel Lewis 58 (21 Feb 1750).
118 Llanelli Public Library plan 1 (1751).
119 Llanelli Public Library plan 3 (Apr 1772).
120 Cawdor (Vaughan) 103/8033 (1758).
121 Cawdor (Vaughan) 53/6139 (1782).
122 Cawdor (Vaughan) 103/8038 (1783).
123 Cawdor (Vaughan) 103/8039 (1788).
124 Llanelli Public Library plan 3 (Apr 1772).
125 Thomas Mainwaring, op.cit.
126 Nevill 19 (24 Oct 1752).
127 Cawdor (Vaughan) 5854, plan of Morfa Bach, 1787, contained in Golden Grove Estate Plan Book IV.
128 Mansel Lewis London Collection 97 — "Lady Mansell's proposals for letting the Shipping Colliery", undated but c1797.

129 Cawdor 2/262 — "Note of Agreement for Part of Morfa Bach" between Lord Cawdor and Alexander Raby (31 May 1804).
130 Llanelli Public Library plan 3 (Apr 1772).
131 Alltycadno Estate correspondence, letters from Wm. Roderick to William Clayton, (2 Dec 1771, 22 Aug 1772).
132 "The Stepneys of Prendergast", Hist. Soc. of West Wales Transactions Vol VII, 1917-18.
133 Alltycadno Estate correspondence, letter from W. Roderick to W. Clayton (14 Aug 1773).
134 Ibid, letter from W. Roderick to W. Clayton (9 Oct 1773).
135 Carmarthenshire lease 140 (20 Oct 1774) (CCL).
136 Alltycadno Estate correspondence, letter from W. Roderick to W. Clayton (26 Nov 1774).
137 Stepney Estate 1110, undated but c1790.
138 Local Collection 465 entitled "Mary Hopkins' Book 1790", but including accounts relating to Llwyn Ifan between 1773 and 1792. It is considered that this was Thomas Jones's account book, later inherited by the Hopkins Family who came to live at Llwyn Ifan.
139 The evidence of Stepney Estate 1110, undated but c1790.
140 Cawdor (Vaughan) 92/7635 (1 Jan 1789).
141 Carmarthenshire lease 126 (20 Feb 1793) (CCL).
142 Stepney Estate 1110, undated but c1790.
143 Cawdor (Vaughan) 87/7350 (1 Oct 1794).
144 Penlle'rgaer A.974, letter from J. Symmons to L. W. Dillwyn (16 Apr 1819).
145 Stepney Estate Office plan 72 — "Map of Upper Pencoid" (CRO).
146 Local Collection 37 (11 May 1824).
147 *The Cambrian* (28 Jan 1804).
148 Penlle'rgaer A.972, letter from J. Symmons to L. W. Dillwyn (8 May 1819).
149 Nevill MS IV, copy letter from C. Nevill to R. A. Daniell (3 Apr 1805).
150 Ibid, copy letter from R. J. Nevill to R. Mitchell (13 Nov 1806) and *The Cambrian* (18 Oct 1806 and 28 Mar 1807).
151 *The Cambrian* (25 Apr 1807).
152 Plan of part of the first field survey for Sheet 37 of the Ordnance Survey 1 inch to 1 mile plan. Surveyed 1813/14 at a scale of 2 inches to 1 mile (NLW).
153 Local Collection 37 (11 May 1824).
154 Llanelli Public Library plan 13, undated but c1824/25; "Plan Book of Llangennech Estate", MS 3.299, undated but c1824/25 (CCL); first edition of Sheet 37 of Ordnance Survey 1 inch to 1 mile plan dated 1830 but surveyed 1825/26.
155 MS. 3.299 — "Map of Penprice", undated but c1824/25 (CCL).
156 Llanelli Public Library plan 13, undated but c1824/25.
157 MS 3.299 — "Map of Penprice", undated but c1824/25 (CCL).
158 Llanelli Public Library plan 13, undated but c1824/25.
159 First edition of Sheet 37 of the 1 inch to 1 mile Ordnance Survey plan (1830).
160 Llanelli Public Library plan 13, undated but c1824/25.
161 Cawdor (Vaughan) 5/118 (1 Sep 1794).
162 Llanelli Public Library plan 10, undated but c1822.
163 Mansel Lewis London Collection 97, letter from A. Raby to Lady Mansell (4 Nov 1797). Mr Walker was probably George Walker who was Raby's manager at this time (Thomas Mainwaring, op.cit.).
164 Cawdor (Vaughan) 109/8292 — Court Leet held at Llanelli on 30 Apr 1801.
165 Carmarthenshire Railway Act (42 Geo III c80).
166 Museum Collection 387, "Report of the Carmarthenshire Railway Company" (9 May 1807).
167 Local Collection 257, "Notes on Llanelly Forge" by R. Craig, quoting papers at the British Transport Corporation Records Office, Ref MPR 81/1.
168 Museum Collection 387, report (9 May 1807).
169 "Some notes on the Mansel Lewis Papers — Part II" by R. Craig, Carmarthen Antiquary 1962.
170 *The Cambrian* (7 Oct 1809).
171 Stepney Estate Office plan 46 (1814) (CRO).
172 "Some notes on the Mansel Lewis Papers — Part II" by R. Craig, Carmarthen Antiquary 1962.
173 Cawdor (Vaughan) 109/8292 — Court Leet held at Llanelli on 30 Apr 1801.
174 "Meeting of tne Subscribers to the Intended Rail Road or Tram-Way from the Flats near Llanelly to Castell-y-Garreg" held at the King's Head Inn at Llannon on Thursday 5 November 1801 (Local Collection 2956).
175 42 Geo III c80 (3 Jun 1802) — "An Act for making and maintaining a Railway or Tramroad from or near a certain place called the Flats, in the Parish of Llanelly in the County of Carmarthen, to or near to certain lime Rocks called Castell-y-Garreg, in the Parish of Llanfihangel-Aberbythick in the said County; and for making and maintaining a Dock or Bason at the Termination of the said Railway or Tramroad at or near the said Place called The Flats".
176 Museum Collection 387 (1802-07).
177 Ibid.
178 Mansel Lewis London Collection 79 (1803).

179 Thomas Mainwaring, op.cit. Mainwaring stated that a price of £9.10.0 per ton was agreed for the tramplates.
180 Copy of "James Barne's Report on the Present State of the Concern (Carmarthenshire Tramroad)" dated 28 Jul 1803, contained in Arthur Mee's copy of "Old Llanelly" by John Innes at CCL.
181 "Canals of South Wales and the Border" by C. Hadfield, quoting the *Hereford Journal* dated 30 Nov 1803.
182 *The Cambrian* (4 Feb 1804). It was stated in Cawdor 2/187 (18 Nov 1844) that Thomas Hill Cox was the owner of the Brondini Colliery.
183 *The Cambrian* (11 Feb 1804).
184 *The Cambrian* (9 Mar 1804).
185 *The Cambrian* (23 Jun 1804).
186 Thomas Mainwaring, op.cit.
187 Charles Hadfield, op.cit.
188 Museum Collection 387 (1802-07).
189 "Some notes on the Mansel Lewis Papers — Part II" by R. Craig, quoting B.T.C. Records Bundle MPR 81/1, Carmarthen Antiquary 1962.
190 Raby's ironworks and collieries at Llanelli used the rail-road and it is likely that he was involved with his brother-in-law, Thomas Cox, in the Brondini Colliery. It has also been stated that he worked an ironstone mine at Mynydd Mawr (Thomas Mainwaring, op.cit.). The first edition of the O.S. 6 inch to 1 mile plan (surveyed 1875-78) showed an "*Old Level (Ironstone)*" exactly at the end of the rail road at Gorslas.
191 R. Craig, op. cit.
192 Museum Collection 387 — "Report of the Carmarthenshire Railway Committee" (9 May 1807).
193 Mansel Lewis London Collection 3, copy of a draft release, assignment and Deed of Composition (10 Jun 1807).
194 *The Cambrian* (4 Jul 1807).
195 *The Cambrian* (11 Jul 1807).
196 *The Cambrian* (12 Sep 1807).
197 "Old Llanelly" by J. Innes, quoting an unseen letter from A. Raby to the Committee of the Carmarthenshire Rail Road Company, dated 9 Mar 1808.
198 *The Cambrian* (29 Jul 1809).
199 *The Cambrian* (25 Aug 1810).
200 *The Cambrian* (5 Oct 1811).
201 *The Cambrian* (6 Apr 1822).
202 Mansel Lewis 2590 — "A Plan of Part of the Carmarthenshire Railway or Tramroad together with part of Sir David Dundas Bart's Estate and other Estates adjoining thereto ...." (Jan 1822).
203 4/5 William IV c70 (27 Jun 1834).
204 Mansel Lewis 2507 — "Prospectus of the Llanelly and Llandilo Railway and Dock Company 1835"
205 *Llanelly Guardian* (12 Apr 1923) quoting an article contained in *The Great Western Railway Magazine.*
206 *The Cambrian* (24 Nov 1804 and 6 Jun 1807). Charles Le Caan was Clerk to the Company from 1806 onwards.
207 Thomas Mainwaring, op. cit.
208 *The Llanelly Guardian* (12 Apr 1923).
209 *The Cambrian* (20 Jul 1811).
210 *The Cambrian* (9 Jun 1810).
211 The Kidwelly and Llanelly Canal Company also attempted to develop the resources of the Gwendraeth Valley from 1812 onwards, overestimated the demand and met with similar financial problems.
212 Cawdor (Vaughan) 21/615 (1 Oct 1794) and 8112 (1 Oct 1794).
213 Cawdor (Vaughan) 43/5847, undated but c1803.
214 Cawdor (Vaughan) 109/8292 — Leet Courts held at Llanelli on 30 Apr and 27 Oct 1801.
215 Cawdor 2/63, Leet Court held at Llanelli on 9 Nov 1803.
216 Llanelli Public Library plan 9 (1814).
217 Mansel Lewis 2578 (c1810).
218 Nevill MS IV, copy letter from C. W. Nevill to R. Michell (15 Nov 1804).
219 Ibid, copy letter from R. J. Nevill to R. Michell (12 Dec 1804).
220 Ibid, copy letter from C. Nevill to W. Hopkin (21 Dec 1804).
221 Ibid, copy letter from C. Nevill to R. Michell (7 Jan 1805).
222 Nevill 650 (26 Jan 1805).
223 *The Cambrian* (31 Aug 1805).
224 Nevill MS IV, copy letter from C. Nevill to R. Michell (1 Aug 1805).
225 "Some notes on the Mansel Lewis Papers — Part II" by R. Craig, Carmarthen Antiquary 1962.

226 Nevill 157 (13 Aug 1805).
227 Nevill MS IV, copy letter from C. Nevill to Jn. Guest (23 Feb 1807).
228 Mansel Lewis 2510 — "A Rough Plan of Part of Morfa Mawr taken March 1808 by Joseph Jones".
229 Llanelli Public Library plan 6 (Jul 1806).
230 Nevill MS XI — "Accounts of Wern Railway Colliery" (Sep 1807 to Mar 1817).
231 "History of Carmarthenshire" edited by J. Lloyd, Vol II p. 369.
232 Llanelli Public Library plan 15 (c1822).
233 Cawdor 2/64, memorandum written by R. B. William (24 Jun 1813).
234 The branch line from the Pembertons' pit at Llwyncyfarthwch, which passed through a tunnel under Marble Hall and joined the railway at the Wern, has been treated as a separate railway and will be considered later.
235 Nevill MS X — it is considered that the "*Pit F*" referred to in the accounts in Feb 1807 was Warde's pit, later to be called the "*Apple Pit*".
236 Ibid.
237 *The Cambrian* (13 Aug 1808).
238 Llanelli Public Library plan 10, undated but c1822; Mansel Lewis 2578 undated but c1810.
239 Nevill MS X, accounts for Sep 1808.
240 Llanelli Public Library plan 10, undated but c1822.
241 Nevill MS X, accounts for 1809.
242 Mansel Lewis 1995 (19 Feb 1811).
243 Nevill MS X, accounts for Feb and Apr 1811.
244 *The Cambrian* (20 Apr 1811).
245 Llanelli Public Library plan 9 (1814) and Stepney Estate Office plan 46 (1814) (CRO).
246 Mansel Lewis 2082, letter (6 Nov 1811).
247 Mansel Lewis 1654 (1810-1822).
248 Llanelli Public Library plan 15 (1822).
249 Mansel Lewis London Collection 108, letter from H. Wright to T. Lewis (17 Mar 1810).
250 *The Cambrian* (5 Nov 1808).
251 *The Cambrian* (7 Oct 1809).
252 Mansel Lewis London Collection 108, letter from H. Wright to T. Lewis (20 Feb 1810).
253 Llanelli Public Library plan 9 (1814).
254 The evidence of Stepney Estate Office plan 46 (1814) (CRO).
255 The evidence of Llanelli Public Library plan 10 (c1822).
256 Nevill 1870-1918 (Jan-Dec 1826) and Nevill 40 (1 May 1827).
257 Nevill 286 (3 Jul 1827).
258 Nevill MS XVIII (Jan 1836).
259 Nevill 294 (8 Jul 1833).
260 Nevill 254 (12 Jul 1856), quoting an unseen transaction and Thomas Mainwaring, op. cit. No documentary evidence relating to the sinking of the pit has been seen.
261 The route was approximately shown on the first edition of the 1 inch to 1 mile Ordnance Survey plan (Sheet 37) but Stepney Estate 1119 (1855), information superimposed on Llanelli Public Library plan 15 (c1822), and Llanelli Public Library plan 23 (1844/45), allow the line of the railway and the position of the tunnel to be accurately plotted.
262 Thomas Mainwaring, op.cit.
263 Local Collection 252 (14 Jan 1828).
264 Diary of L. W. Dillwyn, Vol II, entry for 30 Apr 1828 (NLW).
265 "The Canals of the Gwendraeth Valley (Part 2)" by W. H. Morris and G. R. Jones, Carmarthen Antiquary Vol 8, 1972.
266 "The Canals of South Wales and the Border" by C. Hadfield.
267 Thomas Mainwaring, op.cit.
268 Mansel Lewis 1841, letter from J. Guthrie to T. Lewis (26 Nov 1825). "*Mr Pinkerton*" was James Pinkerton, an engineer to the Kidwelly and Llanelly Canal Company.
269 Ibid, letters from 1825 to 1831.
270 Charles Hadfield, op.cit.
271 *The Cambrian* (16 Oct 1830).
272 Cawdor 2/44 (1833).
273 *Llanelly Guardian* (20 Feb 1896), quoting an article entitled "Railway and Dock of the Llangennech (St David's) Company", contained in the British Almanack 1835.
274 Mansel Lewis 29 (7 Jan 1828).
275 Cawdor 2/70 — a counterpart lease dated 1828 but with no day or month. As only Tunno signed it, it is assumed that it was not executed.
276 Nevill 73 (1 Feb 1826).
277 Mansel Lewis 29 (7 Jan 1828).
278 Nevill 73 (1 Feb 1826); surrender of lease enclosed in document, undated but c16 May 1829.
279 9 Geo IV c91 (19 Jun 1828).
280 Llanelli Public Library plan 3 (Apr 1772).

CHAPTER 6

SHIPPING PLACES

INTRODUCTION

Llanelli emerged as a centre of the coal trade in the 16th and 17th centuries, when land transport was rudimentary, because its coal seams outcropped near the sea-board, necessitating only a short haul from pit-head to ships. Shipping places were needed where vessels could be safely moored in all weather conditions and loaded at all times of the tide. Small vessels sailed up rivers and pills to convenient loading places, where they could safely ground on the sand or mud when the tide receded, and take on coal from barrows, waggons or pack-horses. This simple process seems to have been the only method used at Llanelli until c1749, no evidence having been seen to suggest that purpose-built shipping places with wharfs or quays, linked to pits by canals or railways, were constructed until Squire, Evans and Beynon built their waggon road, from the Engine Fach pit to a shipping place on Spitty Bank, in 1749/50. After this, the process became more complex. Shipping places, railways and canals were provided throughout the area and coal barges and keels were used to load larger ships anchored in the Estuary. The first real dock was not started until 1795, when Roderick and partners formed their shipping place at Seaside but, by 1830, the Llanelli area possessed three tidal docks (the Carmarthenshire Dock, Pembertons' Dock and Llangennech Quay) and one floating dock (the Copperhouse or Copperworks Dock). An Act of Parliament had also authorised the construction of another large floating dock at Machynis Pool (later called New Dock).

This chapter examines the development of the Burry Estuary in terms of the coal shipping trade, gives estimates of the quantities of coal exported from the Llanelli area and considers, in some detail, shipping places, whether rivers, pills, quays or docks, known to have been used to export the coal before 1830.

The Port of Llanelli has not been examined in depth in the physical and administrative senses. Such a study warrants a book to itself and lies outside the scope of this work, which is devoted solely to the coal trade.

(I) THE BURRY ESTUARY BEFORE 1830

Access for sailing ships into the Burry Estuary from Carmarthen Bay or Rhossili Bay was always hazardous because of sandbanks, particularly those between Burry Holmes and Pembrey, known as Burry Bar, and, having negotiated the Bar, a vessel was by no means safe, because shifting sandbanks within the Estuary were a constant threat to navigation. Although coal was exported from the Llanelli area as early as 1566,[1] the first reference seen to the hazard of Burry Bar was a century later, when a John Man at Swansea wrote to Sir Joseph Williamson, a Secretary of State, concerning a Bristol ship, the "*Greyhound*", which had been driven over Burry Bar by a violent storm "*the master and Company not knowing where they were, nor the danger they were in, for if it had not been just on the height of the flood, they had doubtless all perished, it being a most dangerous bar*".[2] Burry Bar, and shifting sandbanks within the Estuary, seriously affected the development of Llanelli's coal industry and continually detracted from the area's other advantages as an industrial centre.

The earliest extant physical evidence relating to the Burry Estuary is a plan, made in 1757, when Sir Thomas Stepney, 7th Bart., was extolling Llanelli's advantages as a trade and industrial centre, during the first phase of its industrialisation. The plan may well have been commissioned by Sir Thomas and was made by William Jones of Loughor who entitled it "*Directions for Ships to come in safe into Burry and to the Several Places of Safety to be within the same*".[3] Although the plan contains surveying errors — for example, the distance between the Llanelli and Gower shorelines is under-estimated — the main channels to, and the shipping channels within, the Burry Estuary are clearly defined. The information on the plan summarises Llanelli's facilities for shipping coal. Safe anchorages are shown in the main channels between The Holmes and Whitford, in the North Pool, off Whitford Point and near Penclawdd. Shipping places, itemised as "*coaleries*", are located at Pool (the present village of Pwll), Dafen Pill, Spitty and Pencoed and also at Penclawdd and Loughor. The information given suggests that ships could sail to these shipping places, or barges or keels could bring coal out to ships moored at the safe anchorages. No dock is shown and no shipping place is marked between Pool and Dafen Pill. (Plate 36)

Apart from the establishment of small shipping places, little change occurred in this pattern until the second phase of Llanelli's industrialisation in the 1790s. This phase was sustained, marking the start of Llanelli's real industrialisation. The consequent increase in industrial activity and the growing volume of coal exports, necessitated the pro-

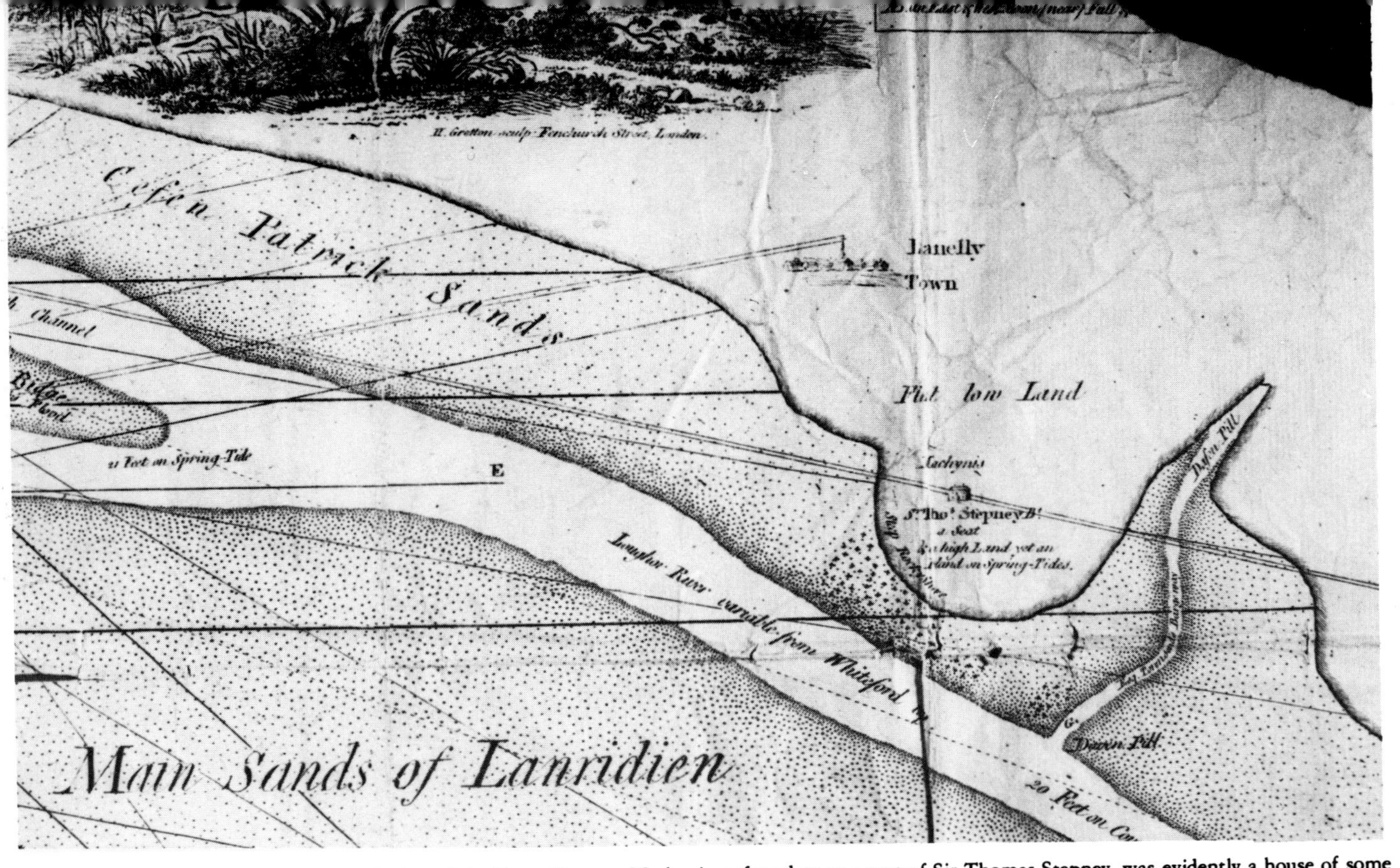

Plate 36 — Part of William Jones's 1757 plan of the Burry Estuary. Machynis, referred to as a seat of Sir Thomas Stepney, was evidently a house of some importance. This tends to be supported by an Eighteenth Century panel painting in Parc Howard Gallery, which shows a large house with well-ordered and extensive grounds.

vision of new, purpose-built shipping places, although many of the old places remained in use. Buoys and pilot facilities were also provided in the Estuary to guide the growing number of ships into harbour or to a safe anchorage. These developments are summarised in the following pages, which are not fully referenced as each shipping place mentioned is dealt with later, in full detail.

In the early 1790s coal shipping methods at Llanelli were much as they had been for the previous 50 years or more. The Mansells may have been using Bank y Llong, Hopkin and Jenkin were shipping from the end of their Llangennech Canal near Pencoed Uchaf and traditional shipping places such as Dafen Pill, Spitty Bank and Pool may have been in use.

This established pattern changed from 1795, however, when Roderick and partners started building Llanelli's first real dock, at the end of their Wern Canal at Seaside. Other industrialists quickly arrived and, over the next 10 years, a number of docks and quays were formed, as they implemented their plans to exploit Llanelli's coal and establish metal industries. By 1799 Raby had connected his various activities, by railways, to a quay he built at Llanelli Flats, near Roderick and partners' dock. This was taken over in 1802, with much of Raby's railways, by the Carmarthenshire Railway Company, who built the Carmarthenshire Dock on the site of Raby's quay. Warde also recommenced shipping from Townsend's Pill in 1802, and, in the late 1790s/early 1800s, John Symmons, the owner of the Llangennech Estate, continued shipping from Llangennech Pill and, perhaps, from Box and Spitty Bank. The Pembertons and the Llanelly Copperworks Company came to Llanelli in 1804 and both possessed an interest in Roderick and partners' dock by 1805, when the Copperworks Company also started to build their own dock at their newly-erected copperworks. John Vancouver and the Earl of Warwick, the new owners of the Llangennech Estate, also probably enlarged the shipping place opposite the tenement of Box to form the Llangennech Quay or Dock, in 1805.

Having provided, or made plans to provide, improved dock and loading facilities, the new industrialists turned their attention to the danger of entry over the Bar and navigational difficulties within the Estuary which, unless improved, would discourage ship's masters from trading at Llanelli. *The Cambrian* announced, in February 1805, that "*A meeting of the Noblemen and Gentlemen interested in the Trade and prosperity of the River Burry is requested at the Falcon Inn, Llanelly on Thursday 21st., at twelve o'clock, to consider the best MODE OF SECURING AND IMPROVING its NAVIGATION, and on several other subjects intimately connected with its welfare*".[4] The statement was signed by

John Rees of Cilymaenllwyd, Charles and Richard Janion Nevill and by John Wedge, who lived in Llanelli and who was surveying the Estuary. The meeting resolved that, as soon as Wedge could produce an accurate drawing, it would be engraved and published for use by the masters of vessels frequenting Llanelli.[5] By September 1805, six buoys had been laid down to mark the greatest depth of water in the main channel between Burry Bar and Penclawdd, with pilots at Penclawdd to take vessels further up the Loughor River, if required. On 19 September 1805, a meeting of "*The Committee for the Improvement of the Navigation of the Burry River*" resolved that printed directions for following the buoys would be circulated until Wedge's plan was finished.[6] It has not been determined when Wedge's chart was published but Warde supplied free copies to ships loading his coal, in October 1808.[7] This version showed Warde's shipping places[8] and it is probable that the original chart was published before this.[9] (Plate 37)

Development of the shipping places continued. The Carmarthenshire Railway Company enlarged the Carmarthenshire Dock, in 1806, by building a 155 yards long western wall; John Symmons announced the opening of his new dock, Llangennech Quay, in 1807; the Pembertons took over Roderick and partners' dock and renamed it Pembertons' Dock; and General Warde established new shipping places at Spitty Bite in 1808 and, perhaps, a new jetty at Dafen Pill in 1809.

The profile of the land/water boundaries also changed when the 1807 Enclosure Act was implemented.[10] By June 1808, work had started on the "*Great Embankment*", from Penrhyngwyn Point to Maesarddafen, to convert the marshes at Machynis to productive land[11] and the River Dafen was diverted from its meandering course, to run into Machynis Pool.[12] This major sea-defence work, the construction of the docks and the formation of slag tips on the foreshore near the Llanelly and Spitty copperworks, must have produced even greater shifts than normal in the main channels and sandbanks in the Burry Estuary, although no documentary confirmation has been seen. Nevertheless, by January 1810, R. J. Nevill, on behalf of the Committee for the Improvement of the Navigation of the Burry River, advertised that Masters of vessels should carefully follow the Burry River buoys.[13] A week later, a letter to *The Cambrian,* signed "*Fairplay*", defended the safety of access into the Burry River. The letter stated that it was being said that "*General Warde's works may obtain a trade for cockleshells*" only and that not even a "*cockleshell*" would attempt to cross Burry Bar in the dark, because there was no depth of water. The writer maintained that the depth of water at Burry Bar was superior to that at entry to Swansea, and that a ship of 115

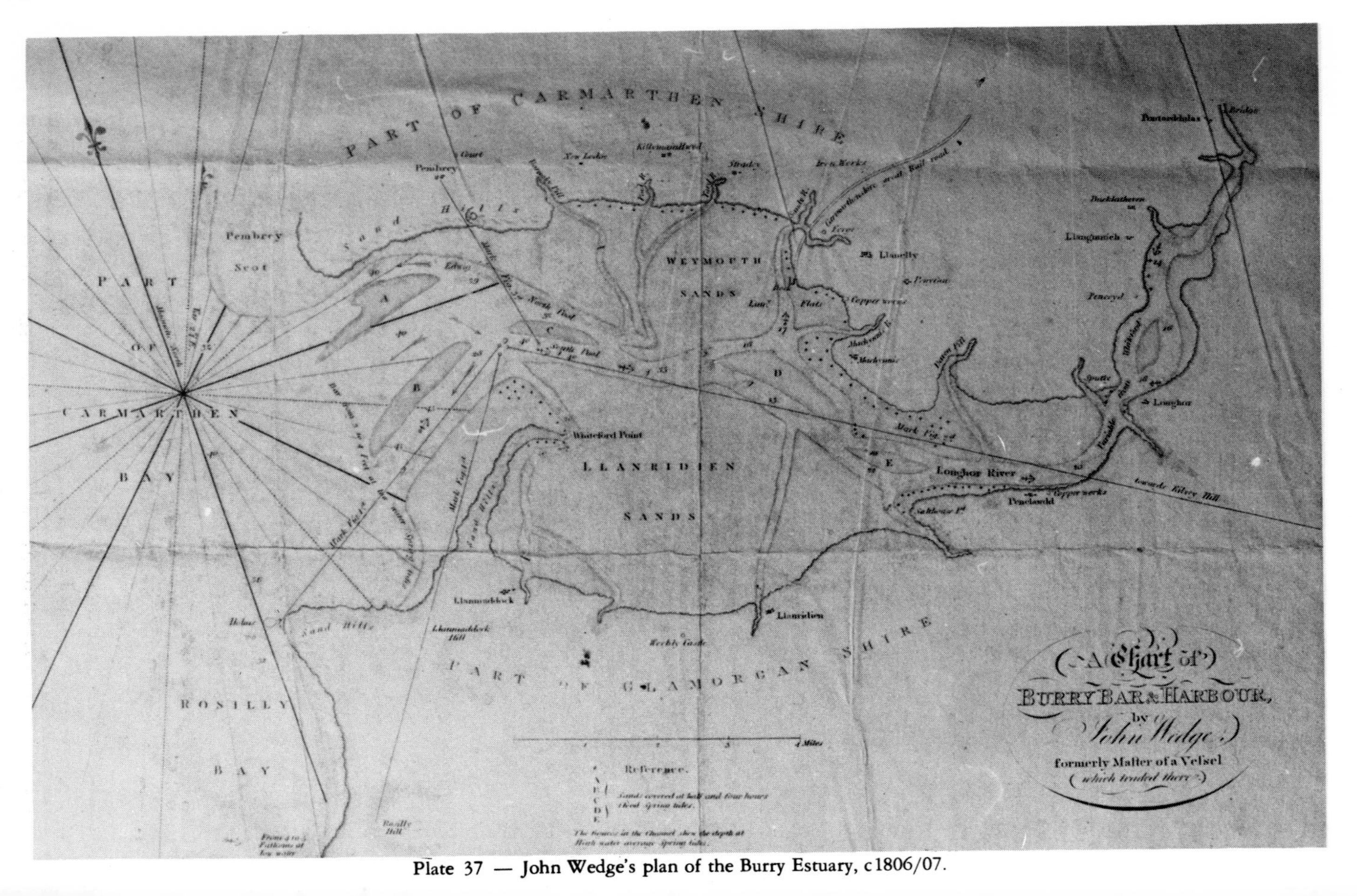

Plate 37 — John Wedge's plan of the Burry Estuary, c1806/07.

tons burthen, the William and Jane (a large trading vessel in 1810), had come safely over Burry Bar some time before high water.[14] Unfortunately, the same ship, outward bound from Llanelly to Cork with coal, foundered on Burry Bar the next year[15] and confidence in the safety of the Burry Estuary must have been shaken. To alleviate this problem, a meeting, on 4 April 1811, of "*Gentlemen interested in the Trade of the Burry River*", resolved to maintain a pilot boat constantly at sea, General Warde offering the use of his yacht until a suitable boat could be found.[16] A further meeting, held on 11 August 1811, resolved to promote a Bill to improve the buoyage and navigation of the Burry River[17] and, anticipating its being passed, John Wedge was asked to report on the Burry River and the cost of improvements.[18] The Act, passed in July 1813,[19] authorised the appointment of fifty nine Harbour Commissioners who were empowered to raise £2,000 on mortgage, against rates and duties to be levied, to effect specified improvements. It was further stipulated that, until the buoyage was improved, vessels need not pay duty to cross Burry Bar. Another clause empowered the Commissioners to direct the deposition of ballast. No ballast was to be placed on the eastern side of the River Lliedi, between Iron Bridge (near Sandy) and Machynis, but deposition was authorised on the western side of the Lliedi, commencing "*at or near a Point Half-way between the entrance to the Carmarthenshire Railway or Tram-road Company's Dock and the Barrel Post* (the present lighthouse remains) *and afterwards to extend the same in opposite directions Northwards and Southwards for the purpose of raising a Shelter or Protection for Ships and Vessels resorting to or frequenting the said Port and Harbour of Llanelly*".* This clause, relating to a breakwater, could have been in the Commissioners' minds since the Bill was proposed, but may have been included because of serious damage to docks, railways and embankments in a violent storm of October 1812.[20] The improvements authorised by the Act were soon implemented. The breakwater was started and the buoyage of the Estuary improved and maintained. By March 1815, the industrialists stated, in coal advertisements, "*the Harbour being now buoyed by Act of Parliament, vessels can at all times come in and out with perfect ease and safety*".[21] The Harbour Commissioners were beginning to come to terms with the peculiar difficulties of navigation, which always had been, and always would be, a feature of the Estuary.

Its safety was still in question, however, and with the impending

---

* This means that the construction of the present breakwater was started from near its middle and not from its land-end outward, as is usually assumed.

opening of Pembrey Harbour, and the possibility that it would take trade from Llanelli, the Commissioners took every opportunity to stress the benefits produced by the 1813 Act. Statistics of the increased number of ships using Llanelli in 1817 were published in *The Cambrian,* in January 1818, when it was said that "*as a proof of the very great increase of trade within Llanelly port in the year 1817, the following statement has been handed to us*"[22] and, in March 1818, R. J. Nevill, advertising for ships to carry freights to ports in Ireland and the West of England, stressed that Cefn Sidan sands was not part of the Harbour of Llanelly. He said "*I am desirous it should be known, that KEFN SHEDAN SANDS, (on which most of the shipwrecks which have occurred in Carmarthen Bay for the last twelve years have happened) do not form any part of the Harbour of Llanelly or of the Bar of Burry; but are situate considerably to the North West of this navigation and appear to have been formed by the junction of the respective rivers of Kidwelly, Carmarthen and Laugharne*".[23] In January 1819 it was stated that "*The River Burry is now in a most excellent state, well buoyed; and the Harbour of Llanelly is deepened, and very materially improved by additional Shipping Places in deeper water*".[24] By 1820 at the latest, a Tide Table was published for the Port of Llanelly, which said "*The Llanelly Flats at present afford good shelter for shipping, which, in the course of a short time, by means of the break-water, which is being daily extended, will be rendered equally as safe, and fully as commodious, as any harbour in the Principality*".[25] An advertise-ment, publicising the Tide Table for 1822, stressed the continuing improvements and stated "*during the late tempestuous weather, unexampled in severity and duration, not the slightest damage was sus-tained by any one of the Vessels in the Harbour; and the Commissioners are still proceeding to give increased access and safety to the Shipping, by deepening the channel and extending the Breakwater*".[26] There can be little doubt that the Commissioners, and Nevill in particular, were doing all they could to improve shipping facilities at Llanelli and the poor image of the Estuary. By June 1823, the Copperworks Company announced their intention of converting their tidal dock to a floating dock[27] and wrote to the Admiralty for formal permission.[28]

As their developments neared completion and their advertising succeeded, the Commissioners must have felt satisfied with the situation. They must also have derived satisfaction from a letter written by a ship's captain, who, explaining why so many ships chose not to load coal at Swansea, said "*but it is notorious that there is more spirit in that little awkward place, Llanelly, than in Swansea*".[29] The shipping facilities at Spitty Bank were improved in the Summer of 1823[30] and the Copperworks Dock became the first floating dock in the Llanelli area, in 1825, when the dock gates were set up.[31] The breakwater was completed in 1828,[32] and the overall improvement in Llanelli's shipping facilities and in its reputation as a port was summarised in *The Cambrian,* which said "*no place affords a more ample scope for gratulation than the Port of Llanelly, the Flats of which, from being a wild shoal, and a dangerous shore, are by a very extensive breakwater become a safe and deep harbour where vessels of burthen, in great numbers, may lie in safety 'to a pack-thread', and so powerful are the different back-waters that the Flats have been scoured away until the ribs of a schooner, wrecked there 40 years ago and never seen since, are become visible for a height of 6 or 7 feet. Vessels of from 40 — 50 tons were formerly considered large for the Flats, while there are now loading there vessels of 450 tons. The convenience of a Wet Dock to load in is also very great and so frequently now do large vessels from the North resort there that the owners who write to insure each other have had a survey and report made, the results of which has been, that they insure to Llanelly, upon the same terms as any other Port*".[33] A further significant step towards the improvement of the Harbour was taken in 1828, when the Llanelly Railway and Dock Company obtained an Act[34] to construct a large floating dock at Machynis Pool, linked by a railway to the St David's pit, which was being sunk at Gelli-gele. This floating dock (the New Dock) was not commenced until 1832, when the St David's pit reached coal, but it was planned in the period considered. It became the first public floating dock in Wales and, for some 20 years, attracted the largest class of vessel to Llanelli.

Despite significant improvements in navigation and in shipping places, it must be acknowledged that the Commissioners were making the best of what was an unsatisfactory location for a harbour. H. R. Palmer, in his report to the Commissioners in 1840, said "*The situation of the Harbour does not appear to have been suggested by any natural circumstances peculiarly favourable to such a purpose, but seems to have been adopted for the service of local purposes previously requiring a work of that nature and which could not, without disadvantage, be removed to a site on which a more perfect Harbour could be formed*" and I. K. Brunel, in a report of 1857, said "*Nature has not done much to fit Llanelly for a Port*".[35] Both were engaged to investigate the silting of the main channel, between the breakwater and the docks, and the blocking of its access by shifting sandbanks in the Estuary. This problem was never solved and was a primary reason for Llanelli's decline as a port, in the late 19th/early 20th centuries.

(II) COAL EXPORTS BEFORE 1830

Before 1830, Llanelli's coal industry was developed chiefly to meet export trade demand. The establishment, in the early 19th century, of iron, copper and lead works using local coal did not alter the situation appreciably and, as late as 1826, well over 90% of the ships left laden with coal.[36] Coal exports, therefore, give a good indication of the development of Llanelli's coal industry up to 1830,[37] and so an attempt has been made to estimate the volume of exports from the Llanelli area. These estimates are approximate, because of several factors: the area covered by this work does not correspond exactly to the administrative area for which export figures have been published; very little information is available for most of the 18th century; much of the data used has been derived from antiquarian works, newspapers and by interpretation of the work of previous researchers; uncertainty exists over weights and measures used at various times by the coal trade in Llanelli. The limitations of the estimates will be discussed later. Accuracy is, therefore, not claimed but it is hoped that a reasonable approximation to the actual figures has been arrived at. More fundamental research into Chancery, Exchequer and Customs records must be carried out to obtain reliable estimates.

Coal exports from the Llanelli area have been studied over four periods, which correspond to the 16th, 17th, 18th centuries and 1800 to 1829. The export figures are given in tons, although most of the early returns for the Llanelli area were in weys or chaldrons, which are measures based on volume and not on weight. The relationship between the modern ton and the old measures can only be determined approximately. This is discussed in Appendix F, which deals with the weights and measures known to have been employed in the Llanelli area between the 16th and mid 19th centuries.

Subject to the errors outlined above, it is estimated that at least 2,500,000 tons of coal were exported from the Llanelli area between 1550 and 1829, of which some 60% was exported between 1800 and 1829.

(a) *Coal exports in the 16th century*

Exchequer records consulted by previous researchers contain information relating to the export of coal from the Llanelli area in the 16th century, but it is difficult to be confident in relation to the actual quantities exported, for the following reasons:-

An Act of 1558[38] specified the main ports and their limits in England and Wales, for fiscal and other related purposes. Cardiff, Milford and Chester were the three head ports for Wales and, from 1559, regular accounts of their foreign and coastal exports were entered into parchment books, which are now known as "*The Welsh Port Books*". The Llanelli area was included in returns made for North Burry, designated as a subsidiary port within the head port of Milford.[39] Previous writers have equated Llanelli to North Burry,[40] but the Pembrey area and the Loughor Estuary between Llangennech and Pontardulais, which lie outside the area covered by this present study, were probably included in North Burry. Coal export returns for North Burry must, therefore, overestimate export figures for the Llanelli area considered in this work. A further major uncertainty also exists about the accuracy of the returns in the Port Books, and it has been suggested that, due to dishonesty and malpractice, they do not always record the actual quantities of merchandise exported.[41] Nevertheless, it is assumed, for the purposes of this work, that the amount of coal exported from the Llanelli area in the 16th century is correctly represented by the returns made for North Burry.

Two published works, by E. A. Lewis[42] and J. U. Nef,[43] contain estimates of coal exports based on examination of the Port Books. Lewis reports the returns for North Burry (and in individual cases for Llanelli itself); Nef presents returns and estimates for "*Llanelly and Burry*", which is taken to mean the entire Burry Estuary, including South Burry on the Glamorganshire side of the river. The quantities given by both authors are listed in Appendix G and are shown in Figure 34. Coal exports between 1551 and 1600 did not exceed 400 tons a year, although there is some evidence of a build-up in trade towards the end of the 16th century. The details of this trade are significant. The first specific reference to the export of coal from Llanelli was in the year 1566 when 2½ weys (10 tons) were shipped to Bideford aboard "*Le Saviour de Bydyford*". This was followed by a number of shippings of coal from "*Burrey*" to France and to Guernsey in 1567 but exports then, apparently, ceased for five years, until 8 weys (32 tons) of "*ring coal*" (bituminous coal) were shipped to France, in 1572. The next entry was in 1586, when coal was shipped to Dartmouth and to Rochelle, followed by an entry for 1593/94, when 48 weys (192 tons) were exported to France from Burry (8 weys specifically listed as from Llanelli). The last entry for the 16th century relates to 1599, when 57 weys (228 tons) were shipped from "*Burry-creek of Carmarthen*" to Rochelle, between 30 April and 11 July.[44] These minor and intermittent coal exports are all that can be positively ascribed to the Llanelli area in the 16th century.

As we know that coal was being mined at Llanelli as early as 1536-39,[45] it must be questioned whether the small quantities of exports listed in the Port Books accurately reflect the quantities of coal shipped in the 16th

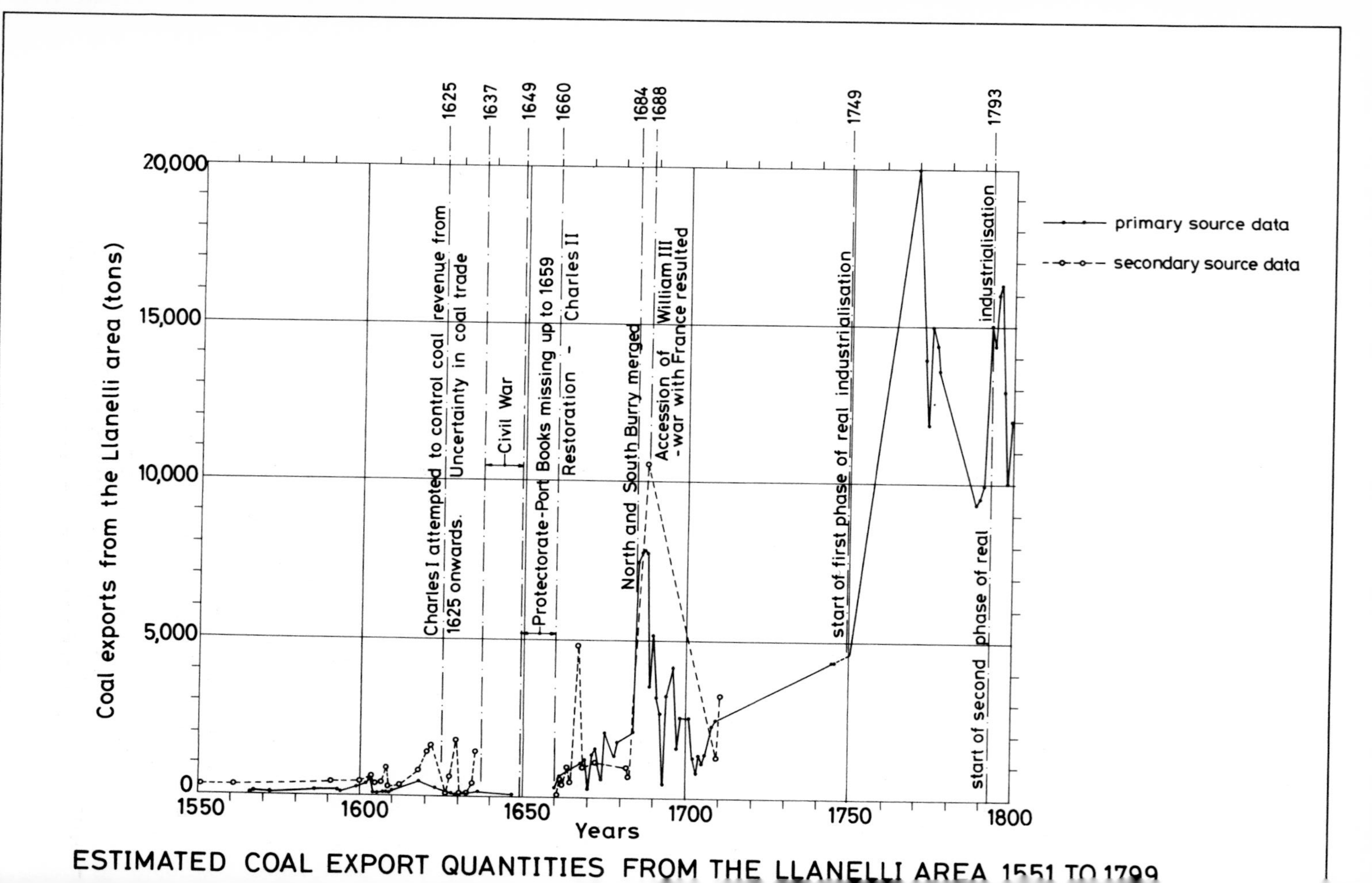

Coal exports from the Llanelli area (tons)
20,000
15,000
10,000
5,000
0
1550
1600
1650
1700
1750
1800
Years
1625
1637
1649
1660
1684
1688
1749
1793
Charles I attempted to control coal revenue from 1625 onwards. Uncertainty in coal trade
Civil War
Protectorate-Port Books missing up to 1659
Restoration - Charles II
North and South Burry merged
Accession of William III - war with France resulted
start of first phase of real industrialisation
start of second phase of real industrialisation
primary source data
secondary source data
ESTIMATED COAL EXPORT QUANTITIES FROM THE LLANELLI AREA 1551 TO 1799

Fig. 34

century. It is possible that all exports were not included in official returns, but piracy in the Bristol Channel and around the South Wales coast might also have prevented the expansion of Llanelli's coal export trade, when the potential of the area as a coal mining centre was just being recognised. Piracy, rife in the Bristol Channel since the 14th century, resurged in the latter half of the 16th century, and the coast of South Wales suffered from pirates, who found shelter and a market for their booty at Cardiff.[46] They operated over the whole of the South Wales coast, goods stolen from as far west as Haverfordwest being publicly sold in Cardiff, in 1576.[47] They would not have been particularly interested in cargoes of coal, but there can be little doubt that the growth of maritime trade generally, in coastal waters around South Wales, was inhibited by piracy in the 16th century. It has been suggested that the first reference in the Port Books to exports from Llanelli occurred in 1566 because Commissioners were appointed, in 1565, to investigate and suppress piracy, their counter measures reducing the activity.[48] It was not eliminated, however, and Commissioners were again appointed in 1575, resulting in naval patrols in the Bristol Channel, which significantly reduced piracy around South Wales by the early 17th century.[49]

Although coal exports from the Llanelli area in the 16th century were small and seemingly intermittent, they were most significant, providing the basis for Llanelli's coal industry and its future as an industrial centre.

(b) *Coal exports in the 17th century*

Lewis, Nef and B. M. Evans used the Welsh Port Books to estimate coal exports from the Llanelli area in the 17th century.[50] The returns are less intermittent than those for the 16th century, but the same uncertainties exist about the area covered by the returns and about their accuracy. The export quantities given by the various authors are listed in Appendix G and are shown in Figure 34. There are discrepancies, probably because Nef considered both North and South Burry together, but the same general picture is obtained, irrespective of the set of quantities taken. Greater detail is available for this century and interpretation of export trends can be attempted.

The increase in exports in the late 16th century continued into the early years of the 17th century, despite the imposition, in 1599, of an export duty of 5 shillings per two tons on coals transported "*overseas*". Although this duty was removed, between 1601 and 1603, on coals transported to Ireland, Scotland, the Isle of Man and the Channel Islands,[51] the quantities shipped from the Llanelli area dropped between 1603 and 1609, until they were as low as they had been in the 1560s. The reason for this interruption in growth is not immediately apparent and it is not known whether it was caused by local or national factors. The 1599 export duty possibly acted against the French trade, which had built up in the 1590s. If this were so, it should have produced a decrease in exports before 1603, but shipments to Rochelle and Brest were regularly listed in the Port Books for these years.[52] The recovery in the export trade, between 1610 and the early 1620s, coincided with an increase in local mining[53] but, thereafter, export quantities, and presumably local coal production, fell back to levels of 50 or 60 years earlier. This decline was probably due to several factors, but chiefly to increases in export duties in 1620 and 1634, which acted against Llanelli's trade with France, and the uncertainty in Britain's coal trade throughout the reign of Charles I, when the coal industry was regarded as a source of income for the Crown.[54] The Civil War and the Protectorate also depressed Llanelli's coal industry between 1637 and 1659. Coal exports were listed in only one year during the Civil War[55] and, although it appears that Port Books were not kept during the Protectorate, it is doubtful if there was a recovery in Llanelli's coal trade during this period. John Vaughan, the major figure in Llanelli's coal industry at this time, was a Royalist and, although only fined after the Civil War,[56] it is unlikely that his interests flourished during the Protectorate.

Vaughan's situation altered completely on the Restoration in 1660 and he began a large scale expansion of the local coal industry, between 1660 and 1665.[57] Coal exports increased, despite interruptions caused by the maritime war with Holland between 1665 and 1667, Vaughan's death in 1669 and a total ban on exports in 1673/74, reaching 2,000 tons in 1683/4. Within a year, however, coal exports increased almost fourfold to an estimated 7,400 tons. It is significant that this large increase coincided with the merging of North and South Burry into the Port of Llanelly for customs purposes.[58] As the 7,400 tons relates to North Burry alone,[59] it must be more than coincidental that coal exports increased so suddenly after a reorganisation of the local customs. South Burry had been administered from Swansea before the reorganisation and, with the whole of the Burry Estuary coming under more vigilant customs supervision, it is probable that evasion of export duties was reduced, with a consequent increase in recorded coal exports. (A similar situation, which supports this opinion, arose in 1847, when it was proposed that the coast between Whitford Point and Loughor should again come under the control of the customs at Swansea. The Collector at Llanelli objected, saying "*It also appears to us that the part of the Coast alluded to is more immediately under the eye of the Officers at Llanelly than at Swansea*").[60]

Coal exports remained fairly constant until the accession of William of Orange in 1688, when resulting hostilities with France, Llanelli's biggest foreign market, virtually destroyed North Burry's foreign export trade. Efforts were made to develop trade with Ireland but, in 1698, total exports were only 2,500 tons.[61] This decline was probably mainly due to the loss of the French trade, in 1689, but Margaret Vaughan's control of the family's coal interests from 1683,[62] when she was over 60, could also have been a cause, in that she neglected to exploit the Estate's coal interests fully.

Although Llanelli's coal industry was in recession by 1700, trading contacts had been made. A regular coastal trade had been established with the West of England (particularly Plymouth, Bideford and Barnstaple), with London and with various parts of Wales. Trade had been established with France (particularly Brest and Rochelle) in times of peace, with Ireland in the latter years of the century and, intermittently, with Portugal (Lisbon and Oporto) and the Channel Islands.[63] These markets survived well into the 19th century and, in some cases, as late as the early 20th century.

(c) *Coal exports in the 18th century*

Coal export returns for the Llanelli area exist only for 1701-10, Port Books for the rest of the century having apparently been destroyed in a fire.[64] Intermittent and non-official evidence has, therefore, been used to estimate the volume of coal exports for the Llanelli area after 1710. The estimates, given in Appendix G and shown in Figure 34, are scanty and probably inaccurate, so there is little scope for interpretation, although some points will be made.

In 1700 little coal was exported but the trade recovered after 1703. This was mainly due to increased coastwise exports, the war with France preventing a recovery of the foreign export trade, and, by 1709, an estimated 2,500 tons of coal were being shipped from the Llanelli area.[65] No information has been seen for 1711 to 1743 but it is likely that the industry steadily recovered, because it was said, in 1724, that "*Llanelly drove a pretty good trade in coals*"[66] and, in 1727, Llanelli was described as a town whose "*inhabitants are principally traders in sea coal*".[67] The number of ships frequenting Llanelli between Michaelmas 1744 and Michaelmas 1746 is known from keelage and layerage returns.[68] These manorial rights belonged to the Stepneys, who had inherited them from the

Vaughans by marriage, but the area over which they extended is not definitely known, although it could have been most of North Burry. (The Stepneys' claim to these rights was disputed in 1802, when evidence was given that "*the Leerage and Keelage of the North side of all Burry River*" belonged to Sir John Stepney, 8th Bart., and had belonged to "*his Family before him*").[69] The returns, made by Edward Dalton, Chief Collector of Customs at Llanelly from 1723 to 1766,[70] showed that 288 vessels, 200 engaged in coastwise and 88 in foreign trade, frequented the Llanelli area between Michaelmas 1744 and 1746. If it is assumed that these vessels averaged 30 tons and were all engaged in the coal trade, a possible coal export quantity of 4,300 tons per annum is obtained for this two year period (See Appendix G).

No information has been seen about ships frequenting Llanelli between 1747 and 1771. Coal exports cannot, therefore, be estimated for a most significant period in the development of Llanelli's coal industry, as John Allen, Squire, Evans and Beynon, Thomas Bowen, Chauncey Townsend and others began the first real industrialisation of the Coalfield, and Sir Thomas Stepney, 7th Bart., attempted to develop the shipping trade of the area. We know, however, that coal was shipped to West Country ports and to Ireland, Brest, Lisbon and Oporto, between 1750 and 1755;[71] that Charles Gwyn mined some 2,600 tons of coal a year, between 1763 and 1770/73, "*which he shipped to Sea and sold to the country*";[72] that large quantities of coal were mined in the Bynea/Pencoed region by Thomas Bowen, Squire, Evans and Beynon and Chauncey Townsend, between 1749 and 1772;[73] and that vessels were waiting for Townsend's coal to be raised from the pits in 1769, when other coal owners had stocks ready for shipping,[74] implying that coal production was more than meeting export trade demand. Coal exports in this period could, therefore, have exceeded all previous totals, with maximum quantities possibly achieved c1769/70, when Chauncey Townsend's colliery at Genwen became operative. This possibility is shown on Figure 34, with an export quantity of 20,000 tons allocated for 1769/70.

The number of ships paying keelage from 1772 to 1777[75] increased considerably over 1746. Improved shipping places had probably attracted a slightly larger class of ship, and coal exports probably averaged some 13,700 tons a year over the 5 year period. The decrease in exports compared to the late 1760s is realistic, because Townsend had died in 1770 and his collieries were little worked; Sir Thomas Stepney had died in 1772 and his son, Sir John, displayed little interest in his father's industrial enterprises; and Charles Gwyn had ceased working coal. The trade was evidently in recession because, in October 1776, it was agreed that Townsend's inheritors would work the Swansea Four Feet Vein at Bynea "*as soon as there shall be a Demand and Sale for the Coal and Culm to be raised therefrom, sufficient to pay the Expences thereof and a reasonable Proffit which there is not at Present, nor (as it is apprehended) will there be for several years to come*".[76] The situation was further aggravated by a scarcity of vessels due to the activities of French ships and fears of a French invasion between 1778 and 1781.[77] It is likely that this recession continued throughout the 1780s. There is little evidence of significant coal mining in this decade and the keelage return for 1788/89 showed a significant decrease, compared to 1777. The estimated coal exports, based on keelage returns from 1788 to 1791,[78] averaged some 9,600 tons a year.

D. Bowen gives, without quoting his source, the number of vessels leaving Llanelli from 1793 to 1800.[79] These have been used to estimate coal exports for this period, when the second phase of Llanelli's real industrialisation commenced. This industrial development coincided, however, with a renewal of hostilities with France, in 1793, and a serious period of inflation, caused by the war and bad harvests. It is, therefore, difficult to decide if fluctuations in exports between 1793 and 1800 were due to national, or predominantly local, factors. The estimates show that exports reached 16,400 tons in 1796, fell back to 10,000 tons in 1798 and then increased to around 12,000 tons in 1799. French ships on trade routes, and fears of a possible invasion after 1793, may well have contributed to the decrease in the export trade[80] but there are, also, convincing local reasons to explain it. The only known working mines in the Llanelli area, in the early 1790s, were those of the Mansells west of the Lliedi, and of William Hopkin and John Jenkin at Llangennech. When the furnace and foundry were established at Cwmddyche, by Givers and Ingman c1793, they took the bulk of the Mansells' output; and the situation did not change when Raby took over the collieries and furnace in 1796/97, Raby writing in November 1797 "*we can scarcely get coals sufficient at present out of the Caerythin Vein for the use of the Furnace*".[81] As Raby's other collieries were not then producing, it is probable that the Mansells' coal was not available for export after 1793/4, because of reduced production and increased local consumption. Hopkin and Jenkin failed to develop the Llangennech collieries; David Hughes and Joseph Jones experienced trouble with their colliery in the Bryn Carnarfon/Baccas region, which they had commenced in 1794; and Roderick and partners, who started mining the Bres/Wern region in 1794/95, raised only small quantities of coal from 1795 to 1800. It is, therefore, likely that coal available for export decreased between 1793 and 1800, because of the reorganisation of the coal industry and because the output of a major colliery was all used by the furnace and foundry.

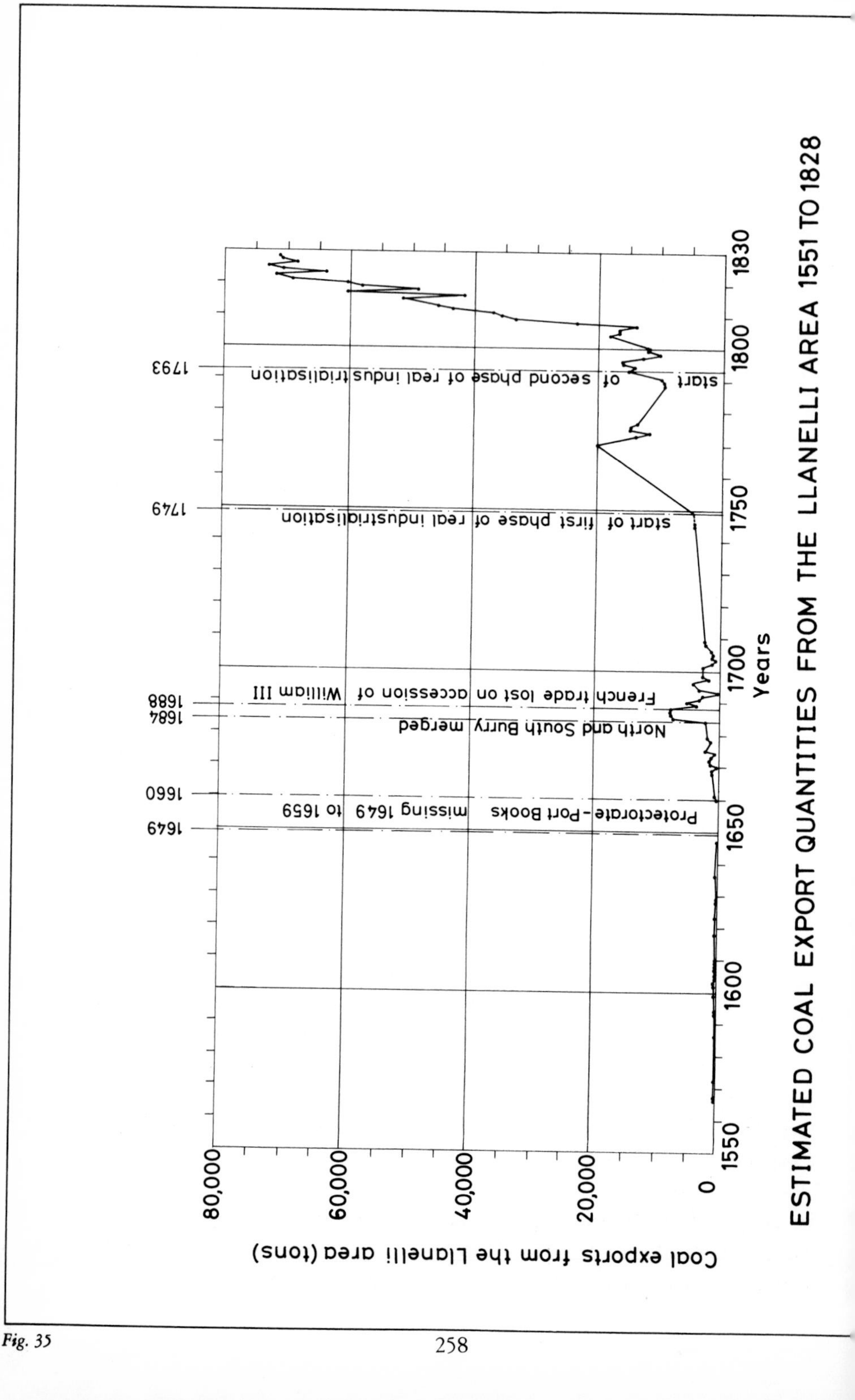

ESTIMATED COAL EXPORT QUANTITIES FROM THE LLANELLI AREA 1551 TO 1828

Coal exports from the Llanelli area (tons)

80,000
60,000
40,000
20,000
0

1550
1600
1650
1700
1750
1800
1830

Years

1649
1660
1684
1688
1749
1793

Protectorate - Port Books missing 1649 to 1659

North and South Burry merged

French trade lost on accession of William III

start of first phase of real industrialisation

start of second phase of real industrialisation

Fig. 35

(d) *Coal exports 1800 to 1829*

Official returns of coal exports from 1816 to 1829 distinguish between coastwise and foreign exports.[82] Shipping returns reported weekly in *The Cambrian* have been used for 1804 to 1816. These are of particular interest, as cargoes of ships sailing to, and cargoes and destinations of ships sailing from, Llanelli were listed. Estimated coal exports, between 1800 and 1829, are given in Appendix G and shown in Figure 35 (which also includes the exports between 1550 and 1800, for comparitive purposes). Hostilities with France, and the reorganisation of the industry, restricted coal exports to some 12,000 tons in 1800, but developments by Raby, Roderick and partners and others were about to result in increased production. Llanelli's first two docks had been formed and the temporary peace with France, between 1802 and 1803, may also have led to increased coal exports, although there is no evidence for this. 18,000 tons were probably exported in 1804 but the improvement was not maintained, exports falling to 14,000 tons in 1807. The renewed war with France in May 1803 and fears of invasion until Trafalgar, in July 1805, may have affected exports,[83] but the main reason was the state of the local coal industry; Roderick and partners, Raby, Vancouver and the Earl of Warwick all experienced financial crises between 1804 and 1807, and Warde's main collieries were not yet producing coal. Consequently, coal production and exports suffered.

Although Britain was still at war, there was an unprecedented growth in Llanelli's coal exports after 1808. The chief reasons were:- Raby opened the Box Colliery; the Pembertons started to work the Bres Pit; the Copperworks Company took over and developed Roderick and partners South Crop Wern Colliery; Symmons took over collieries, developed by the Earl of Warwick and Vancouver, at Llangennech; Warde opened his collieries on a large scale; the Llanelly Copperworks and the Spitty Copperworks started smelting, which required the import of copper ore, with ore ships sailing out laden with coal; a third dock was built by the Copperworks Company; Raby's dock was expanded by the Carmarthenshire Railway Company; Roderick and partners' dock was expanded by the Pembertons; new shipping places were built at Llangennech and Dafen Pill; and transport systems, mainly for the haulage of coal, were laid down throughout the area. With so many developments in so short a time, Llanelli underwent the rapid economic growth which was a feature of early 19th century Britain. Thereafter, single factors, such as the loss of the French trade, became less important to the coal trade within the framework of the new, more complex, industrial structure.

Between 1807 and 1815, estimated coal exports increased from 14,000 tons to 51,600 tons, an average growth rate of some 18% a year. By 1829, coal exports were more than 71,000 tons, despite interruptions in 1816, 1818, 1823 and 1826, having increased almost 7 fold in the 30 years, with an average annual growth of almost 7%. This increase, together with increased consumption of coal by industry and the growing population, created the conditions for the rapid growth of Llanelli's coal industry and laid the foundation for the area's industrial future.

The destinations of ships carrying Llanelli's coal, between 1804 and 1825,[84] show that there were three main markets: other Welsh ports (particularly Carmarthen and Cardigan); West Country ports (particularly St Ives, Barnstaple, Bideford and Plymouth); and Irish ports (particularly Waterford, Wexford, Cork, Kinsale and Wicklow). Ships also sailed to other English ports, the Channel Islands, Copenhagen (until 1816) and, after the Napoleonic Wars, to French ports (particularly Brest) but these constituted only a small percentage of the total trade. Figure 36 shows the extent of Llanelli's coal trade with the various markets from 1804 to 1825. The estimates are based on the number of coal sailings to each market, expressed as a percentage of the total number of coal sailings. If returns which gave the actual quantities of coal exported had been seen, then higher percentages would have resulted for the trade with Irish and foreign ports, because larger ships were used in these markets. Welsh and West Country markets usually took most of Llanelli's coal but, from 1811 to 1819, the Irish market took approximately one-quarter of the vessels, and probably more than that in terms of tonnage.

A number of changes in the distribution of coal exports occurred during the period, which require explanation. From 1809, there was a sudden increase in the Irish trade and a corresponding decrease in the Welsh coastal trade. This followed advertising by the Llanelli coalmasters, to attract larger ships of the Irish market, which could be accommodated due to the improved harbour facilities and navigation of the Estuary. Raby advertised for ships of between 70 and 200 tons, for the Irish trade, in March 1808,[85] and, in August, Warde stressed the ease with which Ireland could be reached from his shipping places.[86] The West Country trade was maintained, because of the profitable two-way trade of copper ore import and coal export. The end of the Napoleonic Wars in 1815 led to increased foreign exports (mainly to Brest and to the Channel Islands) but the most significant change was the increase in the West Country trade from 1819 onwards, which coincided with a steady decrease in the Irish trade. This was mainly because R. J. Nevill, acting for the Llanelly Copperworks

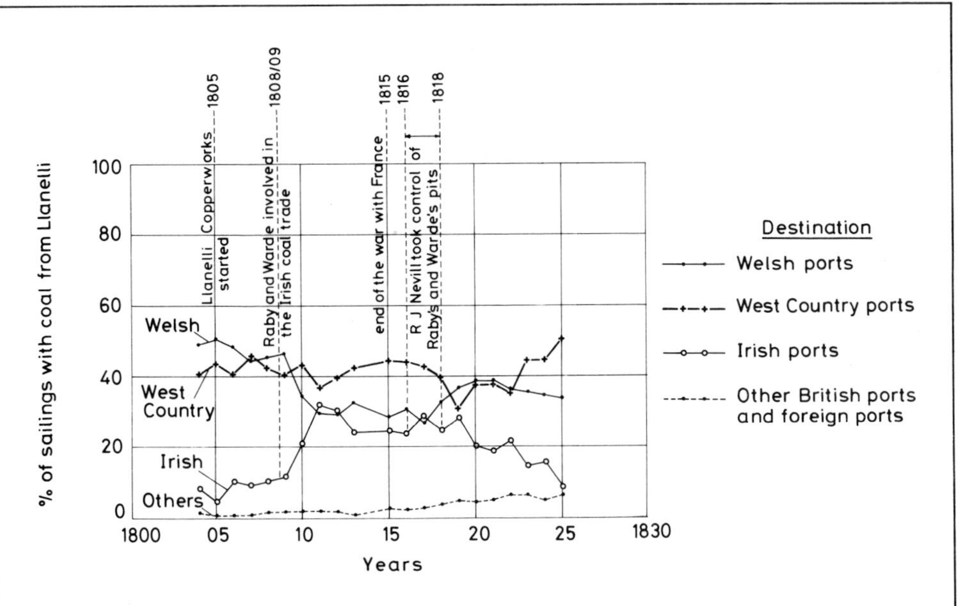

DISTRIBUTION OF COAL SAILINGS BETWEEN 1804 & 1825

36

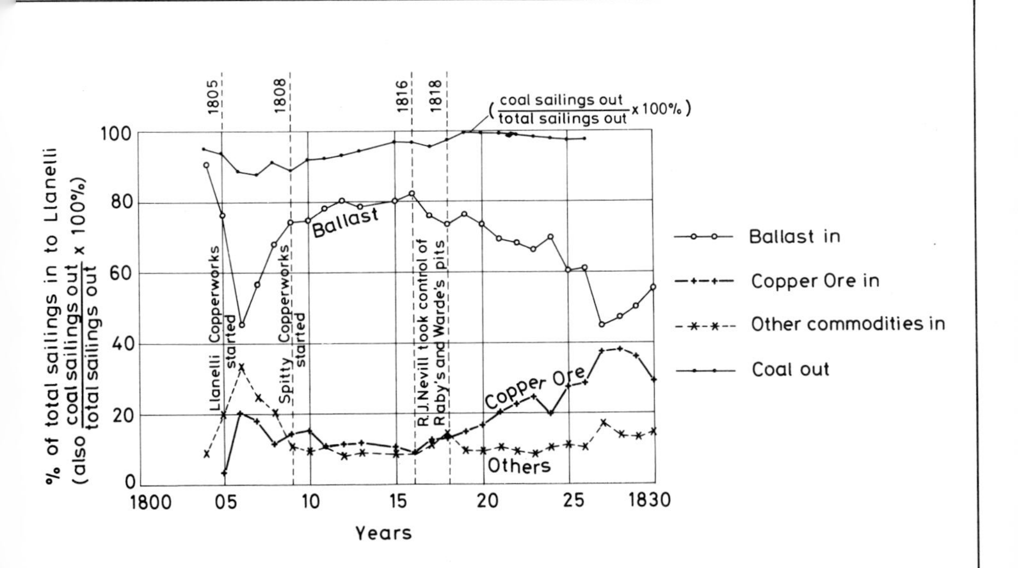

DISTRIBUTION OF IMPORT VESSELS BETWEEN 1804 & 1830
(also variation in % of ships sailing out laden with coal)

7

Company, purchased Raby's collieries in 1816 and took over the management of Warde's collieries in 1818. Nevill concentrated on the West Country copper ore/coal trade, as it yielded greater profit to his Company than Raby's and Warde's established Irish trade.

The returns of ships sailing to Llanelli[87] give further proof of the important part the coal industry played in Llanelli's 19th century industrialisation. Figure 37 shows that over 90% of the vessels entering Llanelli in 1804 were in ballast and, as some 96% left with coal, the export trade relied solely on coal at this time. This situation changed briefly between 1804 and 1808, as materials required for a rapidly expanding industrial area were brought in, and as regular copper ore imports were required for the copperworks at Llanelli and Spitty. In 1806 only 45% of ships entered in ballast, 21% being loaded with copper ore and 34% with other materials. This situation did not last long, and, from 1808, the coal export trade again became the dominating factor. By 1810/11, almost 80% of vessels sailing in were in ballast, the cargoes of the remaining 20% being fairly equally divided between copper ore and general materials. At this time, some 92% of vessels left Llanelli with coal. There was little change until 1817, when R. J. Nevill's influence on the coal industry led to a greater percentage of ships entering with copper ore and a lower percentage in ballast. This trend continued until 1827, when 45% of vessels entered in ballast, 38% with copper ore and 17% with other imports. The influence of the coal industry upon Llanelli's shipping trade had not weakened, however, some 98% of these ships leaving Llanelli with coal. Between 1827 and 1829 the pattern showed signs of changing as the recession of the late 1820s/early 1830s began, and the percentage of ships importing copper ore fell and the percentage of ships sailing in with ballast increased.

(III) THE SHIPPING PLACES USED FOR THE EXPORT OF LLANELLI'S COAL BEFORE 1830

To export large quantities of coal, shipping places were required where vessels could be easily and safely loaded. At least 500,000 tons of coal were exported from the Llanelli area before 1750 but no detailed information has been seen about the shipping places used, although many of the later locations must have been traditional sites used for hundreds of years previously. These traditional sites will be discussed briefly, but this section will be chiefly devoted to shipping places known to have been used between 1750 and 1830. The method of loading known to have been sometimes adopted after 1750, in which keels or barges took coal to large ships anchored in the main channels of the Burry Estuary, will also be considered.

(a) *The probable traditional shipping locations*

The early shipping places would have been on rivers and pills, navigable to small ships at most states of tide, so that coal could be loaded throughout most of the year. All the rivers in the Llanelli area, which led into the Burry and Loughor rivers, could, therefore, have served as channels to shipping places, which would have been sited as close as possible to the collieries. There is no surviving evidence that purpose-built coal stages were provided, and vessels must have been loaded by hand over a period of time. William Jones's 1757 plan and John Wedge's 1808 chart, and the evidence of early leases, suggest that traditional locations could have been at Pool (Pwll), the Dulais (formerly Yard) River and pills around Stradey Marsh, the Lliedi River from Llanelli Flats to the Sandy/Caemain region, Penrhyngwyn, Machynis Pill and Dafen Pill to the region of Cefnymaes, Pill y Ceven and the pill later known as Townsend's Pill, Spitty Bank, Pencoed and Llangennech Pill (Fig. 38). These rivers or pills led directly into the main channels of the Burry or Loughor rivers and, as we know that they were used to ship coal from 1750, it is reasonable to suppose that they were used before that time. No evidence has been seen to confirm this, however.

(b) *Anchorages in the Estuary*

Before docks were built at Llanelli, in the late 1790s/early 1800s, only small ships could sail up the rivers and pills, large ships having no alternative but to wait afloat, at safe anchorages, for coal to be brought out in barges or keels. It is not known when this form of coal shipping was first used in the Llanelli area, but the first documentary evidence was in April 1752, when a ship's captain withdrew from an accepted safe anchorage off Penclawdd back to Whitford, prolonging the loading by many days. The details of this incident summarise the method of shipping employed — On 18 April 1752 the ship "*Matthew*", captained by Isaac Storm, anchored off Penclawdd to take on Bowen's coal. About one third of the cargo was loaded between April 21st and 24th. Storm and his mate then "*ordered all the Keelmen to bring him no more Coals, that if they did, he would not Take it on Board till he Brought his Ship to Whitford*". Coal brought out on the 25th was refused and, on the 3rd and 4th May, the ship fell back down the Burry River to anchor off Whitford, loading not being completed until the 11th or 12th May. On a subsequent visit to the Burry

Estuary, Storm refused to bring his ship beyond Whitford.[88] The evidence of William Jones's 1757 plan[89] strongly suggests that Bowen's coal was brought down Dafen Pill, and the refusal of the ship's captain to anchor off Penclawdd was serious, because it cast doubt on the safety of the Estuary's anchorages. Sir Thomas Stepney, 7th Bart., who was attempting to promote maritime trade and who owned at least two ships and had an interest in several others,[90] was particularly perturbed at the effect this might have on Llanelli's export trade. His representative at Llanelli wrote to a Mr Gibson, stressing that no mention should be made to prospective customers of the number of days loading would take, and gave his reason as "*our Harbour and Craft not being like Newcastle*". This letter also reveals that Storm took a pilot aboard at Dale, to guide him to Penclawdd, "*where large ships have loaded more than once*".[91] As return visits were made, large ships could spend a considerable time loading and still, presumably, find the journey economic. Storm made a third visit before November 1753, when his cargo was loaded within six days.[92] The pilot was probably taken aboard at Dale because of the hazards of Cefn Sidan sands and the Burry Bar.

There is further evidence of the use of anchorages in the Estuary. When Pembrey Harbour was opened, in 1819, it was stated that "*The north side of the Burry River is bold to an extent of nearly 2 miles above the Pembrey Harbour and was formerly well known by the name of the North Pool. There is an excellent Anchorage in this pool in about three fathoms at low water spring tides. It was formerly the station in which SHIPS of 300 tons burden had their cargoes taken off and to them from Barnaby Pill,* by small sloops, for want of a Harbour at Pembrey*".[93] A letter from William Roderick to William Clayton, in 1773, regarding the development of Clayton's coal for the lime-burning trade, said "*I have enquired into the price of Lime Coal per Weigh and find it comes to 21 shillings per weigh put on board the Vessels in the River*". Roderick's breakdown of the on-board price (of which the developer received 2s 10d) included 4/- for barging.[94] His breakdown was:

Working a weigh	5s - 8d	Barging it to the main river	4s - 0d
Carriage to river	6s - 6d or 7d	Land money	1s - 6d
Boarding it	0s - 6d		

As Roderick was referring to coal at the Allt and Brynsheffre, which Thomas Jones wished to work, barges would have brought it from Llangennech Pill to ships in the main channel of the River Loughor. The

* Barnaby Pill lay to the west of Pwll and coal from the present Burry Port area would have been shipped from it.

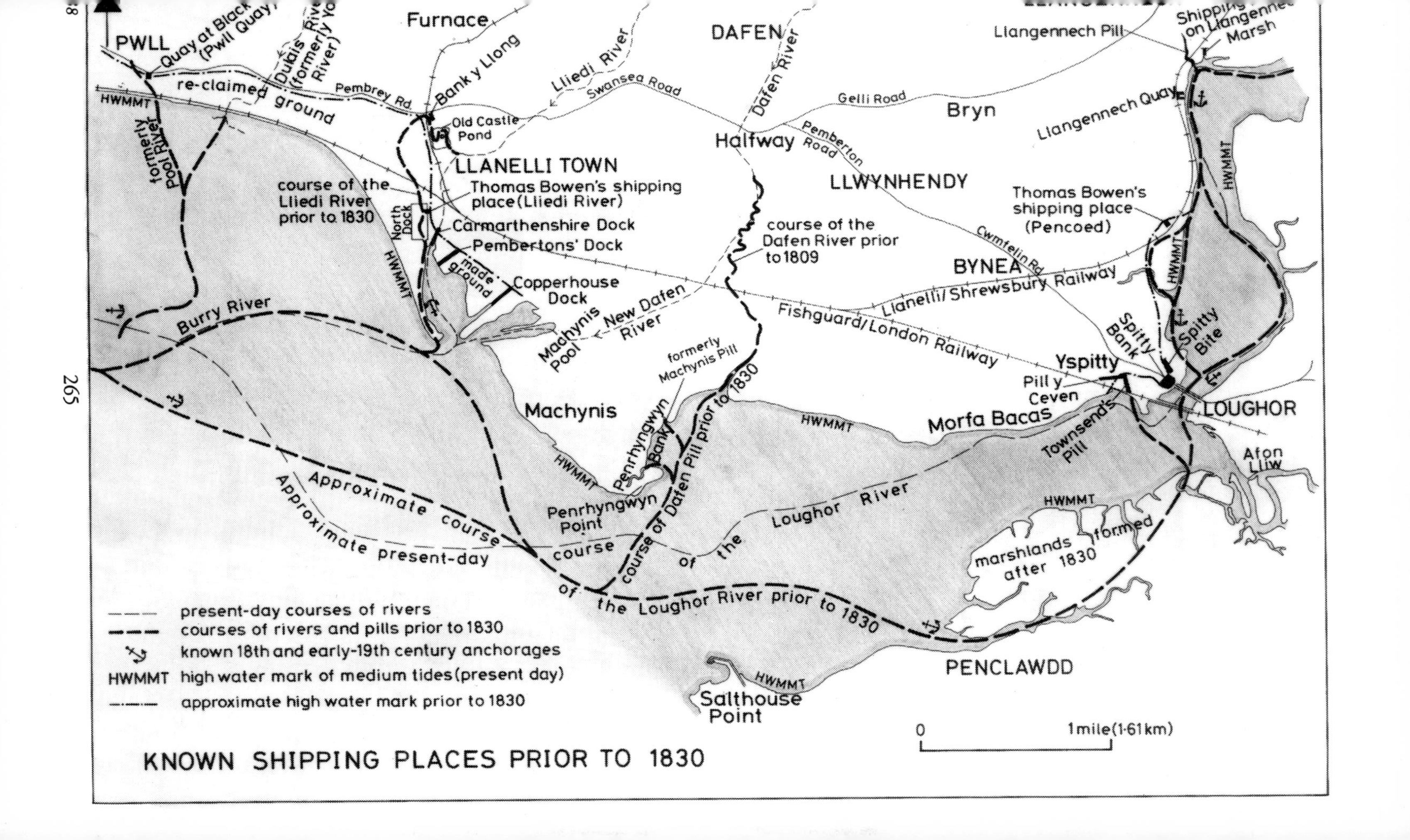

KNOWN SHIPPING PLACES PRIOR TO 1830

relatively high cost of barging probably led Sir Thomas Stepney to consider forming a quay on deep water in the River Loughor, opposite the tenement of Box, where vessels could be loaded directly.[95]

This evidence confirms that large ships at safe anchorages were regularly loaded from barges or keels in the 18th century at Llanelli. We do not know when the method was abandoned, the latest 19th century reference seen being in the 1813 Harbour Act,[96] which specified that vessels used to carry coal from any colliery for loading on ships or vessels in the Rivers Burry, Lliedi and Loughor would not be subject to duty. Known safe anchorages are shown on Figure 38.

(c) *The known Shipping Places*

Shipping places must have been used from the early days of the development of the Llanelli Coalfield, but the earliest known surviving evidence, giving the precise location of a shipping place, is contained in a plan of 1751. This, and all other sites, known from documentary or plan evidence, will be discussed in this section. Their locations are shown on Figure 38.

(1) *Spitty Bank*

Spitty Bank (or Banc y Spitti) was shown on old plans[97] as a plot of land and sand or marsh, covering some 5 acres on the main channel of the River Loughor at Spitty. It had probably been a traditional shipping place but it was first referred to as such on a plan of 1751,[98] which showed a waggon road from Squire, Evans and Beynon's pits in the Bynea/Pencoed region to Spitty Bank (Fig. 26). It is not known if wharfs or staithes were provided but, as their enterprise entailed the provision of the first steam engine and major railway on the Llanelli Coalfield, it is likely that the shipping place was specially constructed to take their almost half-ton coal waggons.[99] By 1772, only David Evans was left in the concern and the shipping place was called "*Dr Evans's Shipping Place*".[100] Evans stopped mining in 1781, when it is likely that the shipping place fell into disuse because, in 1801, Spitty Bank was said to have been "*late in the tenure or occupation of David Evans or his undertenants*".[101] (Plate 38)

In 1801 William Chute Hayton leased Spitty Bank, described as "*All the aforesaid Coal Bank and marsh*", to E. W. R. Shewen (Mansell) at a nominal rent.[102] Mansell later gained possession and sold it to John Symmons, and it became part of the Llangennech Estate in the early 1800s. By March 1804, coal was again being shipped from the Bank, an advertisement for Symmons's Bryn Colliery coal stating "*vessels may load at all*

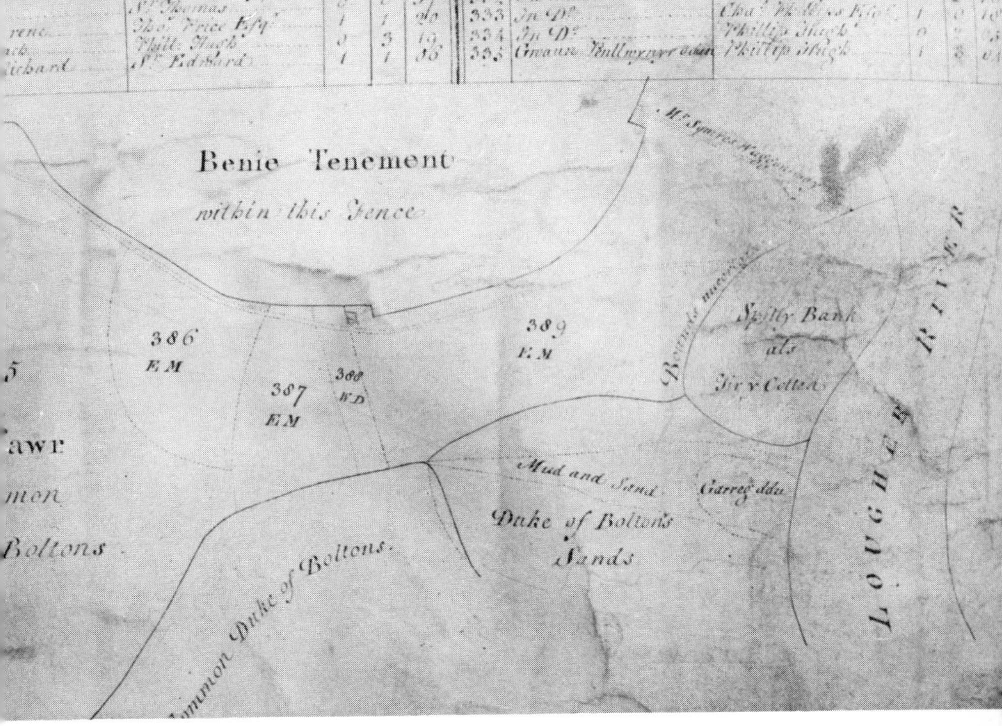

Plate 38 — Part of William Jones's 1751 plan showing Spitty Bank. "*Mr. Squire's Waggonway*" is shown terminating at Spitty Bank.

times of tide at Spitty Bank, opposite Loughor".[103] The Spitty Copperworks was built by Davenport, Morris and Rees,[104] before October 1808,[105] when the shipping facilities were probably improved. Certainly, when the Spitty Copperworks was advertised for sale, in 1814, it was stated that the works were situated at Spitty Bank "*on the most eligible spot of the navigable river Burry, with commodious Quays for shipping*".[106] A new quay and storage area for copper ore were added in 1823[107] and, c1826/27, the Llangennech Coal Company extended the Llangennech Tramroad to Spitty Bank, to bring coal directly to the Copperworks and to ships for loading[108] (Fig. 33). This extension was abandoned before May 1829[109] but, by this time, Spitty Bank only had a limited time left as a main shipping place. After the closure of the Spitty Copperworks in 1831,[110] and the building of the first Loughor Bridge, started in 1832[111] and completed in June 1834,[112] Spitty Bank was no longer used as one of the major shipping places in the Llanelli area.

(2) *Thomas Bowen's Shipping Place at Pencoed*

A plan, dated April 1772,[113] showed "*Mr Bowen's late shipping place*" close to the Pencoed leadworks and stated that Bowen had extensively worked the Swansea Five Feet Vein in this area, from a water-engine pit

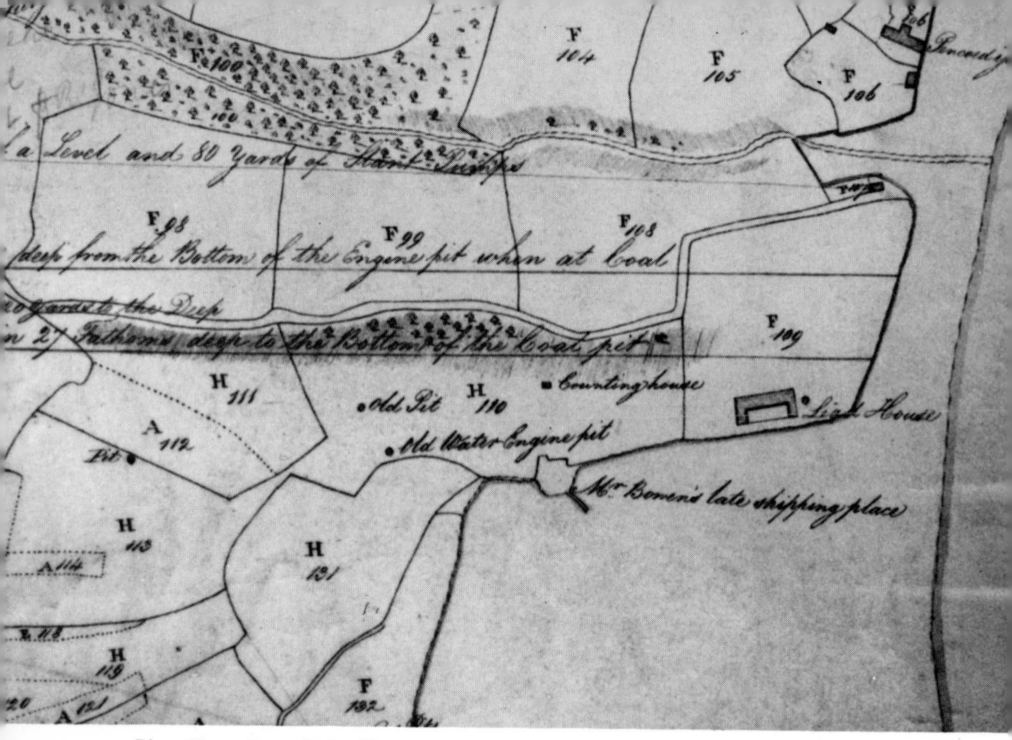

Plate 39 — Part of John Thornton's 1772 plan showing Thomas Bowen's shipping place at Pencoed. The Pencoed Lead House is shown close to the shipping place.

at Ffosfach. Dates were not given, but we know that these activities took place before 1772, and there is documentary evidence that the Bowen mentioned was Thomas Bowen of Llanelli and that the shipping place and colliery were probably established in the early 1750s.

Between 1748 and 1752 Bowen obtained a lease of coal under Pencoed.[114] The leadworks at Pencoed, probably established c1754 with Sir Thomas Stepney's backing, used Stepney's coal,[115] which must have been obtained from Bowen's pits. The leadworks was near the shipping place, so Bowen and the "*Pencoed Company*" probably used it for loading and unloading vessels. Bowen stopped mining in the area before 1770 and the leadworks probably closed about the same time. It is likely that the shipping place was not used after this and the construction of the Llanelly Railway and Dock Company's railway (later the Llanelli/Llandilo railway) to Pencoed, c1836, reclaimed the land between the shipping place and the Loughor River. (Plate 39)

(3) *Thomas Bowen's Shipping Place on the Lliedi River*

In 1802 John Vaughan of Golden Grove and Sir John Stepney, 8th Bart., were in dispute over the Layerage and Keelage rights of the north side of the Burry Estuary. Thomas Bowen supported Stepney, saying in a letter *"The Quay above John Roberts' was erected by me in the year 1756 in order to Ship Coal worked in the Fields East of the Fields late William*

Rich^ds House; for the Trespass of which I paid the late Sir Thos. Stepney Six Pounds a year for 2 years made use of".[116] This evidence confirms that Bowen built a coal shipping quay and used it between 1756 and 1758 but does not give its location. For the following reasons, it is considered that this shipping place was not that at Pencoed, but was probably situated at Old Castle Marsh, on the old course of the Lliedi River:- Bowen was leased coal under the Wern and under a "*waste overflowed by the Sea Tides*", west of the Wern, "*and separated from the Lands of the late Sir Edward Mansell Bart., by a Beach of Stones, commonly called High Water Mark*" by Vaughan, in June 1754.[117] This description allows the waste to be identified as the Old Castle Marsh region. Bowen's coal had previously been leased to John Allen and a plan of 1846[118] located and described pits worked by Allen and Bowen in this region. The plan also showed an "*Old Shipping Place or Stage*", evidently used for shipping coal from these pits, on the main channel of the old course of the River Lliedi (in present day locations, the shipping place lay in the North Dock close to its eastern wall). Bowen must have run into trouble, because Vaughan asked his agent about delays to Bowen's works, in 1756,[119] and rentals of his estate for 1757/58[120] included the entry, "*Mr Thomas Bowen — For the Colliery 0-0-0*", implying that no coal was worked. Although he was also leased coal under Morfa Bach Common (north of Old Castle Marsh) by Vaughan in 1758,[121] Golden Grove Estate rentals up to 1761[122] showed nil returns for Bowen's collieries. Bowen probably failed to develop Allen's old collieries, and the shipping place shown is probably the one he said he had used for two years only. We also know that a John Roberts, who was a shopkeeper and general supplier[123] (probably a ship's chandler) lived in Seaside in the early 19th century and built new houses* near the Carmarthenshire dock.[124] This further supports the conclusion that Bowen's shipping place was on the River Lliedi.

(4) *The Quay at Black Rock (Pwll Quay)*

In November 1755 John Rees of Cilymaenllwyd approached John Vaughan, as Lord of the Manor, for "*leave to build a Wharff near the Pool at the Black Rock in Pembrey Parish*".[125] A year later, the scheme, now involving John and Hector Rees,[126] was described as "*Building a Quay at the Black Rock*".[127] A plan of 1757[128] showed "*coaleries*",** one belonging to Hector Rees being located at Pool, and there is no doubt that this was the quay at Black Rock. The present village of Pwll (formerly Pool) was

* The new houses, built by Roberts, were on the site which became known as Custom House Bank.

** The "*coaleries*" shown on this plan were shipping places where masters could load coal.

once two farms — Penllech and Hendy (Hendu or Tyrhendy or Black Rock)[129] — and the Rees's quay, built c1757, was almost certainly the shipping place, later called Pwll Quay. Although no confirmatory evidence has been seen, it is probable that Edward, Hector Rees's younger son, worked coal at Pwll from 1760. In 1765 he was leased coal under Stepney lands at Pool and, later, coal under Richard Vaughan's lands, in 1768.[130] Mainwaring says that the Pool colliery worked until c1800 and, if this were so, the shipping place must have been used throughout this period.

It was probably not used after 1800, a plan of 1825 referring to it as an "*Old Quay*",[132] and any likelihood of it being redeveloped ended with the construction of the Pool Tramroad in 1826.[133] The site of the Quay became land locked when the South Wales Railway embankment was built in 1852.

(5) *Samuel Townsend's Shipping Place at Maesarddafen*

In January 1760 Sir E. V. Mansell leased lands in the region of Maesarddafen to William Anthony, yeoman, and also a part of the salt marsh, with the exception of the "*Waggonway and Stage erected and built on part of the said demised premises by Samuel Townsend*".[134] It is assumed that the stage was a shipping place and that the waggonway was used to carry a bulk commodity, such as coal. All we know of its location is that it was on the salt marshes in the Maesarddafen/Llwynhendy region, implying that ships or barges using it came up Dafen Pill. Nothing is known of Townsend's activities, coal mining or otherwise, and very little more of him as a person. Sir E. V. Mansell's aunt, Dorothy, married a Samuel Townsend and they had a son, also named Samuel.[135] Father or son could have erected the stage but we have no means of estimating its life, before or after 1760.

(6) *Townsend's Pill*

When Chauncey Townsend developed his colliery at Genwen he shipped coal along the Yspitty Canal to a pill, west of Spitty, which led into the River Loughor, and which was called Townsend's Pill on a plan of 1772[136] (Fig. 21). As he started mining at Genwen in 1766, the pill must have been named about this time. The plan showed a coalbank on the pill, at the end of the canal, but did not show staithes or wharfs. It is, therefore, assumed that canal barges unloaded at the coalbank and the coal was reloaded directly into small ships, or into keels which took the coal out to larger ships at anchor. It must have been used intermittently

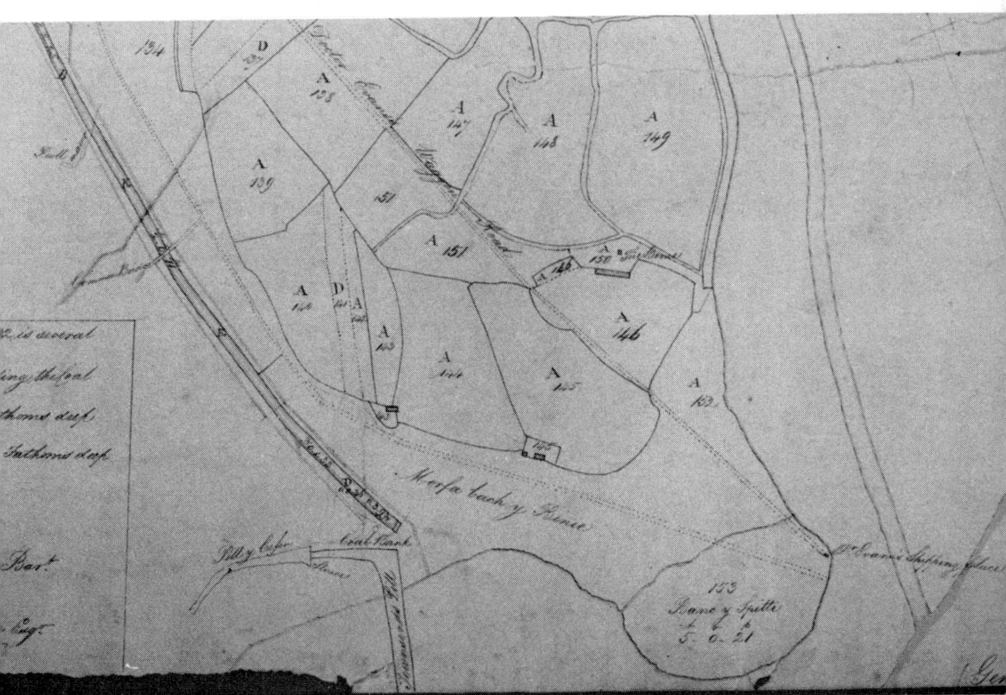

Plate 40 — Townsend's Pill. (Top) Part of John Thornton's 1772 plan showing Townsend's Pill at the end of the Yspitty Canal. (Bottom) Present-day remains of Townsend's Pill — now a drain for the surrounding area.

after Townsend's death, because his son-in-law John Smith continued to work the colliery. Smith was not satisfied with Townsend's Pill as his main shipping place, however, describing it as "*very inconvenient*".[137] On Smith's death, in 1797, his colliery interests and the shipping place passed to his two sons, Charles and Henry, who assigned all their interests in the Llanelli area to Warde, in 1802. Warde reopened some of Townsend's old pits and sank new pits in the Bynea area, and must have used the Pill, although his main shipping place was Spitty Bite, after 1808. Nevertheless, we know that the canal, and presumably the shipping place, were still used in 1810.[138] A plan of 1814 called the pill "*General Warde's Shipping Place*",[139] but it must then have been little used. The construction of the South Wales Railway embankment, c1852, cut off its access to the River Loughor. Townsend's Pill is still identifiable in 1979, serving as a drain for the surrounding area. (Plate 40)

(7) *Bank y Llong*

A plan of 1787 showed "*Bank y Llong*", situated on the old course of the River Lliedi on Morfa Bach Common, near Sandy[140] (On the low ground between the present Sandy Bridge and Old Castle Pond). The name is sufficient authority for assuming that it was a shipping place, but the plan also showed a straight track, from a coal pit to the main road adjacent to Bank y Llong, leaving little doubt that it was a coal shipping site. The old course of the Lliedi was shown as being some 30 feet (9.1m) wide on the plan and would, therefore, have been navigable for small ships as far as Bank y Llong, or for barges to take coal to larger ships at anchor.

We do not know when the Bank was first used for shipping, but it could have been a traditional location. Although it was on common land, it was next to the Mansells' Stradey Estate and this may have been the place Daniel Shewen was referring to in 1762, when he wrote "*we have at this time at least £800 worth of coal on Bank owing to about 7 months Bad Winds for our place of shipping*".[141] It is considered that Charles Gwyn worked the water-engine pit, shown on the 1787 plan, from 1763 onwards, and then Sir E.V. Mansell and Dame Mary Mansell, until c1793. They all must have used Bank y Llong as their shipping place. It is also possible that Givers and Ingman used it to bring in iron ore, when they established their furnace and foundry at Cwmddyche, c1794, and that Raby used it between 1796 and 1799, while building his shipping place at Llanelli Flats. When Raby was negotiating with Dame Mary Mansell, in 1797, for a lease of Stradey Estate coal, and a valuation of the existing colliery was required, one item listed was a "*Coal Barge*". This suggests that coal was possibly loaded into barges at Bank y Llong for reloading into boats at anchor further down the river.[142] The construction of Raby's "*ironbridge*", to carry his railway over the Lliedi, cut off access to Bank y Llong for most shipping, however, and it must have become disused at this time. (See Plate 10)

(8) *Dafen Pill and Penrhyngwyn*

Dafen Pill was undoubtedly one of the traditional shipping places in the Llanelli area. The River Dafen used to run directly into the main channel of the River Loughor and was navigable up to Cefn y Maes. It was certainly used as a bargeway by Townsend and Bowen in 1757[143] and shipping places must have been established along its inland course (the River Dafen), just above Dafen Pill,[144] river and pill being the main shipping route for coal mined at Penyfan, Trostre, Maesarddafen and Llwynhendy.

Before the Great Embankment between Machynis and Maesarddafen was built and the River Dafen was diverted into Machynis Pool in 1808/09,[145] the main course of Dafen Pill ran nearer to Penrhyngwyn Point than it does now.[146] Machynis Pill may also have been navigable, because Penrhyngwyn was considered a suitable site for a shipping place (Fig. 38). Bowen probably used it between 1772 and 1780,[147] and John Smith intended cutting a canal, from the top of Dafen Pill to a large shipping place he planned to build at Penrhyngwyn, in 1785.[148] Smith rented Penrhyngwyn until he died, in 1797,[149] but there is no evidence to suggest that he carried out his plan.

The combination of Dafen Pill and Penrhyngwyn as a shipping place remained a desirable proposition into the early 19th century. When Warde was assigned Smith's coal leases, in 1802, he said he would construct a canal for "*the Navigation of Boats and Barges up to the said Bank called Penrunwen Bank*".[150] Warde later explained that he had intended to finish the Daven pit at Halfway (begun by Smith) and take the coal down a canal to "*Penrhunwen*" to be shipped.[151] When Charles Nevill was examining sites for the Llanelly Copperworks, in 1804, he was also impressed with Penrhyngwyn, and wished to establish the copperworks there and obtain coal from Warde. Nevill wrote to Warde "*Penrunwyn Point I have considered as the only situation that I should be satisfied to build upon*".[152] They could not agree a price for coal, however, and the copperworks was sited at Penrhos, in Seaside. Warde continued to develop the Daven pit but the representation of a shipping place from Penrhyngwyn to the head of Dafen Pill, on Wedge's 1808 plan, represented Warde's intentions only, although it is probable that they were partly implemented, when he built a jetty "*at the mouth of Dafen River*", in 1808/09.[153] The building, in 1808/09, of the Great Embankment from Machynis to Maesarddafen, the diversion of the Dafen River to form the New Dafen River flowing into Machynis Pool, and flooding at the Daven pit, caused Warde to abandon the pit, and the development of Penrhyngwyn as a major shipping place, in 1810.[154] Warde later proposed, in 1821, to extend the Yspitty Canal west for almost a mile, to a point south of Maesarddafen Farm.[155] We do not know his object, but he probably intended to transport coal, raised at Llwynhendy and Genwen, and ship it from Dafen Pill. The scheme was not implemented and it marked the end of plans to develop Dafen Pill and Penrhyngwyn as major shipping places.

(9) *Llangennech Pill (Sir John's Pill)*

The outlet of the Mwrwg brook into the River Loughor, near Llangennech, was formerly known as Llangennech Pill[156] or Sir John's Pill.[157] It was the only pill suitable for shipping coal in the Llangennech area, and may have been used in the early 17th century,[158] although no evidence has been seen to confirm this.

Thomas Jones may have used Llangennech Pill in the mid 1770s[159] and, certainly, when Hopkin and Jenkin began to work the Allt and Brynsheffre coal in the early 1790s,[160] a disused canal ran from Llangennech into Llangennech Pill.[161] Hopkin and Jenkin probably used a new shipping place, opposite Box Tenement, but, when Symmons first bought the Llangennech Estate in the early 1800s, he must have recommenced shipping from Llangennech Pill, because land bordering it was described on a plan[162] as "*This piece sold to John Symmons, Esq., for a Coal Bank and Shipping Place*". When Symmons sold the Estate to the Earl of Warwick and Vancouver, in 1804, they adopted the shipping place opposite Box Tenement and commenced building a dock, later called Llangennech Quay. When Symmons re-possessed the Estate, in 1806, he completed their dock, and Llangennech Pill must then have fallen into disuse as a shipping place (Fig. 23). It still exists, but the Llanelli/Shrewsbury railway embankment has restricted the tidal flow, and it is narrow and silted compared to the waterway that was once used to ship coal. (Plates 41 & 42)

(10) *The Shipping Place on Llangennech Marsh*

Plans of c1825 and 1826[163] showed an "*Old Shipping Place*" on Llangennech Marsh, between Llangennech Pill and Viney, adjacent to the

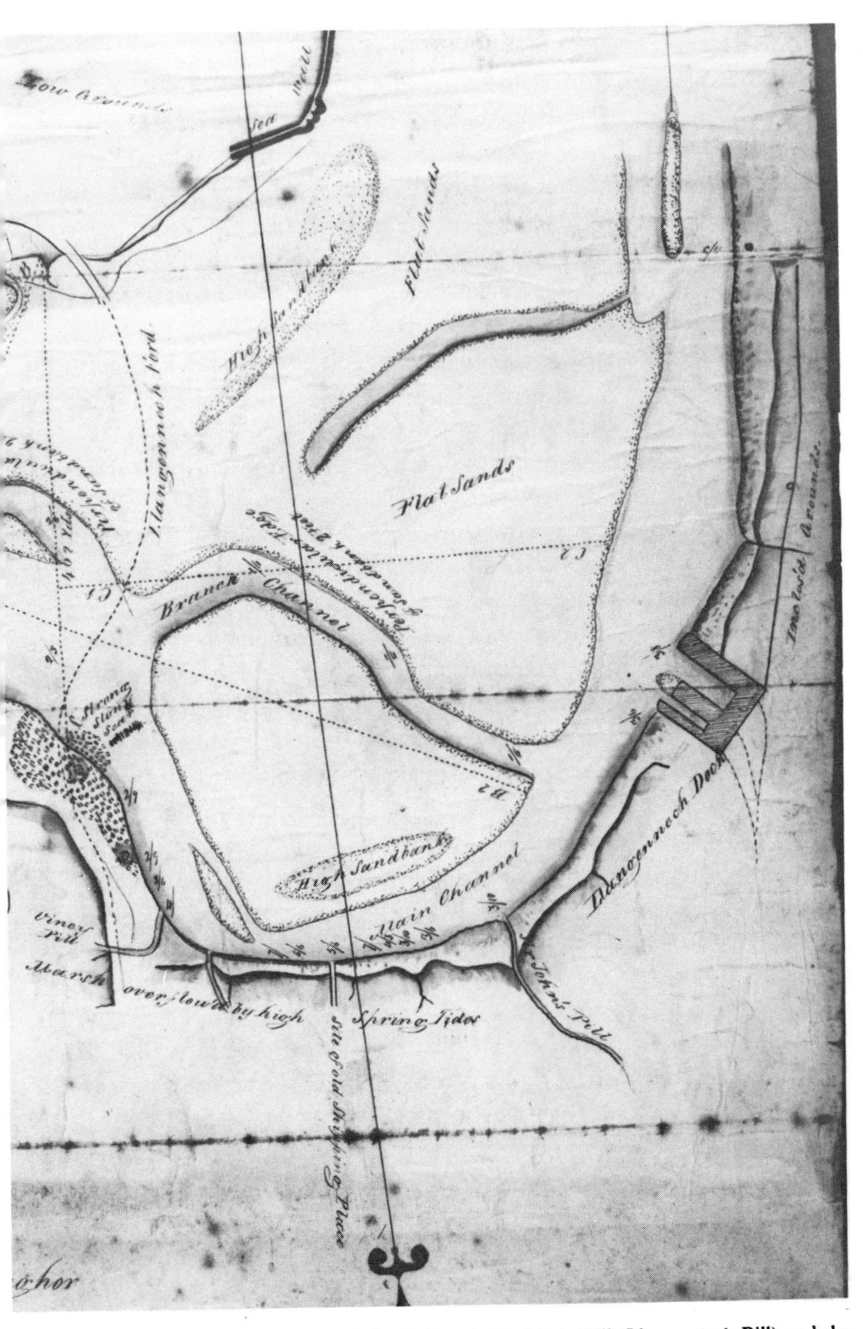

Plate 41 — Plan showing Llangennech Dock (Quay), Sir John's Pill (Llangennech Pill) and the shipping place on Llangennech Marsh in 1826.

Loughor River. It is not known who formed it but there is evidence to suggest that it was Thomas Jones or Hopkin and Jenkin. (Plate 41)

Jones was leased coal at Brynsheffre, in 1774,[164] when he also obtained a sub-lease of coal at the Allt.[165] Jones intended shipping from a point opposite Box tenement, but available evidence suggests that he shipped from Llangennech Pill. Nevertheless, an account book for Llwyn Ifan confirms that waggon roads were made from Brynsheffre, in 1777, an entry in May also recording "*paid the workmen for work done on the new road on Llangennech Marsh and on the Sea Bank about the Shipping Bank*".[166] The reference to a road (probably a waggon road) over Llangennech Marsh, suggests that the shipping bank could have been the old shipping place shown on the 1825 plan.

In January 1789, Hopkin and Jenkin took over Jones's Allt coal[167] and, in February 1793, the Brynsheffre (and Talyclyn) coal.[168] The 1789 lease gave Hopkin and Jenkin permission "*to pass over part of Llangennech Marsh that is situate lying and being between a tenement, the property of Sir John Stepney called Viney and a certain Pill called Sir John's pill for the purpose of conveying the said coal or culm to any ship*". This description makes it virtually certain that Hopkin and Jenkin originally intended shipping from the "*Old Shipping Place*" shown on the 1825 and 1826 plans. As they had taken over Jones's interests, it is very possible that it already existed, as the lease made no reference to Hopkin's and Jenkin's intention to form it. As already seen, Hopkin probably extended the Llangennech Canal, and shipped from the Llangennech Quay location, and may only have used Llangennech Marsh while the Quay was being built.

There is more uncertainty, because an un-dated note in the Golden Grove records states "*Mr Jones of Duffryn has erected a Shipping Bank on Llangennech Common from whence he shippes his Coal*".[169] The schedule dates the note to c1790 but there is no evidence that this is correct and it could justifiably be dated to c1780. We do not know if "*Mr Jones of Duffryn*" was Thomas Jones of Llwyn Ifan.

The evidence relating to the shipping place on Llangennech Marsh is not conclusive, but it was probably formed by Thomas Jones and subsequently used by Hopkin and Jenkin.

(11) *The Shipping Place opposite Box Tenement (Llangennech Quay or Dock)*

The third shipping place located for the Llangennech area was south of the Pill and Marsh sites and opposite the tenement of Box (Figs. 23 and

Plate 42 — (Top) Present-day remains of Llangennech Quay. Two side walls and the sedimented dock basin are still visible. (Bottom) Present-day remains of Llangennech Pill.

38). Again, the surviving evidence does not give us a full picture of its history. Thomas Jones, Hopkin and Jenkin and Symmons were all involved in its early life and it is possible that they all used the three Llangennech shipping places.

We know that there was no constructed shipping place opposite the tenement of Box in August 1772, when Sir Thomas Stepney, 7th Bart., proposed mining at Brynsheffre and building a waggon road to a shipping place on the Loughor River, near the tenement of Box. Roderick, Clayton's agent, confirmed the originality of the scheme when he wrote "*This scheme of taking part of Box Tenemt. is newly come into a plan. If you grant him a term of years on that spot, he will create a key to load the vessels, the Water by the side of that bank being deeper than anywhere higher up*".[170] Stepney died in October 1772, before his plans could be realised,[171] and Thomas Jones took over Stepney's coal interests[172] and was granted the Brynsheffre lease, in October 1774. Roderick confirmed his intention of forming the shipping place opposite Box tenement in a letter of November 1774, which stated "(Jones) *wishes to have a spot of Land by the water in the Tenement of Box for laying down the Coles ready for shipping*".[173] Jones probably shipped coal from Llangennech Pill and Llangennech Marsh but no evidence has been seen to confirm that he shipped from opposite Box Tenement, despite his intention of doing so.

A plan of c1790[174] shows that an old canal existed, from Llangennech to Llangennech Pill, when Hopkin and Jenkin took over Thomas Jones's coal interests, between 1789 and 1793. Hopkin undoubtedly extended this canal south, reference being made in 1824 to "*the little canal which Mr Hopkin's formerly cut*".[175] It was shown following the course of the old canal as far as Llangennech Pill, with a southerly extension to a point opposite the tenement of Box,[176] where they must have shipped coal and formed a shipping place. No details have survived to show if this was merely a coalbank, or whether staithes were erected, or a dock basin formed.

The period over which Hopkin and Jenkin worked is not known but Symmons purchased Stepney's Llangennech lands, probably in the early 1800s, and took over the working collieries and railroads on them.[177] There is evidence that Symmons shipped from Llangennech Pill, as he purchased land adjacent to it as a "*Coal Bank and Shipping Place*",[178] but he might also have used the spot opposite the tenement of Box. When Warwick and Vancouver purchased Symmons's Estate, after it was advertised in 1804,[179] they started to develop and modernise its coal resources and transport system. No conclusive evidence has been seen, but it is considered that they started to build a dock or quay opposite the tenement of

Box, or improved existing facilities there. When the Estate was regained by Symmons, he continued to develop its resources, advertising, in April 1807, that his collieries were in full work and that coal was ready for shipping at the "*New Dock on the Loughor or Burry River*".[180] There is, therefore, little doubt that a substantial dock or quay had been formed by this time. By agreement or lease, Richard Davenport, George Morris and William Rees then took over the collieries on the Estate and established a copperworks on Symmons's land at Spitty Bank,[181] the Spitty Copperworks being built before October 1808.[182] They must also have used the dock or quay opposite the tenement of Box to ship coal. Their enterprise failed and, in February 1814, an auction sale of all their stock-in-trade was advertised.[183] One of the auction sites was called "*Llangennech Collieries and Quay*", the earliest reference seen to the name "*Llangennech Quay*", although it must have been used before this. The available evidence suggests that Symmons worked his Estate's industrial interests until he sold them to the Llangennech Coal Company, and it is likely that coal was loaded into barges at Llangennech Quay, which were then towed by steam tug to the Spitty Copperworks and to the docks at Llanelli.[184] Evidence exists to suggest that the Llangennech Coal Company improved Llangennech Quay in 1825 or 1826. A plan of 11 May 1824,[185] although of small scale (3 inches to 1 mile), clearly shows "*Llangennech Dock*" as a narrow dock channel, with two quay walls each supplied by a rail track, whereas a plan of 1826, at 20 inches to 1 mile,[186] shows two dock channels, with three quay walls and three rail tracks. The Llangennech Coal Company used Llangennech Quay or Dock as their main shipping place until 1834, when the New Dock (at Machynis Pool) was opened.

The remains of two quay walls and one dock basin still exist. They are tangible reminders of the shipping place, proposed by Sir Thomas Stepney in 1772 and abandoned for coal shipping purposes in 1834, although probably used by small ships for some time afterwards.[187] (Plates 41 & 42)

(12) *Pill y Ceven*

"*Pill y Ceven*", shown on a plan of 1772,[188] led into Townsend's Pill, which then led into the River Loughor. When the Rev. David Hughes and Joseph Jones were leased coal under "*Morfa Baccas*" in 1794,[189] Townsend's canal and Townsend's Pill were disused. They took coal in barges by canal to Pill y Ceven, and then into Townsend's Pill, where the coal was loaded into small ships, or the barges took it out to larger ships at anchor. Pill y Ceven, therefore, was a link between the Baccas Canal and Townsend's Pill (Fig. 24). The period over which the Pill was used is not known but it was probably short-lived, as no reference has been seen to Hughes and Jones's activities after 1797.

Pill y Ceven still exists but the building of the South Wales Railway Embankment, in 1851, cut off its direct access to the River Loughor, and what was once a navigable pill is now only a small drain.

(13) *Roderick, Bowen and Griffiths' Dock (Pembertons' Dock)*[190]

Some time after October 1794 Roderick and partners constructed a canal from their pits at the Bres and Wern to a shipping quay at Seaside (near St John's Church) (Fig. 25). It was probably little more than a wharf or staithe at first,[191] but it was converted into a dock before 1804, although no details of this development have been seen. Certainly, in late 1803 or early 1804, when the partnership was in financial trouble and Symmons was trying to persuade Vaughan to transfer the Wern coal lease to him,[192] William Roderick described the dock and canal, in a letter to John Vaughan, thus "*Dock made from the Channel of the River up to high water mark, 300 yards in length, 18 feet deep at the upper end, 40 feet wide with Three Jetty heads for Shipping of Coals. A Barge on the River to compleat the filling of large vessels on low tides out in the river. A Cannal from the Pitts to the Water Edge being about ¾ of a mile in length with two Boats 48 feet in length thereon to carry down the coal*".[193] Coal was, therefore, brought down the Wern canal, unloaded at a coal-bank and then reloaded into ships which were moored at jetty heads in the dock. If the tide was too low to allow large ships to enter the dock, a barge took coal out to them at anchor in deeper water.

Roderick and partners assigned the North Crop coal to John Pemberton in 1804.[194] The agreement has not been seen, but there is evidence that Pemberton was also assigned a share in the canal and one side of the dock. At the same time, the Nevills, who wanted a shipping place to import ore for the proposed copperworks, also took a one-third share in the dock, although they were considering constructing a dock of their own.[195] It appears that Pemberton did not honour his bargain[196] and did not use the dock, but the Llanelly Copperworks Company had to use it, because they started smelting copper ore on 20 September 1805, before their own dock was completed.[197] Copper ore was regularly imported from August 1805, (with a single cargo, probably for trial purposes, in April 1805).[198] This must have been unloaded at Roderick and partners' dock until the Company's own dock came into use, in December 1805 or January 1806.[199]

In 1807 Roderick and partners disposed of their concern to the

Pembertons and the Copperworks Company. The 1804 assignment of North Crop coal to Pemberton was legally finalised on 16 February 1807,[200] and the South Crop coal, and all the colliery plant, the Wern Canal and the dock, were assigned to the Copperworks Company, on 11 May 1807, the assignment being subject to such rights as Pemberton had by virtue of his 1804 and 1807 assignments.[201] Undoubtedly, the use of the dock was one of these rights.

After this, the dock's history is uncertain. The Copperworks Company owned it but they probably did not use it to any extent, as their Copperhouse (or Copperworks) Dock was built by December 1805. Roderick requested permission to continue to make use of it, immediately after the assignment and sale of the South Crop Colliery and dock in May 1807,[202] and, although no agreement has been seen, it is assumed that this was allowed until he was made bankrupt, in May 1808. The Pembertons must have used it after they reached coal at the Bres pit and, as their engine did not start working until January 1808,[203] it is probable that they shipped from late 1808 or early 1809, after Roderick's bankruptcy. The Pembertons, therefore, were probably the only users of the dock at this time, although it was owned by the Copperworks Company. By 1814, at the latest, the dock was called "*Messrs. Pembertons' Dock*".[204] It is possible that the Copperworks Company sold the dock to the Pembertons but no bill of sale has been seen. By 1817, R. S. Pemberton was applying for 5000 tons of shipping to carry coal to coastal ports and to Ireland,[205] and a steady trade must have been established at the dock.

The Pembertons' enterprises did not prosper, however. In 1828 they proposed a scheme to link the St David's pit to their railway at Llwyncyfarthwch,[206] to allow the Llangennech Coal Company to ship at Pembertons' Dock, and opposed a Bill for the construction of a new floating dock at Machynis Pool with a railway up to Gelly-gille, where the St David's pit was being sunk.[207] They failed to block the Bill and the Pembertons took steps to withdraw from coal mining in Llanelli, in 1829, when they advertised their collieries for sale or letting,[208] George Bruin becoming their undertenant[209] when he took over their collieries and the dock, in 1830. Although the Pembertons left Llanelli at this time, the name "*Pembertons' Dock*" was used for the remainder of the dock's life.[210] It was subsequently filled in and built over and today no trace remains of what was Llanelli's first real dock. (Plate 43)

(14) *Alexander Raby's Dock (The Carmarthenshire Dock)*[211]

When Raby was leased coal under the Stradey Estate, in 1797, it was

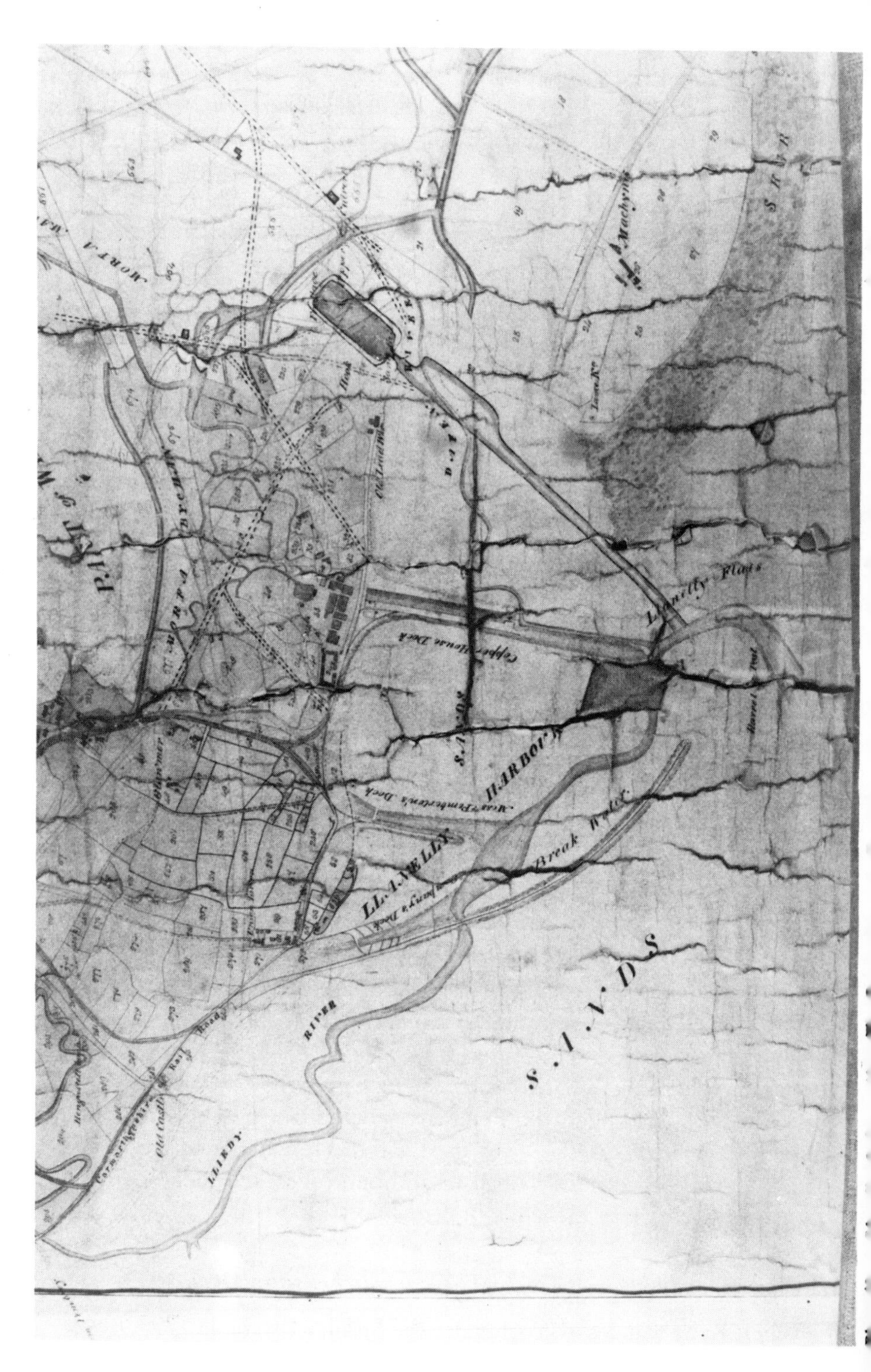
LLANELLY HARBOUR
SANDS
Copper House Dock
Llanelly Flats
Break Water
RIVER
LLIEDY
Old Castle
Carmarthenshire Rail Road

stipulated that he should construct a transport system from the iron furnace at Cwmddyche to "*communicate with the sea*"[212] and it was stressed that he should be "*particularly bound to make a proper Shipping Place*" [213] Raby probably started to build his shipping place at Llanelli Flats, close to Roderick, Bowen and Griffiths's dock, in 1797/98. No detail of its construction has been seen, but it was on the site of the eastern quay wall of the later Carmarthenshire Dock[214] and probably became operational c1799[215] (Figs. 9 and 28).

The Act of 1802, authorising the construction of a railway or tramroad from Llanelli Flats to Castell-y-Garreg, stipulated that Raby's existing railway system and his dock should be purchased by the Company. In relation to Raby's dock, it said "*and a Dock hath been made at the Termination of the said Railroad, which Dock is hereby intended to be made a Part of the Dock and Bason by this Act directed to be made.*"[216] The Carmarthenshire Railway Company paid Raby £3,117 for his railway system and dock[217] and proceeded with the improvements to the dock. These probably entailed the deepening and widening of the basin and the provision of more extensive berthing and loading facilities at the quay, which, Mainwaring says, were done by two contractors, named Thomas Leyson and William Rees, and completed in 1805.[218] A second phase of improvements was started, in May 1806, when tenders were invited for the excavation of a further 20,000 cubic yards from the dock basin and for the completion of the masonry to the dock walls,[219] and, in October, the foundation stone of a new western pier, 155 yards (142m) long, was laid.[220] It was announced, in April 1807 "*the extension of the Dock belonging to the Carmarthenshire Rail-Road Company*" was going ahead with all possible facility and that "*additional Jetty-heads with mooring posts are already completed*",[221] and the improvements to the dock were undoubtedly proceeding at a rapid rate. Raby's financial crisis broke at this time, however, and, in June 1807, he convened a meeting of his creditors and assigned his leasehold interests and personal estates to trustees to repay his debts.[222] It transpired that Raby had misused the Carmarthenshire Railway Company's funds, to construct branch lines to his own works and, as he was also the major user of the railway, a financial crisis resulted. By July 1807 the dock dues and register tonnage were advertised for letting by auction[223] and work on dock improvements probably stopped, because Raby wrote to the Company, in March 1808, complaining about the safety of the dock and saying that the railway and dock would not succeed if they were not properly finished.[224] The dock dues and register tonnage were re-advertised for letting in July 1809.[225]

However, the dock was now used increasingly for the export of coal. Raby's Stradey Estate collieries were still working and, in November 1808, he opened the Box Colliery, which was linked to the Carmarthenshire Dock by the Box railroad. Warde also began to ship coal from his Old Castle Colliery at the dock in April 1811,[226] and, in October, *The Cambrian* reported that the Carmarthenshire Railway Company had instructed their engineer (James Barnes) to survey and make estimates for "*a considerable extension of their dock and shipping places*".[227] At this time, a Bill was promoted to allow the formation of the Kidwelly and Llanelly Canal Company, one clause relating to the construction of a junction canal, from Pembrey to the Carmarthenshire Railway near Old Castle House, with the formation of additional shipping places and perhaps other docks to the west of the Company's Dock. A General Assembly of the Carmarthenshire Railway Company considered the Bill,[228] the Company and the Copperworks Company opposing it, in April 1812.[229] The Bill was passed in June 1812 and the Kidwelly and Llanelly Canal Company was authorised to form shipping places or docks at the west side of the Carmarthenshire Dock.[230] The Carmarthenshire Railway Company had opposed the Bill because of fears that the extension of a canal system up the Gwendraeth Valley would take the Carmarthenshire hinterland trade from their railway and, also, because they probably considered that the Kidwelly and Llanelly Canal Company's shipping places and dock would prove superior to their own dock and would take most of the local trade. In November 1813 it was announced that "*the long projected improvements at the Carmarthenshire Rail Road Dock commenced this week, by which considerable accommodation will be afforded to the trade at Llanelly*".[231] It is not known what these improvements were, but they were probably carried out because of rivalry between the two Companies. The Kidwelly and Llanelly Canal Company did not build their junction canal to Llanelli, however, and the shipping places were not constructed. By 1814 the Carmarthenshire Dock probably had nine loading points[232] but it is unlikely that the facilities were expanded after this.[233] The Carmarthenshire Railway Company's intention of developing the mineral resources of the Carmarthenshire hinterland had not been realised, and it was only the local Llanelli trade, chiefly of coal, which kept the dock and the lower reaches of the railway in operation. The building of the breakwater, after 1813, provided shelter from the tides and the westerly winds, but must also have contributed to silting of the Carmarthenshire Dock, which was at the top of the developing harbour. The linking of the Pool Colliery to the Carmarthenshire Railway, in 1826, did not necessarily increase trade at the Carmarthen-

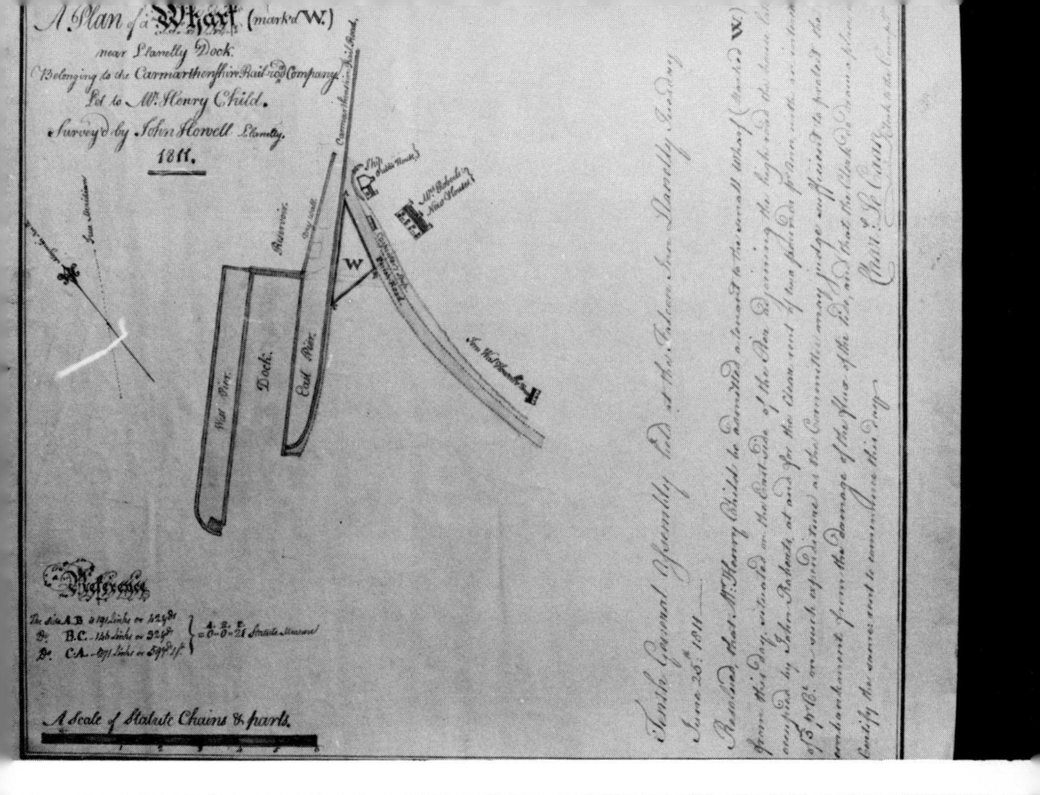

Plate 44 — The Carmarthenshire Dock. (Top) Plan of the dock in 1811. (Bottom) The dock today (Centre of photograph).

shire Dock, because the railway was joined to the other docks on Llanelli Flats and the provision of gates at the Copperworks Dock must have led to it being favoured by the ships trading at Llanelli.

In 1829 the Carmarthenshire Dock was still used but it had not become the major dock the Company planned when Raby's shipping place was purchased in 1802. The diversion of the River Lliedi through the head of the Dock in 1845[234] ensured its continued existence, however, and the eastern quay wall is still used today by boats of a local sand-dredging company, some 180 years after Raby started shipping there. (Plates 43 & 44)

(15) *The Copper-House (Copperworks) Dock*[235]

In December 1804 the Nevills concluded that, although they had obtained the use of part of Roderick, Bowen and Griffiths' Dock, they should make their own shipping place. R. J. Nevill wrote to the clerk to the Copperworks Company "*We judge it will be most eligible and not be attended with any extra expense to make a small dock for our own purposes separate from the one third part of R & B*".[236] Surveys were immediately carried out to estimate costs[237] and construction must have started in 1805, because Charles Nevill reported in August "*The Dock proceeds with considerable rapidity, the workmen promised to have the work completed in a month. I have no doubt but they will accomplish this in that time . . .*"[238] Nevill was too optimistic, however, and, in November 1805, two months after smelting had commenced at the copperworks, the dock was still unfinished and copper ore must have been unloaded at Roderick, Bowen and Griffiths' dock.[239] The Copperhouse Dock probably became operative in December 1805 or January 1806, because Charles Nevill stated, in early February 1806, "*Have begun to scour our Dock by means of the Back Water and am happy to say it answers every expectation I have formed of it*"[240] and it was still being developed in November 1806, when R. J. Nevill reported "*The Dock is very considerably extended and in about 5 weeks the whole length will, I trust, be completed*".[241] Although no plan has been seen to show the extent of construction at this time, it is probable that a dock basin and channel, of some 700 yards (640m) combined length, had been excavated, from the copperworks to the confluence of the Lliedi and Burry Rivers. Shipping quays had been built on the western side of the basin only, served by a single railway line (Fig. 30).[242] The "*Back Water*" referred to in February 1806 was obtained from a scouring reservoir, at the head of the dock, which was fed by tidal water and by water from the River Lliedi, which was diverted by a weir built close to Llanelly Mill. (This diverted water course had originally been built to operate Roderick and partners' water engines at the Wern and for the Wern Canal, but was used by the Company to scour their dock basin. The Llanelli Burgesses repeatedly, but unsuccessfully, tried to compel the Company to restore the waters of the Lliedi to their natural course, after 1813).[243] Further scouring action was provided, in 1809, by the New Dafen River. The Company undertook the excavation of its new course[244] and led it into the channel joining the Copperhouse Dock to the sea.[245]

The Dock had, therefore, been significantly developed in five years, but the launching of the Kidwelly and Llanelly Canal Company, in June 1812, provided a further stimulus for the expansion of facilities. In July 1812, the Copperworks Company wrote to the Admiralty informing them of its intention to erect two additional quays, wharfs or jetties[246] and, in 1813, announced that it was intended to promote a Bill to authorise the construction of a canal or railway, from the Copperhouse Dock to Gwilly Bridge in Llanedy parish.[247] It was also stated that the Company would consider allowing its dock to become public.[248] The Dock became available for public use for a trial period of two years, in November 1814,[249] but the Bill was not pursued. Nevertheless, in September 1817, the Company informed the Admiralty of its intention to erect additional quays and revived the canal or tramroad scheme[250] but, again, it did not promote the necessary Bill.

We do not know if the Dock was used publicly after the trial period, but the Company still had ambitions to make it the premier dock in Llanelli. They must have decided to make it a floating dock when they started to develop Warde's Box Colliery in 1819, and, when the new pits were opened in June 1823, *The Cambrian* stated "*it has been determined immediately to finish the already advanced work of a floating basin, by which vessels may load at all hours of day and night*".[251] On 24 June 1823 the Company informed the Admiralty that they intended to "*erect floodgates with proper side walls and embankments across a Canal or Dock formed by us for the use of these works for the purpose of forming a floating dock*".[252] The work must have been put in hand at this time, and the Dock became the first floating dock in the Llanelli area, probably in 1824 or 1825.[253] As ships could now load afloat at all states of tide, masters were undoubtedly attracted to the dock and its trade must have flourished.

The Dock had undoubtedly been successful during its first 25 years, due, in no small measure, to the two-way trade of copper ore import and coal export. This two-way trade was established when most of the ships frequenting the other shipping places in the Llanelli area entered in ballast, and its advantages to the Company were considerable. In 1804, some 91%

of the ships trading at Llanelli entered in ballast, compared to only 47% in 1828. In the same period, ships entering Llanelli loaded with copper ore increased from none to 39%. The bulk of this copper ore trade was concentrated at the Copperhouse Dock.[254]

The Dock remained in constant use for 122 years after 1829 but it has now been filled in and only part of the entry channel remains.

(16) *General Warde's Shipping Place at Spitty Bite*

Spitty Bite lay just up the Loughor River from Spitty Bank, shielded from the westerly winds by the promontory that formed Spitty Bank. Its suitability as a shipping place was recognised as early as 1778, when John Smith, referring to "*the Bank between Spitty Bank and Binie Lands*", stated "*I apprehend it will make a much better shipping place than that at the end of our Canal*",[255] but he did not develop it. When Warde gained Townsend's and Smith's coal leases, in 1802, he inherited only Townsend's Pill, at the end of the Yspitty Canal, for shipping, so his plans included the building of a new shipping place at Spitty Bite. Warde started his developments in late 1806/early 1807, when work on the new shipping place probably also started, as it was completed in time to take ships when Warde's Carmarthenshire Collieries were formally opened in August 1808. An advertisement, issued for the opening, set out the advantages of Spitty Bite as, "*the vessels lying out of the race of the tide — the good ground to lie upon — the being to the windward side of the river, by which vessels may almost always be able to stand out — and in particular to the Irish Trade, from the being enabled to reach the Irish channel by the ebb and the Bay of Dublin with the next flood*".[256] Three quays or jetty heads appear to have been provided, because the Apple Pit Tramroad ended in three separate branches at Spitty Bite[257] (Fig. 31).

It was constantly used until 1816, when Warde's financial troubles caused him temporarily to suspend workings in the Bynea region,[258] but it was used again in late 1817/early 1818, when R. J. Nevill became manager of Warde's collieries.[259] It probably stayed in use until the early 1820s, when the opening of the deep pits of the Box Colliery, in 1823, probably led to the closure of workings in Bynea. It may then have fallen into disuse although, in January 1829, when Warde objected to the siting of a bridge at Loughor, he stated that his shipping places lay above Spitty and were frequented by "*the largest coasting colliers navigating the Bristol Channel*".[260] Warde's death in 1830, and the construction of the first Loughor Bridge in 1833/34, marked the effective end of coal exports from Spitty Bite. Subsequent land reclamation has completely covered the site of what was once a major coal shipping place.

CHAPTER 6— REFERENCES

1 "Welsh Port Books 1550-1603" by E. A. Lewis (1927).
2 "Trade of Llanelly" by Hopkin Morgan, quoting Calendar of State Papers Domestic 1675-76, p. 529 at PRO: Carmarthenshire Local History Magazine Vol. II (1962).
3 Brodie Collection 69 (1757).
4 *The Cambrian* (9 Feb 1805).
5 Ibid, (23 Feb 1805).
6 Ibid, (28 Sep 1805).
7 Ibid, (1 Oct 1808).
8 Brodie Collection 68 (1808). This chart was entitled "A Chart of Burry Bar & Harbour engraved for the use of Genl. Warde's Colliery and given gratis to Vessels loading at it, 1808".
9 Llanelli Public Library plan 291, undated but c1808. This chart was entitled "A Chart of Burry Bar & Harbour by John Wedge formerly Master of a Vessel which traded there" and was undated.
10 "An Act for inclosing lands in Llanelly in the County of Carmarthen and for leasing part of the said lands and applying the Rents thereof in improving the Town and Port of Llanelly in the said County", 8 Aug 1807 (47 Geo III c107).
11 *The Cambrian* (7 May, 4 Jun 1808).
12 Cawdor (Vaughan) Box 228 — Report on the overflow of the River Dafen by D. H. Jones (24 Jan 1877).
13 *The Cambrian* (27 Jan 1810).
14 Ibid, (3 Feb 1810).
15 Ibid, (27 Apr 1811).
16 Ibid, (6 Apr 1811).
17 Ibid, (31 Aug 1811).
18 Ibid, (16 Nov 1811).
19 "An Act for the Improvement of the Navigation of the Rivers Bury, Loughor and Lliedi in the Counties of Carmarthen and Glamorgan", 2 Jul 1813 (53 Geo III c183). R. J. Nevill was a main participant in the implementation of the Act and drew up the provisional list of Commissioners — See Cawdor (Vaughan) Box 228, document entitled "Llanelly Harbour Bill — 31 March 1813".
20 *The Cambrian* (24 Oct 1812).
21 Ibid, (4 Mar 1815).
22 Ibid, (17 Jan 1818).
23 Ibid, (14 Mar 1818).
24 Ibid, (9 Jan 1819).
25 Local Collection 253 (1820). It is not known if this was Llanelli's first tide-table but no earlie one has been seen.
26 *The Cambrian* (5 Jan 1822).
27 Ibid, (14 Jun 1823).
28 Nevill 243 (24 Jun 1823).
29 *The Cambrian* (4 Oct 1823).
30 Local Collection 37 (11 May 1824).
31 "Old Llanelly" by J. Innes, quoting an un-identified source. No documentary evidence has been seen to confirm this statement.
32 Ibid.
33 *The Cambrian* (20 Sep 1828).
34 9 Geo IV c91 (19 Jun 1828).
35 Mansel Lewis 1855 — "Port of Llanelly, Llanelly Harbour Improvements". Reports to the Commissioners by Henry R. Palmer (15 Oct 1840) and by Isambard K. Brunel (23 Nov 1857).
36 "*Ship News*" in *The Cambrian* between 1804 and 1826.
37 This statement may not be strictly true for the latter 20 years of the period considered when iron, copper and lead works and an increasing population used large quantities of local coal. Unfortunately, coal production figures have not survived but, even as late as 1864/65, some 80-83% of the total coal production of the Llanelli area was being exported by ship (evidence of "Report of the Commissioners appointed to inquire into the Several Matters relating to Coal in the United Kingdom 1871").
38 I Eliz I c2, quoted in "Sources for the history of ports" by R. C. Jarvis, Journal of Transport History, pp 76-93, Vol III (1957).
39 "Welsh Coal Trade during the Stuart Period 1603-1709" by B. M. Evans, M. A. Thesis, Un. of Wales, Aberystwyth (1928).
40 "Trade of Llanelly" by Hopkin Morgan, Carmarthenshire Local History Magazine, Vol II (1962); B. M. Evans, op.cit.; "The Rise of the British Coal Industry" by J. U. Nef. (1932).
41 R. C. Jarvis, op.cit.; J. U. Nef, op.cit.
42 E. A. Lewis, op.cit.

43 J. U. Nef, op.cit.
44 E. A. Lewis, op.cit.
45 "The Itinerary in Wales of John Leland in or about the year 1536-1539" edited by L. Toulmin Smith (1906).
46 "Cardiff — A History of the City" by W. Rees, (1969) pp 161-163.
47 "The Economic History of South Wales prior to 1800" by D. J. Davies, (1933) p.51.
48 Hopkin Morgan, op.cit.
49 W. Rees, op.cit., D. J. Davies, op.cit.
50 E. A. Lewis, op.cit., J. U. Nef, op.cit., B. M. Evans, op.cit.
51 J. U. Nef, op.cit., Vol II pp 219-221.
52 B. M. Evans, op.cit.
53 A number of coal leases were granted in this period and Walter Vaughan, who came to live at Llanelli c1616, probably helped produce an expansion in the local coal industry (See Chapter 2).
54 J. U. Nef, op.cit., Vol II pp 222-223.
55 B. M. Evans, op.cit.
56 "Cadets of Golden Grove" by F. Jones, Transactions of the Honourable Society of Cymmrodorion, Part II, (1971).
57 Cilymaenllwyd 77 (15 Nov 1660); Derwydd 687 (13 Sep 1662); Cwrt Mawr 979 (20 Sep 1663); Penller'gaer B. 15.15 (14 May 1665).
58 B. M. Evans, op.cit.
59 The Welsh Port Books continued to record coastwise coal exports separately for North Burry and South Burry after 1684. Only the foreign coal exports were merged under North Burry (B. M. Evans, op.cit.). As coastwise exports significantly exceeded foreign exports, a realistic estimation of the separate quantities exported from North Burry can be made for the years after 1684 — See Appendix G for details. J. U. Nef's quantity of over 10,000 tons in 1685 must relate to the combined coal exports for North Burry and South Burry.
60 Customs 74/3, Llanelly, Collector to Board, letter (19 Jul 1847) (H. M. Customs and Excise Library).
61 B. M. Evans, op.cit.
62 John Vaughan's son, Walter, died in 1683. The Vaughans' Llanelly Estate and its coal interest then passed to John Vaughan's widow, Margarett, to hold in trust for her four daughters. Margarett Vaughan was 67 years old when the French export trade was lost and she remained in control of the estate until her death in 1703, aged 81.
63 B. M. Evans, op.cit.
64 The evidence of B. M. Evans, op.cit., H. Morgan, op.cit., J. U. Nef, op.cit.
65 B. M. Evans, op.cit.
66 "The South Wales Coal Trade" by C. Wilkins, p. 38, quoting an unidentified source.
67 "A New present State of England" by D. Defoe, p. 309.
68 Cawdor (Vaughan) 21/614 — "An Acco^{t} of Keelage money received at Llanelly from Michaelmas 1744 to Michaelmas 1746".
69 Cawdor (Vaughan) 41/5812, letter from Morgan Thomas to John Vaughan dated 8 May 1802, quoting evidence given by Thomas Bowen.
70 Hopkin Morgan, op.cit.
71 Stepney Estate 1101, copy letters between 1750 and 1755.
72 Mansel Lewis London Collection 10 — "Stradey Collieries and Binie — Queries for Mr Rees Jones dated February 1828".
73 Evidence of Llanelli Public Library plan 3 (1772).
74 Mansel Lewis London Collection 98, letter from D. Shewen to Joseph Shewen (14 Apr 1769).
75 Cawdor (Vaughan) 21/614 — "An account of Keelage money received by Evan Griffiths, Esq., from 5 July 1772 to 5 July 1777".
76 Mansel Lewis 127 (9 Oct 1776).
77 Carmarthenshire document 67 (6 Oct 1780) (CCL).
78 Cawdor (Vaughan) 21/614 — "An Acc^{t} of the Keelage of Vessels cleared on the Lordship of Sir John Stepney, Bart., in this Port for the 3 last years ending the 5th day of July 1791".
79 "Hanes Llanelli" by D. Bowen (1856).
80 The Coal Commission Report of 1871 listed the quantities of coal exported from Llanelli to foreign countries between 1790 and 1799. A significant decrease after 1793 was maintained for the rest of the century.
81 Mansel Lewis London Collection 97, letter from A. Raby to Dame Mary Mansell (4 Nov 1797).
82 "Report of the Commissioners appointed to inquire into the Several Matters relating to coal in the United Kingdom" Vol III, Report of Committee E (1871).
83 In February 1804, it was reported that the gentlemen at Llanelli were contemplating erecting a battery of cannons on Machynis Point "*to protect the navigation of the Burry River*", and real fears must have been held respecting the presence of French ships (*The Cambrian* dated 24 February 1804).
84 *The Cambrian* (1804 to 1825). Detailed listings of sailings with their destinations were given

in these years. After 1822 the listing of sailings became intermittent although full coverage of ships entering Llanelli was maintained until 1847.

85 *The Cambrian* (5 Mar 1808).
86 Ibid, (13 Aug and 1 Oct 1808).
87 Ibid, (1804 to 1829) — Ship News relating to Llanelli.
88 Cilymaenllwyd 135 (28 Feb 1753).
89 Brodie Collection 69 (1757).
90 Stepney Estate 1101, copy letter from Sir Thomas Stepney to an un-named person (18 Aug 1750).
91 Stepney Estate 1101, copy letter from an un-named representative of Sir Thomas Stepney to a person named Gibson (22 Nov 1753).
92 Ibid.
93 *The Cambrian* (22 May 1819).
94 Alltycadno Estate correspondence, letter from William Roderick to William Clayton (9 Oct 1773).
95 Ibid, letter (22 Aug 1772). The formation of a dock or harbour at Llanelli, c1750, has been generally attributed to Sir Thomas Stepney, because of an extract from a letter, no longer extant, written to Sir Thomas by a Thomas Cole. This extract, quoted in "Some Notices of the Stepney Family" by F. Harrison, read, "*If Capt. Biggin likes your coals he will undertake to vend you 50 to 60 thousand chaldrons a year. At the same time he very much likes the last account you sent of your harbour*". It is likely that Stepney's "*harbour*" was, in fact, the Burry Estuary.
96 53 Geo III c183 (2 Jul 1813).
97 Llanelli Public Library plans 1 and 3 (1751 and 1772 respectively).
98 Llanelli Public Library plan 1 (1751).
99 Nevill 19 (24 Oct 1752).
100 Llanelli Public Library plan 3 (1772).
101 Mansel Lewis 1872 (1801).
102 Coleman 329 (27 Oct 1801) (NLW).
103 *The Cambrian* (31 Mar 1804).
104 "The Industrial Development of South Wales" by A. H. John, quoting Public Record Office C/114/136; the evidence of *The Cambrian* (12 Feb 1814).
105 *The Cambrian* (1 Oct 1808); Brodie Collection 68 (1808).
106 *The Cambrian* (23 Apr 1814).
107 Local Collection 37 (11 May 1824).
108 Mansel Lewis 29 (7 Jan 1828).
109 Nevill 73 (1 Feb 1826). Surrender of lease, undated but c16 May 1829, enclosed in document.
110 "*Y Diwygiwr*", No XII, November 1847 p. 366, stated that the Spitty Copperworks had lain idle for 16 years. I am indebted to Mr R. O. Roberts of University College, Swansea for bringing this reference to my attention.
111 *The Cambrian* (17 Mar 1832).
112 Ibid, (7 Jun 1834).
113 Llanelli Public Library plan 3 (Apr 1772).
114 Nevill 19 (24 Oct 1752) quoting an unseen lease.
115 Stepney Estate 1101, copy letters between 1753 and 1755.
116 Cawdor (Vaughan) 41/5812, letter from Morgan Thomas to John Vaughan (8 May 1802). Copy of Bowen's letter enclosed.
117 Cawdor (Vaughan) 21/623 (28 Jun 1754).
118 Mansel Lewis London Collection 87 — "A tracing of Stradey, Kille and Old Castle Marshes" (8 May 1846).
119 Cawdor (Vaughan) 102/8029, letter from John Vaughan to Philip Lloyd (6 Mar 1756).
120 Cawdor (Vaughan) 103/8033 (Michaelmas 1758).
121 Cawdor (Vaughan) 21/624 (22 Apr 1758).
122 Cawdor (Vaughan) 102/8028, rentals (1759 to 1761).
123 Nevill 284 (24 Jun 1825).
124 Castell Gorfod plan 24 (1808) and 1 (1811).
125 Cawdor (Vaughan) 102/8029, letter (22 Nov 1755).
126 Cawdor 2/187 (18 Nov 1844) shows that Hector Rees's father, eldest son and grandson were all named John Rees. Hector died in 1760 and both his father and son pre-deceased him. It was probably Hector's son who initially approached John Vaughan in 1755.
127 Cawdor (Vaughan) 102/8029, letter (27 Nov 1756).
128 Brodie Collection 69 (1757).
129 Stepney Estate 772 (22 May 1832). The entire Pwll area was termed "*Black Rock*" on a plan of 1775 ("Carmarthen Bay on the South Coast of Wales" by M. Mackenzie, Llanelli Public Library plan 1214, 25 Nov 1775).
130 Stepney Estate 1094 (28 Jul 1765) and Cawdor (Vaughan) 114/8430 (11 Jan 1768).
131 Thomas Mainwaring's Commonplace Book (LPL).

132 Mansel Lewis London Collection 44, Plan of Penllech Farm (1825).
133 "Canals of South Wales and the Border" by C. Hadfield.
134 Mansel Lewis 181-182 (19 Jan 1760).
135 Castell Gorfod 149 (1808).
136 Llanelli Public Library plan 3 (1772).
137 Mansel Lewis London Collection 114, letter (25 Sep 1778).
138 Nevill MS X, entries in 1810.
139 Stepney Estate Office plan 74 — Enclosure Awards 1810 to 1843. (CRO).
140 Cawdor (Vaughan) 5854, Golden Grove Estate Book IV (1787). "Bank y Llong" could be translated as Shipping Bank.
141 Mansel Lewis London Collection 98, letter (14 Feb 1762).
142 Mansel Lewis London Collection 97, "Lady Mansell's proposals for letting the Shipping Colliery", undated but almost certainly late 1797. It is considered that the canal from "*the Forge to Carmarthenshire Dock*", referred to by J. Innes in "Old Llanelly", and the Vauxhall Canal, referred to by C. Hadfield in "Canals of South Wales and the Border", were, in fact, the navigable course of the River Lliedi from Llanelli Flats up to Bank y Llong.
143 Brodie Collection 69 (1757).
144 Cawdor (Vaughan) 5854, plan of "Trawstre in Llanelly" (1785).
145 *The Cambrian* (7 May, 4 Jun 1808); Cawdor (Vaughan) 109/8292 — Court Leet held at Llanelli on 27 Oct 1809.
146 Plan of part of the first field survey for Sheet 37 of the Ordnance Survey 1 inch to 1 mile plan, surveyed 1813/14 at a scale of 2 inches to 1 mile (NLW). No accurate plan showing the route of Dafen Pill prior to the building of the Great Embankment, in 1808/09, has been seen but it is likely that the river was still close to its original route in 1813/14.
147 Stepney Estate 1099, rentals between 1773 and 1780.
148 Nevill 27 (27 Jun 1785).
149 Stepney Estate 1099, rental for 1797.
150 Un-numbered Carmarthenshire lease (2 Oct 1802) (CCL).
151 Thomas Mainwaring's Commonplace Book, extract from "General Warde's letter of 21 Mar 1814 to Mr Goodeve." (LPL).
152 Nevill MS IV, copy letter from C. Nevill to Gen. Warde (31 Oct 1804).
153 Nevill 170 (30 Sep 1809).
154 Nevill MS X, entries between 1809 and 1812.
155 Llanelli Public Library plan 11 — "A plan for the continuation of General Warde's Canal 1821."
156 Stepney Estate 1110, undated but c1790.
157 Information superimposed, at a later date, on Stepney Estate Office plan 72 — "Map of Upper Pencoid", 1761 (CRO). It is assumed that "*Sir John*" was Sir John Stepney, 8th Bart.
158 Coal was being mined at the Allt as early as c1611 — See Chapter 2.
159 See previous discussion on Thomas Jones — Chapter 3.
160 Cawdor (Vaughan) 92/7635 (1 Jan 1789).
161 Stepney Estate 1110 (c1790).
162 Stepney Estate Office plan 72 — "Map of Upper Pencoid" (1761) (CRO).
163 Llanelli Public Library plans 14 (undated but c1825) and 933 (1826).
164 Carmarthenshire lease 140 (20 Oct 1774) (CCL).
165 Alltycadno Estate, correspondence of Wm. Roderick, agent, letters (9 Oct 1773 and 7 Jan 1774).
166 Local Collection 465, accounts for Llwyn Ifan from 1773 to 1792.
167 Cawdor (Vaughan) 92/7635 (1 Jan 1789).
168 Carmarthenshire lease 126 (20 Feb 1793) (CCL).
169 Cawdor (Vaughan) 21/641 (undated but given as c1790 in the schedule).
170 Alltycadno Estate, letter from William Roderick to William Clayton (22 Aug 1772).
171 "The Stepneys of Prendergast", Hist. Soc. of West Wales Transactions Vol VII (1917-18).
172 Alltycadno Estate, letters from W. Roderick to W. Clayton (14 Aug, 9 Oct 1773 and 7 Jan 1774).
173 Ibid, letter (26 Nov 1774).
174 Stepney Estate 1110 — "Plan of Mr Hopkin's intended canal" (undated but c1790).
175 Local Collection 37 (11 May 1824).
176 "Plan of Llangennech and Talyclun Marshes and adjacent lands", undated but contained within Local Collection 37 (11 May 1824).
177 Penlle'rgaer A.974 (16 Apr 1819).
178 Stepney Estate Office plan 72 — "Map of Upper Pencoid" (1761) (CRO).
179 *The Cambrian* (28 Jan 1804).
180 Ibid, (25 Apr 1807).
181 No legal agreement or lease between Symmons and Davenport, Morris and Rees has been seen.
182 *The Cambrian* (1 Oct 1808).
183 Ibid, (12 Feb 1814).
184 *Llanelly Guardian* (20 Feb 1896) quoting an article entitled "Railway and Dock of the Llan-

gennech (St David's) Company", contained in the British Almanack for 1835.

185 "Plan of Llangennech and Talyclun Marshes and adjacent lands", contained in local Collection 37 (11 May 1824).

186 Llanelli Public Library plan 933 — "Chart of A Part of the River Loughor 1826".

187 "Old Llanelly" by J. Innes.

188 Llanelli Public Library plan 3 (1772).

189 Cawdor (Vaughan) 5/118 (1 Sep 1794).

190 The dock was also sometimes referred to as "*Middle Dock*" during the period considered. It would later be termed "*Lead Works Dock*" or "*Cambrian Dock*" but the name "*Pembertons' Dock*" was extensively used from the early 19th century onwards.

191 Nevill 800 (1848-1852) showed what could have been the representation of a single shipping stage at the top end of the dock and termed it "*Roderick & Bowen's Shipping Place*".

192 Cawdor (Vaughan) 14/450 letter from J. Symmons to J. Vaughan (26 Dec 1803).

193 Cawdor (Vaughan) 43/5847, entitled "Mr Roderick's Proposal" (undated but late 1803 or early 1804).

194 Cawdor 2/133 — brief of a Bill in Chancery, R. A. Daniell and others against Lord Cawdor and others (1814).

195 Nevill MS IV, copy letter from R. J. Nevill to R. Michell (12 Dec 1804).

196 Cawdor 2/133, Brief of a Bill in Chancery (1814).

197 Nevill MS IV, copy letter from C. Nevill to Messrs. Guest & Savill (23 Sep 1805).

198 "*Ship News*" in *The Cambrian* during 1805.

199 Nevill MS IV, copy letter from C. Nevill to R. Michell (15 Nov 1805).

200 Nevill 411 (29 Apr 1825) and Nevill MS IV, letter from C. Nevill to J. Guest (23 Feb 1807). The actual assignment document does not appear to have survived.

201 Cawdor 2/133 — Instructions for Assignment of Leases of Collieries etc., between Messrs. Roderick & Co and Messrs. Daniell & Co (1807). The assignment document does not appear to have survived.

202 Nevill MS IV, copy letter from Savill, Guest, Daniell and Michell to Wm. Roderick (21 May 1807). When Roderick was declared a bankrupt in May 1808 he was described as a "*Chapman and dealer*", and probably wished to retain the use of the dock in 1807 in order to continue this trade.

203 *The Cambrian* (16 Jan 1808).

204 Stepney Estate Office plan 46 (1814) (CRO).

205 *The Cambrian* (16 Sep 1817).

206 Local Collection 252 (14 Jan 1828).

207 Calendar of the Diary of Lewis Weston Dillwyn, Vol II, entry for 30 Apr 1828 (NLW).

208 *The Cambrian* (28 Feb 1829).

209 Thomas Mainwaring, op.cit.

210 The name "*Pemberton Dock*" was used on the first 6" to 1 mile scale Ordnance Survey plan of the Llanelli area published in 1891.

211 There is evidence that the shipping place was originally known as "*Mr Raby's Dock*" (Mansel Lewis London Collection 79 dated 1803) or as "*Squire's Dock*" (Old Llanelly by J. Innes). Between 1802 and 1830 it was generally referred to as "*The Carmarthenshire Railroad Dock*" but was also referred to, in the plans and documents of the period, by the several names of "*Railroad Dock*", "*Railway Company's Dock*", "*Llanelly Dock*", "*Upper Dock*" and "*North Dock*". The first reference seen to the name "*Carmarthenshire Dock*" was dated 1811 (*The Cambrian*, 16 Nov 1811), but this does not appear to have been the usual terminology prior to 1830.

212 Mansel Lewis 480 (25 Mar 1797).

213 Mansel Lewis London Collection 97, undated document entitled "The description of the lands for the line of the Railway and Canal etc . . .".

214 The Carmarthenshire Railway Act (42 Geo III c80), dated 1802, authorised the purchase of Raby's dock and it is known that only an eastern quay wall existed at that time — See *The Cambrian* (9 Jun 1804).

215 "History of Carmarthenshire", ed. J. Lloyd, Vol II, quoting an unidentified source.

216 42 Geo III c80 (3 Jun 1802).

217 Museum Collection 387 (1802-07).

218 Thomas Mainwaring, op.cit. (LPL).

219 *The Cambrian* (3 May 1806).

220 *The Cambrian* (11 Oct 1806). The provision of a western dock wall to give shelter from the sea and winds had been proposed as early as 1804, when it was suggested that the "*stone-ships*", constructed for the purpose of blocking-up French ports, could be used as "*a western wall to the great dock now forming on Llanelly Flats*" (*The Cambrian* 9 Jun 1804).

221 *The Cambrian* (25 Apr 1807).

222 Mansel Lewis London Collection 3, copy of a draft release, assignment and deed of composition dated 10 June 1807.

223 *The Cambrian* (11 Jul 1807).

224 "Old Llanelly" by J. Innes, quoting a letter, no longer extant, from A. Raby to the Committee of the Carmarthenshire Rail Road Company, dated 9 March 1808.
225 *The Cambrian* (29 Jul 1809).
226 Ibid, (20 Apr 1811); Nevill MS X, entries during 1811.
227 Ibid, (5 Oct 1811).
228 Ibid, (19 Oct 1811).
229 Ibid, (18 Apr 1812).
230 52 Geo III c173 (20 Jun 1812).
231 *The Cambrian* (13 Nov 1813).
232 Llanelly Public Library plan 9 (1814) showed four individual rail turn-outs to the eastern quay wall and five to the western wall. It is assumed that each turn-out led to a different loading point.
233 Llanelli Public Library plans 10 (c1822) and 15 (c1822) showed the dock with onlv eight turn-outs: four to each quay wall.
234 Local Collection 569 — Harbour Master's monthly reports from 1842 to 1854.
235 This dock was generally referred to as the "*Copperhouse Dock*" on the plans and documents of the period, although the name "*Copperworks Dock*" was sometimes used — See Cawdor (Vaughan) 109/8292 — Court Leet held at Llanelly on 19 October 1813. The later-used name, "*Nevill's Dock*", has not been encountered in any pre-1830 references.
236 Nevill MS IV, copy letter (12 Dec 1804).
237 Ibid, copy letter (14 Dec 1804).
238 Ibid, copy letter (1 Aug 1805).
239 Ibid, copy letter (15 Nov 1805).
240 Ibid, copy letter (4 Feb 1806).
241 Ibid, copy letter (13 Nov 1806).
242 The evidence of Llanelli Public Library plan 9 (1814).
243 Cawdor (Vaughan) 109/8292 — Court Leet held at Llanelly on 19 Oct 1813 and Cawdor 2/63 — Court Leets held at Llanelly between 1819 and 1823.
244 Cawdor (Vaughan) 109/8292 — Court Leet held at Llanelly on 27 Oct 1809.
245 The evidence of Llanelli Public Library plan 15 (c1822).
246 Nevill 240 (Jul 1812).
247 *The Cambrian* (11 Sep 1813).
248 Cawdor (Vaughan) Box 228 — Bundle headed "Papers on the subject of a proposed Canal in Carmarthenshire". Accounts of a meeting held on 30 Apr 1813.
249 Nevill 636 (24 Nov 1814).
250 *Carmarthen Journal* (5 Sep 1817).
251 *The Cambrian* (14 Jun 1823).
252 Nevill 243 (24 Jun 1823).
253 No document relating to the installation of the dock gates has been seen, although we know that Llanelli had a floating dock before September 1828 (*The Cambrian* 20 Sep 1828). J. Innes, op.cit., stated that the gates were erected in 1825, but did not quote his source, and R. Craig stated that the Copperworks Dock was a floating dock with lock gates by 1824 or earlier, ("The Emergence of a Shipowning Community at Llanelly 1800-1850", The Carmarthen Antiquary, Vol III, Part I 1959).
254 "*Ship News*" in *The Cambrian* between 1804 and 1828. Copper ore would have been unloaded at Spitty Bank, where the Spitty Copperworks was sited, but the majority of the ore would have been smelted at the Llanelly Copperworks, which was a much larger concern and did not experience the intermittent working that marked the life of the Spitty Copperworks. It is unlikely that much copper ore would have been unloaded at Llanelli's other shipping places.
255 Mansel Lewis London Collection 114, letter from John Smith to Sir E. V. Mansell (25 Sep 1778).
256 *The Cambrian* (13 Aug 1808).
257 Mansel Lewis 2578 (undated but c1813).
258 Penll'rgaer B.18.64 (1817/18) and B.18.66 (25 Nov 1817).
259 Nevill 817 (1 Jan 1818) and an endorsement, dated 20 Apr 1818, on Nevill 813 (22 Oct 1814).
260 *The Cambrian* (24 Jan 1829).

APPENDIX A

COAL SEAM NAMES USED BEFORE 1836

All the coal seam names known to have been employed in the Llanelli Coalfield before 1836 are listed in Table II, Chapter 1. Correlation of these names to known seams necessitated interpretation of available data in both the historical mining and geological contexts, and a brief explanation of the reasoning behind the correlation is presented in this Appendix.

The Gelli Group of coal seams

Surviving evidence does not allow the seams of the Gelli Group, worked in the Bryn/Llangennech region before 1830, to be positively identified, but the thicker bottom seam of the Group, and probably one or more of the thin seams above it, were exploited. Early working took place near Gelli-gele farm, reference being made to the Gelly-Gilly Vein in 1803,[1] to the Gelly Little Vein in 1824[2] and to the Gellygele Vein c1825.[3] These seams were also worked in the Bryn region, reference being made to the Gellicilau or Bryn Vein[4] and to the Gellywhiad or Small Vein[5] (after the farm of Gellywhiad) in 1836.

Un-named seam(s), Penyscallen Vein, un-named seam(s) (Grovesend Beds)

Considerable uncertainty exists about workings in the Penyscallen Vein and in the 3 or 4 thin un-named seams, above and below it. These have never been fully defined, and the situation is further complicated because they outcrop close to each other, east of the Gors Fault and, probably, immediately east of the Box Fault, between the northern and southern outcrops of the Swansea Four Feet Vein. Surviving evidence is not sufficient to allow us to allocate known names to individual seams, but the thicker Penyscallen Vein was probably the most exploited.

The Little Vein in the Box/Capel region was referred to in 1805[6] and the Genwen Small Vein (after Genwen Farm) at Cwmfelin in 1812.[7] The Little Vein and the Small Vein were also referred to in the Box/Capel region, in 1822[8] and 1825[9] respectively, and the outcrop of the Penyscallen Vein, in the Penprys region, was considered to be a continuation of the outcrop of the Swansea Four Feet Vein and was termed, in error, the Penprys Vein in 1825.[10] The lack of knowledge of the correlation of the seams in the Llanelli Coalfield is illustrated by the fact that, as late as 1835/36, the Penyscallen Vein in the Cwmfelin/Pencoed region was considered to be the Rosy Vein, because it was the topmost known and exploited seam, and was identified with the topmost seam west of the Box Fault.[11]

The Swansea Four Feet Vein

Although the Swansea Four Feet Vein only exists east of the Box Fault, it was extensively exploited before 1830 because of its great thickness. The inability to correlate the seam occurrences led to different names being given to the same seam at different locations in Llanelli. It was also thought to be the Rosy Vein or the Fiery Vein, and confused with the Swansea Five Feet Vein, the only other seam of comparable thickness in the Llanelli Coalfield. At least 17 different names are known to have been allocated to it by 1836, a number of them in use at any one time. This confused situation can be clarified by considering the seam names used at different locations in the Coalfield.

In the region between the Box and Gors Faults, workings in the Swansea Four Feet Vein, near to the Box Fault, were referred to as workings in the Penyfigwm or the Rosey or the Great Vein in 1762.[12] The name Penyfigwm was used because the northern outcrop of the Swansea Four Feet Vein occurred at Penyfigwm Farm (north of Capel Uchaf),[13] and the name Great related to the great thickness of the seam. We do not know whether the name Rosey was an alternative name in this region, or whether it was confused with the Upper Swansea Six Feet Vein at Llanelli, because both were the topmost known and exploited seams each side of the Box Fault.[14] For the latter explanation to be correct, the Upper Swansea Six Feet Vein west of the Box Fault must have been called the Rosey before 1762, but no evidence has been seen to confirm this. Certainly, as late as 1758, the Upper Swansea Six Feet Vein at Llanelli was known as Thomas David's or Ca-main Vein (See later). The Penyfigwm or Great Vein, east of the Box Fault, was again referred to in 1776,[15] and the name Penyfigwm Vein was again used in the Box/Bryngwyn area in 1805,[16] and in 1808, when the Box Colliery was opened.[17] Thereafter, no reference to the name has been seen, the Swansea Four Feet Vein at the Box Colliery being called the Box Vein in 1813[18] (after the farm of Box on which the Box Colliery was situated). In 1825 it was called the Great Vein[19] and, by 1836, it was called the Box Great[20] or the Nine Feet Vein in the Box/Llandafen area.[21]

In the region along the southern outcrop of the Swansea Four Feet

Vein, between the Gors and Plas Isaf Faults, a plan of the Cwmfelin to Pencoed area, dated 1772,[22] clearly shows that the seam was known as the Fiery Vein. The allocation of the name Fiery was undoubtedly due to its being identified with the Lower Swansea Six Feet Vein west of the Box Fault (where the Swansea Four Feet Vein is absent), which had been called Fiery as early as 1754. This confusion in the minds of the early miners arose because they thought that the second known seam below ground at Cwmfelin (the Swansea Four Feet Vein) must be the same as the second known seam west of the Box Fault (the Fiery Vein), the effects of faulting not being understood. The name was used again in the Cwmfelin/Pencoed area in 1802.[23] In 1807 the Swansea Four Feet Vein, in the same area, was referred to as the Old Penyfan or Daven Pill Fiery Vein[24] and, in 1808, as the Daven Pill Burry River Fiery Vein.[25] Although these names were quoted in advertisements in *The Cambrian* directed at the coal-shipping trade (advertisements which often used names designed to attract the buyer, as opposed to the normal seam name), they show that it was assumed, in error, that the Swansea Four Feet Vein at Cwmfelin corresponded to the Lower Swansea Six Feet Vein, worked some 40 years earlier in the Llwynwhilwg/Penyfan region and shipped from Dafen Pill. The seam remained known as the Fiery Vein in this region up to 1836,[26] although it was referred to as the Old Fiery Vein in 1809,[27] and the True Fiery Vein in 1813.[28]

In the region of the northern outcrop of the Swansea Four Feet Vein, between the Gors and Plas Isaf Faults from Clyngwernen Isaf to Glanmwrwg, the seam was called the Penprys Vein between Clyngwernen and Penprys, and the Glanmurrog Vein at Glanmwrwg, in 1803.[29] Both names remained in use, and the seam was termed the Glanmwrwg Great Vein,[30] and also the Penprys Vein[31] in the Clyngwernen/Penprys area, in 1825. The opening of the St David's pit, in 1832, led to the seam being called the St David's Vein at Gelly Farm in 1836,[32] when reference was also made to the Penprys or St David's Vein[33] and to the "*Llangennech Vein 6ft thick now worked at the St David's Colliery*".[34]

Un-named seam between the Swansea Four and Five Feet Veins

There is evidence that a thin, un-named seam occurs between the Swansea Four Feet and Swansea Five Feet Veins in the Cwmfelin/Pencoed region, although no representation of this seam has been given on geological plans. The evidence is in a report of 1836[35] on the Genwen Coal District, which referred to a thin seam, 1ft 8in thick, lying between the

Cwmfelin Vein (Swansea Five Feet) and the Fiery Vein (Swansea Four Feet), as the Golden Vein, a name formerly used for the Swansea Five Feet Vein in this region. The thickness of each of the named seams was given in the report, and, even allowing for correlation inadequacies, it seems that a thin seam, lying between the Swansea Four Feet and Five Feet Veins at Cwmfelin, was termed the Golden Vein in 1835/36.

The Swansea Five Feet Vein

The Swansea Five Feet Vein was the other seam of great thickness in the Coalfield, and it was exploited wherever it occurred, east of the Box Fault, before 1830. As with the Swansea Four Feet Vein, considerable confusion existed in relation to its correlation and, because of this, it was known by a number of regional names. It was also thought to be either the Fiery Vein or the Golden Vein, and was also confused with the Swansea Four Feet Vein, which was of comparable thickness. An examination of the known names used at different locations clarifies the confused situation.

In the region between the Box and Gors Faults, the seam, at its outcrop at Penygaer, was probably known as the Stradey Vein in 1767. This is deduced from a document of that year, which stated that a seam of coal at Penygaer was "*under lease as Stradey Vein*".[36] No other reference to this name has been seen and, by 1805, the Swansea Five Feet Vein, immediately east of the Box Fault, was termed "*The Llanerch Vein otherwise called the Fiery Vein*".[37] The Swansea Five Feet Vein had doubtless been equated to the Fiery Vein at Llanelli, because both were the topmost but one important seams, immediately east and west of the Box Fault respectively. Figure 3 also shows that the early colliery viewers probably assumed that the outcrop of the Swansea Five Feet Vein, at Penygaer, was a direct continuation of the outcrop of the Fiery Vein at Llanerch, the Box Fault being regarded as a zone of disturbed ground interrupting continuity of workings. In 1813, the seam was referred to as the Pinygare[38] or Penygare Vein[39] (after the farm of Pen-y-gaer Uchaf where the seam outcrops), but developments at the Box Colliery led to a change of name in this region in 1823. The Box Colliery pits were sunk from the Swansea Four Feet Vein (at 192 feet; 58m) to the Swansea Five Feet Vein (at 420 feet; 128m), between 1819 and 1823 and, when the "*new coal*" was first shipped from Llanelli, in June 1823, the seam was called "*Warde's Fiery Vein*".[40] Reference was made to "*Warde's Fiery Vein or the Penygare Vein*" in 1827,[41] but Warde's name was not identified with the seam after his death, in 1830. In 1835 the seam at the

Box Colliery was simply referred to as the Fiery Vein,[42] and, in 1836, as the Fiery Vein and as the Box Fiery Vein.[43]

In the region along the southern outcrop of the Swansea Five Feet Vein, between the Gors and Plas Isaf Faults, it is suspected that the seam's outcrop in the Ffosfach area was called the Marsh Vein between 1755 and 1761[44] (after the marshland east of Ffosfach), but it was known as the Great or Golden Vein, in the area between Cwmfelin and Pencoed, by 1772.[45] The seam was called "*Great*" because it was nine feet (2.7m) thick in this region, but it is not known if the name Golden was an alternative name, or whether it was confused with the Swansea Three Feet Vein at Llanelli, because both seams were the topmost but two known, exploited seams in their regions. (The first reference seen to the use of the name Golden Vein for the Swansea Three Feet Vein at Llanelli dates to 1794, however).[46] It was called the Great Vein in 1802,[47] the Gwndwn Mawr Golden Vein in 1807,[48] the Old Gundunmaur or Golden Vein in 1808,[49] the Great or Old Golden Vein in 1812[50] and the Old Golden Vein in 1813.[51] Workings in the Swansea Five Feet Vein, between Cwmfelin and Pencoed, were temporarily discontinued before 1820, the next reference seen to the seam here being in 1835, when the seam was referred to as the Great Fiery Vein or the Cwmfelin Vein.[52] In 1836 it was called both the Fiery Vein[53] and the Cwmfelin Vein.[54] We do not know why the name Fiery was substituted for Golden when mining recommenced in this region, but it may have been because it was realised that the seam corresponded to the Swansea Five Feet Vein at the Box Colliery, where it had been called Fiery since 1823.

In the region of the northern outcrop of the Swansea Five Feet Vein between the Gors and Plas Isaf Faults, reference was made to "*Alt Cole*" (Allt Coal) in 1773/74[55] and, because the Swansea Five Feet Vein outcrops on the Allt, the term "*Alt Cole*" may be synonymous with the name "*Allt Vein*". No reference to the actual word "*vein*" has been seen, however. In 1803, in the Clyngwernen to Brynsheffre area, the seam was shown to be two separate seams, termed the Glyngwernen Upper and Lower Veins.[56] The name Glyngwernen Vein* was still used in this region in 1825[57] but, by 1836, the seam was being called the Cornhwrdd Vein,[58] Cornhwrdd being a farm on the northern outcrop of the Swansea Five Feet Vein, between Clyngwernen and the Allt.

The Upper Swansea Six Feet or Rosy Vein

It is considered that the Upper Swansea Six Feet Vein, west of the Box

* Edward Martin used "*Glyngwernen*" instead of "*Clyngwernen*" in his 1803 report.

Fault, was known as Thomas David's Vein or the Ca-Main Vein in the 1750s. Coal leases of 1754 and 1758[59] made reference to both Thomas David's Vein and the Fiery Vein along the southern outcrop of the coal seams between the Box and Stradey Faults, and to Thomas David's or Ca-Main Vein, and the Fiery or Ca-Plump Vein under Morfa Bach (the Sandy area), at the northern outcrop. Both leases stated that Thomas David's Vein was 2 feet (0.61m) thick and the Fiery Vein 4 feet (1.22m) thick. We know that the Upper Swansea Six Feet Vein outcrops at Ca-Main, and the Lower Swansea Six Feet Vein outcrops at Ca-Plump, just north of Morfa Bach Common.[60] This evidence leaves little doubt that the Upper Swansea Six Feet Vein was called Thomas David's Vein (probably after some now-forgotten early miner), or the Ca-Main Vein, between 1754 and 1758.

We are not certain when the seam was first called the Rosy Vein. The Swansea Four Feet Vein, immediately east of the Box Fault, was called the Penyfigwm or Rosey or Great Vein in 1762,[61] but we do not know if Rosy was an alternative for Penyfigwm in this region, or whether it had been so called because it was wrongly equated to the Upper Swansea Six Feet Vein, west of the Box Fault, as both were the topmost known seams each side of the Fault. If the latter explanation is correct, the Upper Swansea Six Feet Vein west of the Box Fault, must have been called Rosy (or Rosey) by 1762, at the latest.[62] No confirmatory evidence of this has been seen.

However, in 1788, when the Upper Swansea Six Feet Vein was encountered in a new pit on the northern limb of the Llanelli Syncline west of the Box Fault, it was thought that a new seam had been discovered. The sulphurous smell of the extracted coal led to the seam being called the Stinking Vein,[63] but no further reference to this name has been seen and, by 1798, the seam west of the Box Fault was called the Rosy Vein.[64] The origin of the name is not clear but, after 1798, the seam was called the Rosy,[65] Rosey[66] or Rose Vein[67] in all documents, plans and correspondence, except for a brief period, in 1805/06, when the seam was confused with the Swansea Four Feet Vein and called the Penyfigwm or Great Vein.[68] Attempts were still being made to equate the topmost seams, each side of the Box Fault, as late as 1836, and the seam was called, in error, the Little Vein,[69] on the assumption that it corresponded to the topmost seam worked at the Box Colliery.

One further name was used for the seam before 1830. It was encountered south of the Carnarfon Fault in the Baccas or Carnarfon Colliery, but lack of correlative knowledge did not allow the colliery viewers to even suspect that they were dealing with the Rosy Vein, and they gave it the regional name, Carnarfon Upper Vein.[70]

The Lower Swansea Six Feet or Fiery Vein

A reference to Llanelli's coal, in a letter written in 1749, said "*one sort here is called the Fiery Vein*".[71] Although it is possible that the seam referred to was the Swansea Four Feet Vein at Cwmfelin, it is considered that this is the first use seen of the name Fiery for the Lower Swansea Six Feet Vein, west of the Box Fault. The origin of the name Fiery is not known. Certainly, in 1754, the name Fiery Vein was used in the Bigyn Hill and Bryn (Seaside) areas[72] and, in 1758, reference was made to the Fiery or Ca-Plump Vein under Morfa Bach[73] (a common between the present Old Castle Pond and Pembrey Road). As the Lower Swansea Six Feet Vein outcrops under a field once termed Ca-Plump[74] (just north of Pembrey Road), it is undoubtedly the seam known as the Fiery or Ca-Plump Vein in the 1750s. It is also considered that the seam along its southern outcrop, immediately east of the Box Fault in the Penyfan/Llwynwhilwg region, was called the Llwynwilog Vein in 1754, after Llwynwilwg-fawr, where the seam outcrops.[75]

The seam was known as the Fiery Vein until after 1830, although variations were introduced into advertisements aimed at the shipping trade. Reference was made to Bowen's Real Old Fiery Vein in 1808,[76] to the Old Fiery Vein in 1811, and to Lady Mansell's or Raby's Old Fiery Vein in 1819.[77] The continuing attempts to correlate the seams each side of the Box Fault led to the suggestion that the seam was the Rosy Vein, but this view was not generally held.[78] One other name is known to have been used for the seam before 1830. The Baccas or Carnarfon Colliery was sunk to the seam in the late 1790s or early 1800s, but lack of correlative knowledge led to its being called the Carnarfon Lower Vein.[79]

The Swansea Three Feet or Golden Vein

An agreement for working coal under Bigyn Hill, in 1729, referred to the Bwysva Vein.[80] It was stated that the coal lay under Caswddy Hill (Bigyn Hill), Bwysva (not located) and Coed Hirion (the old name for Ty-Isaf), allowing location of the seam to the Bigyn/Penyfan area. It was also stated that the seam lay "*in and under*" landshares called Cae'r Bwysva and Dorgraig. Caer Bwysva has not been located, but Dorgraig was a land-share, just south of the present Bigyn School, on the outcrop of the Swansea Three Feet Vein.[81] The Bwysva Vein was, therefore, an early name for the Swansea Three Feet Vein, and is also the earliest known name for any coal seam in the Llanelli area. It was still used in 1754, when the seam was called the Boisva Vein,[82] but its use may have lapsed thereafter. In 1758/59 the seam under Stradey Estate lands was called the Parkycrydd or Killyveig Vein[83] (after two land-shares situated along the northern outcrop of the seam west of the Box Fault) but, before 1776, it was also called the Country Coal and Dray Pitt Vein, after an unlocated pit, and as the Engine Vein,[84] after a water-engine pit situated between the present People's Park and Pentip School.

We do not know when the name Golden Vein was first used for the seam. The Swansea Five Feet Vein at Cwmfelin was known by this name in 1772,[85] when the Swansea Three Feet Vein at Llanelli was variously called the Parkycrydd or Killyveig or Country Coal and Dray Pitt or Engine Vein. It is, therefore, possible that the seam west of the Box Fault was named after the Swansea Five Feet Vein at Cwmfelin, because both were the third known seams below ground level. The earliest reference seen referring to the seam at Llanelli as the Golden Vein dates from 1794, when reference was made to the Old Golden Vein, west of the Box Fault along the northern outcrop of the seam.[86] After this, the seam was always referred to as the Golden Vein in this region.

One regional name was given to it c1825, when the seam, at its northern outcrop between the Gors and Acorn Faults near Porthdafen, was called the Trosserch Vein[87] and, in 1836, uncertainty caused by attempts to correlate the seams throughout the Llanelli Coalfield, led to the seam being called the Fiery or Golden Vein, west of the Box Fault.[88]

The Swansea Two Feet or Bushy Vein

The Swansea Two Feet Vein was possibly called the Biggin or Hill Vein, in the Bigyn/Penyfan region, in 1754, from the evidence of a coal lease,[89] which included the coal in the Biggin or Hill Vein, the Boisva Vein and the Llwynwhilog Vein, but excluded the coal in Thomas David's Vein and the Fiery Vein. Thomas David's Vein was the Upper Swansea Six Feet, the Fiery Vein was the Lower Swansea Six Feet, the Boisva Vein was the Swansea Three Feet and the Llwynwilog Vein was the Lower Swansea Six Feet east of the Box Fault in the Penyfan/Llwynwhilwg region, leaving only the Swansea Two Feet Vein, in the Bigyn/Penyfan region, to qualify for the name of the Biggin or Hill Vein.

By 1762, the Swansea Two Feet Vein along its northern outcrop west of the Box Fault was called the Cwm Vein,[90] probably because the seam outcropped along the south side of the glacial valley, running from the village of Furnace to Felinfoel. Although Cwm was still used in 1799,[91] the name Caereithin Vein had, by this time, virtually superseded it in all documentation and correspondence. The first use seen of the name Caereithin Vein was in 1794,[92] and it was given because the seam outcropped across fields known as Caereithin mawr and Caereithin bach, situated between Old Road and New Road, south of Furnace.[93] By 1806, however, the seam at its southern and northern outcrops, west of the Box Fault, was called the Bushy Vein,[94] the new name being generally used thereafter, although Caereithin (or Caerythen) was still occasionally used up to 1836.[95] The origin of the name Bushy is not known with any certainty, but it is tempting to speculate that the English entrepreneurs, who came to Llanelli in the late 18th/early 19 centuries, modified their translation of Caereithin (field of gorse) to field of gorse bushes, and this led to the name Bushy.

In 1819, reference was made to the Lower or Caithen Vein[96] west of the Box Fault and, in 1836, the problems of seam correlation led to the name Golden or Bushy being given to the seam at the Bres Colliery.[97] One regional name is known to have been temporarily used, just before 1835/36. The seam was discovered near its northern outcrop, between the Box and Gors Faults, and lack of correlative knowledge led to the assumption that it was a new seam. As it was on Llanlliedi Farm lands, it was called the Llanlliedi Vein.[98]

Un-named seam(s), Cille Nos. 1, 2 and 3 Veins

A plan of 1822[99] showed a "*Brick Manufactory*" at Trebeddod, with an adit adjacent to it called "*Brickyard Level in the Brickyard Seam*". As its position coincides with the known outcrop of the Cille No. 2 Vein, it appears that Brickyard Seam was the Cille No. 2 Vein. No other reference to the Cille Veins or the un-named seams above them has been seen before 1836.

The Pwll Little Vein

No reference has been seen to the Pwll Little Vein before 1830 but, when the Stradey Level (immediately west of the Stradey Fault) was opened in 1832, the seam was called the Black Rock Vein.[100] It was named after the farm of Hendy (or Tyrhendy), sometimes called Blackrock, situated in the present Pwll area.[101]

The Hughes or Pwll Big Vein

Leases of 1765[102] and 1768[103] referred to the Pool Vein at Hendy and Penllech (Pwll). The seam was named after the North Pool, where large vessels were loaded with coal from small sloops or keels.[104] It may have been so named many years before, a plan of 1757 itemising "*Coaleries*" within the Burry Estuary, and stating that one was "*at the Pool*".[105] The colliery was abandoned, staying unworked until 1825. The next reference seen to the Pool Vein was in 1836, when it was called the Pwll Vein.[106] There is no confirmation that the Pool Vein of the 1760s and the Pwll Vein of the 1830s corresponded to the Hughes or Pwll Big Vein but, when the Pool Colliery was finally abandoned in the early 1880s, workings were in that seam.[107] It is, therefore, assumed that the Hughes or Pwll Big Vein, between the Stradey and Moreb Faults, was known as the Pool or Pwll Vein before 1830.

*The Cilmaenllwyd Vein**

Reference was made, in 1762, to the Stradey Vein.[108] There is insufficient detail to identify the seam, but it was probably given the name because it outcropped in the Stradey region. Reference was made, in 1767, to the Stradey Vein at Penygaer,[109] where the Swansea Five Feet Vein outcrops and, as already seen, it is likely that this name was given because both seams were the topmost but one known seams of consequence in their particular regions. If this were so, it is possible that the Cilmaenllwyd Vein, immediately west of the Stradey Fault, was known as the Stradey Vein from the 1750s to 1800.

In 1822, the Cilmaenllwyd Vein between the Stradey and Box Faults, north of Trebeddod, was called the Cwtta Vein[110] (after the farm of Cwta where the seam outcrops), but no other reference to this name has been seen.

Contradictory evidence regarding the Cilmaenllwyd Vein was provided in 1836,[111] when it was said that the Cilmaenllwyd Vein outcropped at Newgate (once a house near Cilymaenllwyd Mansion), confirming that the name now used had been given to the seam by that time. However, the same document referred to the Pwll and Stradey Veins, outcropping between Pwll and Stradey, and an accompanying plan showed representations of the outcrops of only the Hughes and Cilmaenllwyd Veins, immediately west of the Stradey Fault, implying that the Pwll and Stradey Veins were the Hughes and Cilmaenllwyd Veins respectively.

* This is the spelling of Cilymaenllwyd adopted on Geological Survey plans.

The Cwmmawr Vein

The Cwmmawr Vein between the Stradey and Box Faults north west of Felinfoel was known as the Hengoed Vein in 1822[112] (after the farms of Hengoed-fawr and Hengoed-fach near which it outcrops). No other reference to the Cwmmawr Vein has been seen before 1830, although we know that the seam was exploited along its outcrop, immediately north of Hengoed-fawr farm, before 1744.[113]

Un-named seam(s) between the Cwmmawr and Goodig Veins

One of the thin un-named seams between the Cwmmawr and Goodig Veins, north-west of Felinfoel between the Stradey and Box Faults, was known as the Pantylliedy Fawr Vein in 1822[114] (after the farm of Panty-lludu-fawr where it outcrops).

The Goodig Vein

No reference has been seen to the Goodig Vein before 1830.

The Gwscwm Vein

The Gwscwm Vein, north-west of Felinfoel between the Stradey and Box Faults, was known as the Trevenna Seam or Vein in 1822[115] (after the farms of Trefanau Uchaf and Trefanau Isaf where it outcrops).

REFERENCES — APPENDIX A

1 "Report on the Llangennech Collieries by Edward Martin dated 21 Sept 1803", contained in Local Collection 37 (11 May 1824).
2 Catalogue of Plans of Abandoned Mines, Vol IV, 1930, referring to a working plan of the Gelly-Hwyad Colliery, no longer extant.
3 Llanelli Public Library plan 13, undated but c1825.
4 GS 171/218 (1836 to 1842) (IGS).
5 Nevill 1783 (6 Aug 1836).
6 *The Cambrian* (2 Nov 1805).
7 Llanelli Public Library plan 8 (1 Jul 1812).
8 Catalogue of Plans of Abandoned Mines, Vol IV, 1930, referring to a working plan of the Box Colliery, no longer extant.
9 Nevill 1827-1869, weekly reports of the Box Colliery between March and December 1825.
10 Llanelli Public Library plan 13, undated but c1825.
11 Nevill MS XVIII (Jan 1836) and GS 171/218, op.cit.
12 Nevill 705 (2 Aug 1762).
13 Mansel Lewis London Collection plan 63 (1820).
14 In 1805/06, a legal dispute took place between two industrialists at Llanelli and it was stated that the Rosy and Penyfigwm Veins were one and the same, merely because they were the top veins on each side of the Box Fault (Mansel Lewis London Collection 34 — Supplemental Case on Leases of Coal Mines under the Stradey Estate 1806).
15 Mansel Lewis 127 (9 Oct 1776).
16 *The Cambrian* (2 Nov 1805).
17 Ibid, (5 Nov 1808).
18 Ibid, (15 May 1813).
19 Nevill 1827-1869, Box Colliery weekly reports Mar-Dec 1825.
20 Nevill MS XVIII (Jan 1836).
21 GS171/218, op.cit.
22 Llanelli Public Library plan 3 (Apr 1772).
23 Carmarthenshire lease (2 Oct 1802) (CCL).
24 *The Cambrian* (14 Nov 1807).
25 Ibid, (13 Aug 1808).
26 *The Cambrian* (3 Apr 1813); Catalogue of Plans of Abandoned Mines, Vol IV, 1930, referring to an 1815 plan of the Llwynhendy and Genwen Collieries, no longer extant; Nevill MS XVIII (Jan 1836).
27 Nevill MS X, entry for 7 Jan 1809.
28 *The Cambrian* (15 May 1813).
29 "Reports of the Llangennech Collieries etc... " 1803, contained in Local Collection 37 (11 May 1824).
30 MS 3.299 (CCL) — Plan Book of Llangennech Estate, undated but c1825.
31 Llanelli Public Library plan 13, undated but c1825.
32 Nevill 412 (6 Dec 1836).
33 GS171/218, op.cit.
34 Nevill MS XVIII (Jan 1836).
35 Nevill MS XVIII, report on the Genwen Coal District (Jan 1836).
36 Mansel Lewis London Collection 4, printed sheet entitled "Lands proposed by Sir Edward Vaughan Mansel, Bart., to be sold by Auction" 1767. It was probably assumed that the Swansea Five Feet Vein (the topmost but one known seam immediately east of the Box Fault) corresponded to the Cilmaenllwyd Vein (the topmost but one known seam immediately west of the Stradey Fault). The Cilmaenllwyd Vein in that region was known as the Stradey Vein — See GS171/218, op.cit.
37 *The Cambrian* (2 Nov 1805).
38 Ibid, (15 May 1813).
39 Mansel Lewis 2578, undated but c1813.
40 *The Cambrian* (14 Jun 1823). The seam was named after General George Warde, the proprietor of the Box Colliery. The Swansea Four Feet Vein was identified with Warde's Fiery Vein later in the 19th century, but there is no doubt that this name was initially given to the Swansea Five Feet Vein at the Box Colliery.
41 Nevill 819 (26 Sep 1827).
42 Thomas Mainwaring's Commonplace Book (LPL).
43 Nevill MS XVIII, reports on the Llandafen Coal District and on the Pencoed Coal District, both dated Jan 1836.
44 Cawdor (Vaughan) 102/8029 letters (22 Nov 1755, 6 Mar and 17 Apr 1756 and 9 July 1761). John Vaughan of Golden Grove was attempting to let the Marsh Vein to entrepreneurs, who were already working the Swansea Four and Five Feet Veins between Cwmfelin and Pencoed.

As Lord of the Manor, Vaughan was entitled to custodianship of the seashore between high and low water marks and, in terms of this consideration, the Swansea Five Feet Vein just east of Ffosfach is the only realistic equivalent to the Marsh Vein.

45 Llanelli Public Library plan 3 (Apr 1772).
46 It is possible that the Swansea Three Feet Vein at Llanelli was called the Golden Vein after the name had first been given to the Swansea Five Feet Vein at Cwmfelin. The first use of the name seen is certainly that at Cwmfelin, in 1772, and the Swansea Three Feet Vein at Llanelli was being called the Parkycrydd or Killyveig Vein about this time.
47 Carmarthenshire lease (2 Oct 1802) (CCL).
48 *The Cambrian* (14 Nov 1807). Gwndwn Mawr was a farm situated on the outcrop of the Swansea Five Feet Vein between Cwmfelin and Pencoed.
49 *The Cambrian* (13 Aug 1808).
50 Llanelli Public Library plan 8 (1 Jul 1812).
51 *The Cambrian* (15 May 1813).
52 Thomas Mainwaring, op.cit.
53 GS171/218, op.cit.
54 Nevill MS XVIII, reports on the Llandafen Coal District and the Box Coal District, both dated Jan 1836.
55 Alltycadno Estate correspondence between 9 Oct 1773 and 26 Nov 1774.
56 Local Collection 37, op.cit.
57 Llanelli Public Library plan 14, undated but c1825.
58 GS171/218, op.cit.
59 Cawdor (Vaughan) 21/623 and 21/624 (28 Jun 1754 and 22 Apr 1758 respectively).
60 Mansel Lewis 2538, "Map of Killey" by M. Williams (1805).
61 Nevill 705 (2 Aug 1762).
62 Mansel Lewis London Collection 34 — Supplemental Case on Leases of Coal Mines under the Stradey Estate 1806 — shows that mis-identification, on the basis of both being the topmost known seams each side of the Box Fault, was the reason for the Rosy and Penyfigwm Veins being considered the same seam in 1806.
63 Ibid.
64 Mansel Lewis 49 (20 Jan 1798).
65 Nevill 650 (26 Jan 1805); *The Cambrian* (6 Aug 1808); Nevill 682 (30 Mar 1815).
66 *The Cambrian* (6 Jul 1811).
67 Ibid, (9 Aug 1806).
68 Mansel Lewis 380, letters in 1805/06; 382 to 384 (c1806).
69 GS171/218, op.cit.
70 Ibid.
71 Stepney Estate 1101, letter from Sir Thomas Stepney to Mr Shepard (7 Sep 1749).
72 Cawdor (Vaughan) 21/623 (28 Jun 1754).
73 Cawdor (Vaughan) 21/624 (22 Apr 1758).
74 Mansel Lewis 2538, "Map of Killey" by M. Williams (1805).
75 Cawdor (Vaughan) 21/623 (28 Jun 1754).
76 *The Cambrian* (6 Aug 1808).
77 Ibid, (6 Jul 1811, 9 Jan 1819).
78 GS171/218, op.cit.
79 Ibid.
80 Stepney Estate 1100 (1 Aug 1729).
81 Llanelli Public Library plan 9 (1814).
82 Cawdor (Vaughan) 21/623 (28 Jun 1754).
83 Mansel Lewis London Collection 34 — Case on Leases of Coal Mines under the Stradey Estate, Mr Bell's opinion 1805.
84 Ibid. and Mansel Lewis 127 (9 Oct 1776).
85 Llanelli Public Library plan 3 (Apr 1772).
86 Mansel Lewis 189 (1794).
87 Llanelli Public Library plan 13, undated but c1825.
88 GS171/218 op.cit.
89 Cawdor (Vaughan) 21/623 (28 Jun 1754).
90 Nevill 705 (2 Aug 1762) and the evidence of Mansel Lewis London Collection 34 (1806).
91 Nevill 155 (3 Apr 1799).
92 Mansel Lewis 189 (1794).
93 Llanelli Public Library plan 9 (1814).
94 Llanelli Public Library plan 7 (1806) and *The Cambrian* (9 Aug 1806).
95 GS171/218, op.cit.
96 Nevill 410 (18 May 1819).
97 GS171/218, op.cit.
98 Nevill MS XVIII, report on the Box Coal District (Jan 1836).

99 Mansel Lewis 2590 (Jan 1822).
100 Museum Collection 181 (1827-1834).
101 Stepney Estate 1104 (4 Aug 1840).
102 Stepney Estate 1094 (28 Jul 1765).
103 Cawdor (Vaughan) 114/8430 (11 Jan 1768).
104 *The Cambrian* (22 May 1819).
105 Brodie Collection 69 (1757).
106 GS171/218, op.cit.
107 "Geology of the Gwendraeth Valley and adjoining area", Special Memoir of the Geological Survey of Great Britain (1968).
108 Nevill 705 (2 Aug 1762) and the evidence of Mansel Lewis London Collection 34 (1805).
109 Mansel Lewis London Collection 4, printed sheet entitled "Lands proposed by Sir Edward Vaughan Mansel, Bart., to be sold by Auction" 1767.
110 Mansel Lewis 2590 (Jan 1822).
111 GS171/218, op.cit.
112 Mansel Lewis 2590 (Jan 1822).
113 "A Map of Hengoed belonging to Geo. Thomas, Gent, Surveyed by Wm. Jones, 1744" (NLW) showed "*Old Collierys*" along the seam outcrop.
114 Mansel Lewis 2590 (Jan 1822).
115 Ibid.

APPENDIX B

MINING METHODS AT LLANELLI BEFORE 1830

The progress in mining technology in the Llanelli Coalfield before 1830 has to be examined briefly, so that certain mining methods referred to in the text can be understood. The information that has survived for Llanelli in this respect is surprisingly meagre, and it has been difficult to compile an accurate summary. Because of this, comparisons have had to be made with known developments in other coalfields.

(1) THE METHODS USED TO GAIN ACCESS TO AND WORK THE COAL

(a) *Access to the coal*

When mining technology was primitive, small quantities of coal were extracted from outcrop workings and from bell pits (Fig. 39). In outcrop workings, access was gained where a coal seam outcropped under a minor thickness of superficial deposit, and the coal was extracted for a limited distance down its dip, with no support to the roof of the working, scouring taking place for a short distance along the seam. In bell pits, shallow shafts were sunk just down from the outcrop of the seam and workings effected radially from the bottom of the shaft. The roof of the working was unsupported and it was abandoned when it became unsafe. Evidence has survived to show that both these methods were used in the Llanelli area. In 1866, the mineral and general land agent to the Stepney Estate, in relation to a dispute concerning coal royalties due to the Estate, stated "*with regard to the outcrop workings in particular I say that it is common through all the coalfields of South Wales to find the traces of old workings or scourings whenever coal crops to the surface of the land... but I do not consider such workings or grubbings constitute open mines or collieries in the ordinary sense of the terms*".[1] Possible evidence of the working of bell pits was given, on a plan of 1772,[2] which showed four "*Old Coal Pits*" within a total distance of only 30 yards (27m), just down from the outcrop of the Swansea Five Feet Vein at Bynea. Both types of working would have been effected on higher ground, above the natural ground water level. These mining methods were probably employed at Llanelli throughout the 16th and 17th centuries, and may have been used in the 18th and 19th centuries by small operators, such as farmers extracting small quantities of coal for lime-burning purposes.

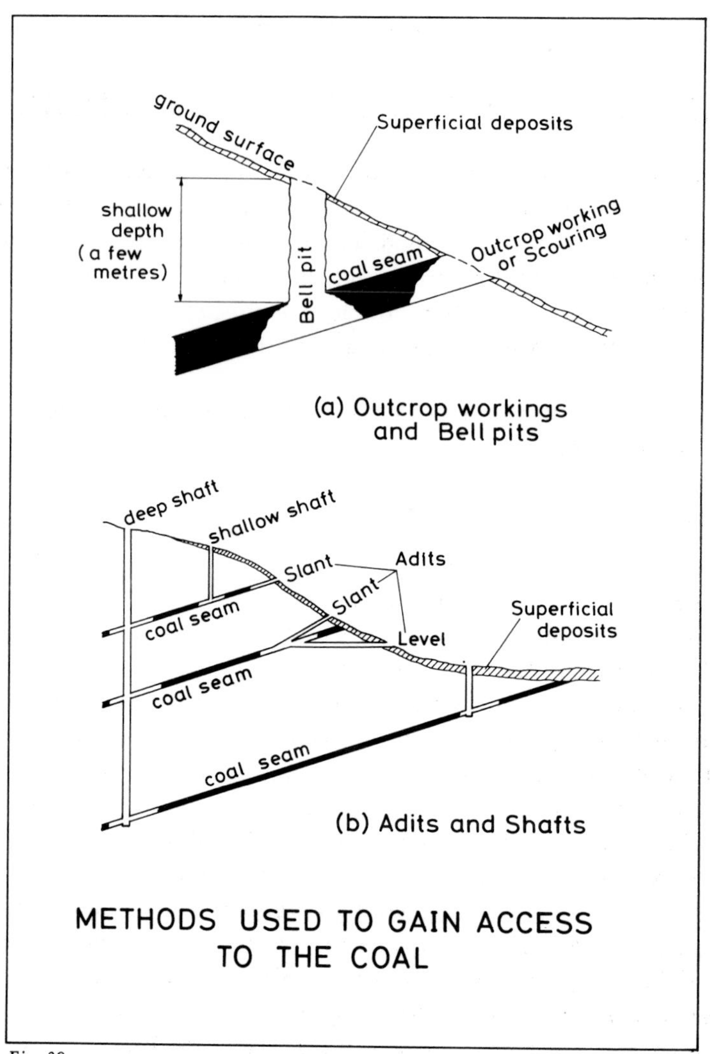

Fig. 39

It is likely that the limited output from outcrop workings and bell pits was inadequate to meet the export trade demand as early as the 16th century, and the driving of adits and the sinking of shafts or pits, which gave access to much larger quantities of coal, were adopted (Fig. 39). Adits, called slants or levels, depending on whether they sloped or were virtually horizontal, were driven into hillsides to follow or meet the coal seams, and workings were extended as far as possible from these points of access. The roof was supported by propping, and winnings of coal were effected over a considerable period of time. Shafts or pits were sunk to

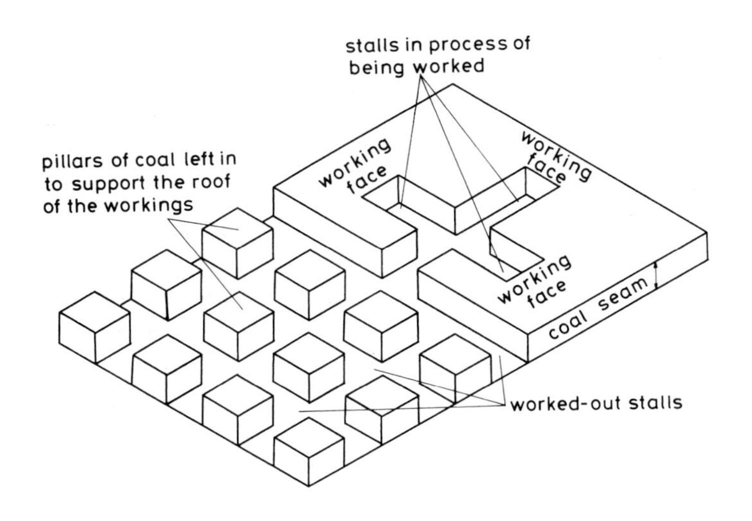

PILLAR AND STALL WORKING (IDEALISED)

40

meet the coal seams, and workings were extended from these points of access in a similar manner to that for adits. Shafts enabled coal to be mined at greater depth, shafts and adits often being worked together to achieve an increased winning. There is evidence that shafts were being sunk at Llanelli in the early 1600s[3] and, although their depth is not known, it is likely that they were shallow — less than 100 feet (30m). The first evidence of deeper shafts in the Llanelli Coalfield relates to 1750 to 1770, when Squire, Evans and Beynon and Chauncey Townsend were sinking to 200 to 240 feet (61 to 73m) in the Bynea region.[4] By c1820, the Pembertons had sunk the Bres pit to about 550 feet (168m),[5] and the Llangennech Coal Company took the St David's pit down to 660 feet (201m), between 1825 and 1832.[6]

(b) *Working the coal*

In outcrop and bell pit workings, all available coal within reach of the workmen was extracted but, when the larger collieries began to use shaft and adit access, it became standard practice to leave coal uncut to support the roof (Fig. 40). This method, termed "*pillar and stall*" working, was probably adopted at Llanelli from the very beginnings of an organised industry, but no confirmation of this has been seen in plans or contemporary references for the 16th and 17th centuries and for most of the 18th century. The first proof is contained in early 19th century plans of workings at Roderick and partners' South Crop Wern Colliery, between c1797 and 1806,[7] which show that approximately two-thirds of the coal was extracted, leaving one-third behind to support the roof. Later, in 1836, Richard Janion Nevill stated that this was the method adopted in relatively shallow workings at the Bryn Colliery, although he also confirmed that back-working of the pillars, with a resulting increased take, had become standard procedure. He wrote "*The seam is 2'-6" thick generally. The pillars are 4 yards square and all the openings 4 yards wide. The first working which is very regular thereby taking out ⅔ of the coal at first & then ½ of the remaining ⅓ by splitting the pillars*".[8] This method yielded an 83% take and it was probably this practice, and not longwall working,* which Nevill referred to, in 1837, when he offered to lend Sir John Morris two or three colliers "*who are used to long work*", to instruct Morris's workmen.[9] Little evidence has survived to suggest that "*pillar robbing*" was prevalent at Llanelli before 1830, although it must have occurred. Later, it became standard procedure for large companies, working in the manner described by Nevill, and for small concerns, operating under sub-lease from the larger companies or from the landowners. (Plate 45)

(2) THE METHODS USED TO DE-WATER THE WORKINGS AND RAISE THE COAL

(a) *De-watering the workings*

The drainage of workings has always been a major problem in coal mining, and Llanelli's early miners soon found that de-watering was necessary if the industry was to advance beyond minor outcrop and bell pit extraction, above the levels of the natural ground water tables. It is probable that the drainage adit or level was the first de-watering method employed at Llanelli. In this method, workings in the region of high ground were drained by means of an adit, inclined slightly downwards, which allowed the water to run away to lower ground (Fig. 41). It was probably used at Llanelli in the 16th century, although no confirmatory evidence has been seen, but it was certainly employed in the Bynea region just before 1627.[10] A century later, collieries in the Bigyn Hill region employed only this method for drainage,[11] and it is likely that it was still used in 1829.[12]

---

* There is no evidence to suggest that longwall working, in which all the coal is removed in a single advancing face with the waste being packed behind, was used in the Llanelli Coalfield before 1830.

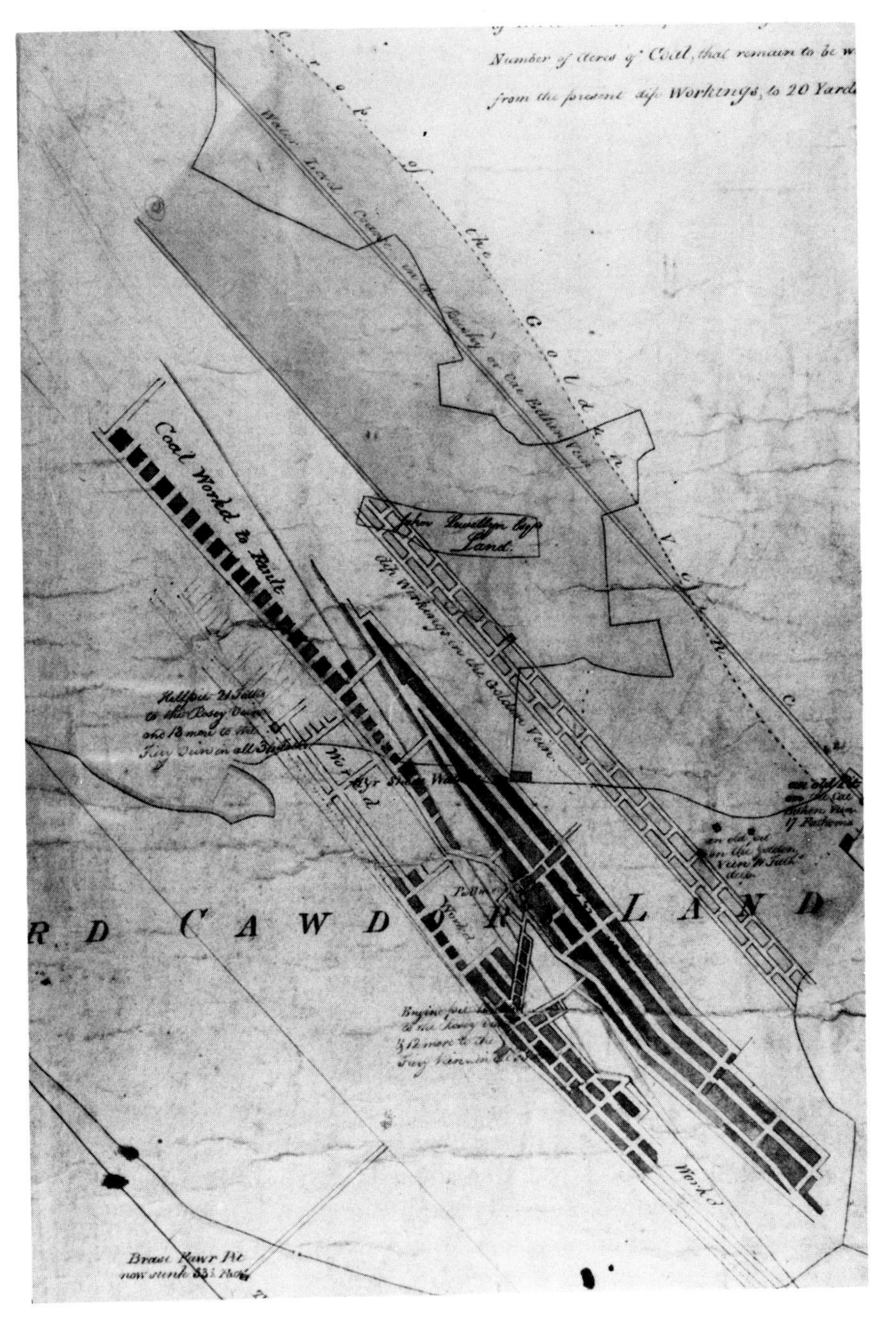

Plate 45 — Plan showing workings in the Golden Vein at the South Crop Wern Colliery between c1797 and 1806. Regular pillar and stall workings are shown. The pillars were actually square but were drawn as rectangles because of dimensional foreshortening, in plan view, in the direction of dip of the seam.

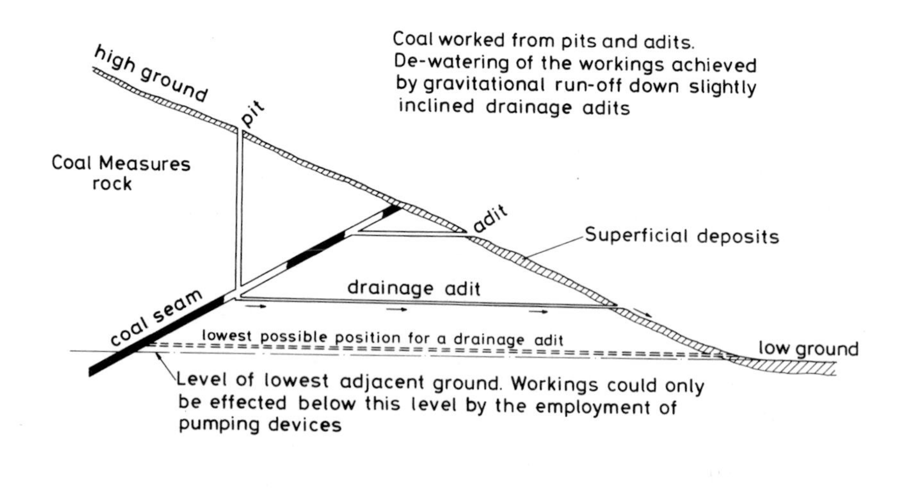

Fig. 41

Drainage adits could not be used if workings were at a lower level than the surrounding ground surface and, to work seams at depth, mechanical means had to be used to raise the water out of the pits. In the early days of the British coal industry, this mechanical effort was provided by wheels impelled by water power (water engines), or gins impelled by horse power (horse engines), which operated chains of buckets or barrels, and later pump rods,* but no description of the methods employed at Llanelli has been seen, and we do not know when these devices were actually introduced. The fact that skilled mining viewers were being brought to Llanelli to supervise the sinking of pits in the early 17th century[13] suggests, however, that winnings were being effected from below the water table at this time, and water engines and horse engines were certainly used in the Coalfield throughout the 18th century.[14] The water engine may still have been occasionally used for this purpose as late as 1829.[15] One ingenious combination of the water and/or horse engine and the drainage adit was employed by Squire, Evans and Beynon, between 1749 and c1780, to dewater workings in the Swansea Four Feet Vein at Bynea. The coal winning pit was some 140 feet (43m) deep, with coal

* The reader should refer to "Annals of Coal Mining and the Coal Trade" by R. L. Galloway, 1st series, or to modern texts on Industrial Archaelogy for a full explanation of water and horse engine construction and operation.

mined down the dip of the seam to a depth of about 200 feet (61m). The water or horse engine was not powerful enough to raise water from this depth, so it was raised part of the way up the coal winning pit to a drainage adit, which took it to a second, shallower pit. Another water or horse engine again raised the water part of the way up the second pit to a second drainage adit, which took it to a third, even shallower, pit. The process was repeated until the water was finally raised to the top of the last pit, or part of the way up it, to a drainage adit which took the water to low ground. This system was termed "*A Water Level drove to the Fiery Vein by Dr Evans*" on a plan of 1772.[16] A diagrammatic representation of its principle of operation is given in Fig. 42.[17]

The water lifting and pumping power of water and horse engines was limited, however, even when used in the manner of Squire, Evans and Beynon's water level. The water engines also needed a steady water supply, which often diminished in the Summer when coal working was at its peak, and the horses (or oxen) which powered the horse engines were expensive to keep. The introduction of Thomas Newcomen's atmospheric steam pumping engines, in 1712, therefore, heralded a new era in Britain's coal industry because they provided the power, reliability and

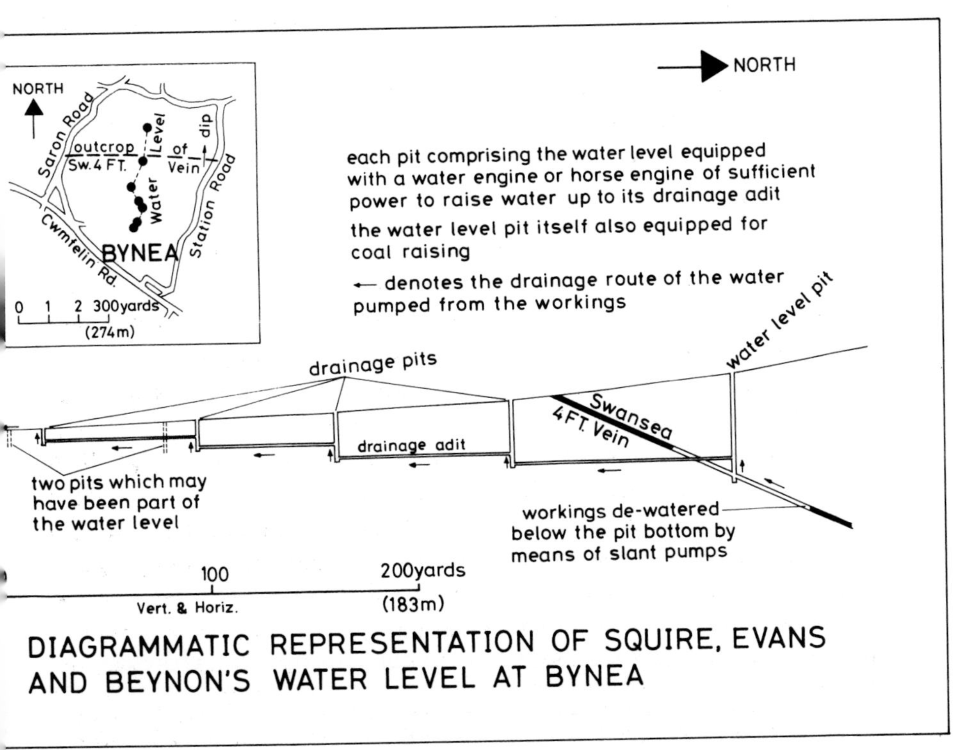

DIAGRAMMATIC REPRESENTATION OF SQUIRE, EVANS AND BEYNON'S WATER LEVEL AT BYNEA

42

economy of operation, which had previously been lacking. Newcomen's steam engines quickly came into general use in Britain's coalfields after 1712, but Llanelli did not immediately adopt them. This was probably because Llanelli's coal industry was controlled by local people from 1700 to 1750, and they could not meet the high cost of installing a steam engine. Consequently, they were not used in the Llanelli Coalfield until c1750, when Squire, Evans and Beynon installed a small pumping engine at the 204 feet (62m) deep Engine Fach pit, at Bynea.[18] Two more steam engines were installed in 1767/69, one at Bynea (by Chauncey Townsend)[19] and the other at Pwll (by Edward Rees),[20] but the failure of Llanelli's first phase of industrialisation prevented its general use in the area between 1770 and 1795. When the sustained second industrialisation was about to commence in 1793/94, at least two of the three early engines were standing idle. No information has been seen about the types of steam engine at Bynea and Pwll but, as they were installed between 1750 and 1769, it is assumed that they were Newcomen atmospheric engines. The steam engine came into general use in Llanelli's coal industry between 1794 and 1829, when at least 12 engines were installed for pumping purposes. By this time, James Watt's condensing engines and Richard Trevithick's high pressure engines were available, and Newcomen's atmospheric engine no longer enjoyed its previous monopoly. All three types of engine were installed to dewater coal workings at Llanelli. Few details of their size or power are known, but the meagre information which has survived is summarised in Table III.

(b) *Raising the coal*

In the early days of Llanelli's coal industry, coal was probably carried out of the pits on workmens' backs, or was raised by simple, manually-operated windlasses, but these methods could only have been used in very shallow pits. When deeper pits were sunk, it became impossible to raise coal manually, so horse engines and water engines were used, these methods probably being introduced in the Coalfield in the 17th century. Coal was raised by water and horse engines throughout the 18th century,[29] both methods carrying over into the 19th century[30] (and even later*), with the steam engine being used mainly for pumping, although the water lifted by the steam engines was probably used to impel coal-raising water wheels. Additionally, some of the steam engines listed in Table III may have been used for both pumping and winding purposes, although no evidence has been seen to confirm this. Nevertheless, the

---

* A horse engine was used for coal raising at the Pant Colliery, in the region of Stradey Woods, as late as 1910. (John Francis Papers, report by S. E. Bowser on the Pant Colliery dated 4 Nov 1910 CR0).

TABLE III — DETAILS OF STEAM PUMPING ENGINES EMPLOYED IN THE LLANELLI COALFIELD BETWEEN c1750 AND 1829

Coalmaster(s)	Colliery	Years of Installation	Surviving details of steam engine
Henry Squire, David Evans and John Beynon	Engine Fach (Bynea)	c1750	None
Chauncey Townsend	Genwen	c1767/69	None
Edward Rees	Pwll	c1768/69	None
David Hughes and Joseph Jones	Baccas	c1795	None
Alexander Raby	Caemain	1798/99	"50 inch cylinder condensing"[21]
Alexander Raby	Caebad	1798/99	"40 horse power atmospheric"[22]
Alexander Raby	Engine Fach (?) (near Cilfig)	1804	4 to 6 horsepower Trevithick. 7 inch cylinder operating at 25lb/in^2 but capable of being operated at 50lb/in^2 [23]
Alexander Raby	Box	1808	"Atmospheric"[24]
Roderick, Bowen and Griffiths	Wern	1805	None
George Warde	Genwen	1806	52 inch cylinder single acting Boulton and Watt operating at 8lb/in^2 [25]
George Warde	Llandaven	1806	4 horsepower[26]
George Warde	Old Castle (?)	1810	None
The Pembertons	Bres	1807/08	"A double-power steam engine of 45 inches diameter"[27]
The Pembertons	Llwyncyfarthwch	c1820/22	None
John Symmons or John Vancouver or Messrs. Davenport, Morris and Rees	A colliery on the Llangennech Estate	prior to 1819	None
Richard Janion Nevill acting for George Warde	Box	1819/23	"52 inch pumping Double Engine"[28]
Martyn John Roberts	Bryn (?)	c1820/21	None
Martyn John Roberts	Pwll	c1825/26	None
Llangennech Coal Company	St David's	c1825/32	None

general picture which emerges is that coal was raised mainly by water engines right up to 1830, with only two known instances of steam engines being used solely for winding purposes. Alexander Raby installed a "*Cylinder Steam Engine*"[31] at the original Box Colliery winding pit (97) in 1807/08, and R. J. Nevill and General Warde installed a "*25 inch Winding Engine*" at the new Box Colliery winding pit (100) in 1822/23.[32]

The water-balance or balance-tub technique for raising coal may also have been used at Llanelli. This method used two cisterns, one at the top of the shaft carrying water, and the other at the bottom of the shaft carrying coal. When the weight of the water cistern exceeded the weight of the coal cistern plus the ropes or chains and the friction of the pulleys and guides, the water cistern descended and the coal cistern was raised to ground level, where it was unloaded. The water was then drained from the water cistern at the shaft bottom, and the system was ready to bring up another load of coal. This method was probably used at one of the pits comprising the South Crop Wern Colliery, because a valuation of its plant and buildings, in 1807, referred to a "*New Water Engine pit, 28 fathoms to the Fiery Vein,*" and said there were "*Two cisterns in the said pit*".[33]

REFERENCES — APPENDIX B

1 Stepney Estate 1105, Bill of Complaint in Chancery, Sir John Stepney Cowell Stepney and others v William Chambers the younger, first filed 1862.
2 Llanelli Public Library plan 3 (Apr 1772).
3 "The Rise of the British Coal Industry" by J. U. Nef, Vol I, p. 417, quoting Star Chamber Proceedings, James I, 155/5.
4 The evidence of Llanelli Public Library plan 3 (Apr 1772) and Mansel Lewis 2565 (1772).
5 We are not certain when the Pembertons reached the Bushy Vein at c550 feet (168m) in the Bres pit. It had only been sunk to the Fiery Vein by 1813 (Thomas Mainwaring, op.cit.), but had been taken down to the Bushy Vein by the time the Pembertons left Llanelli in 1829 (J. Innes, op.cit.).
6 *The Cambrian* (16 Jun 1832).
7 Llanelli Public Library plans 4 — 7 (July 1806).
8 Nevill 1780/81 (6 Aug 1836).
9 "The South Wales Coal Industry" by J. H. Morris and L. J. Williams, p. 60, quoting Nevill MS VIII, copy letter dated 12 Dec 1837.
10 Derwydd 233 (27 Apr 1627) and Carmarthenshire document 397 (14 Jun 1627) (CCL).
11 Stepney Estate 1100 (1 Aug 1729).
12 No specific evidence relating to the use of drainage adits has been seen for the 1800 to 1829 period, but they must have been employed in areas such as Llwynhendy and Bynea where unworked coal occurred in high ground regions.
13 "The Rise of the British Coal Industry" by J. U. Nef, Vol I, p. 417, quoting Star Chamber Proceedings, James I 155/5.
14 Refer to Chapter 3 for the numerous instances of the use of water engines in the 18th century. Llanelli Public Library plan 3 (Apr 1772), provides visual evidence of the large number of these engines employed in a relatively small mining area.
15 A number of the Llangennech Estate Collieries would have needed drainage provisions in 1829, but there was only one steam engine operating in addition to that used to sink the St David's pit.
16 Llanelli Public Library plan 3 (Apr 1772).
17 The representation is based upon the details given in Llanelli Public Library plan 3, on the geological and topographical information on Geological Survey 1 : 10560 sheet SS59NW, on the description of a water level given in "Annals of Coal Mining and the Coal Trade" by R. L. Galloway, First Series pp. 158-9 and by comparison with details of an actual water level operating at Blaenavon in 1824, kindly supplied by Dr Gerwyn Thomas of the Department of Industry at the National Museum of Wales.
18 The date of c1750 was supplied by Benjamin Jones in 1835 (see Thomas Mainwaring, op.cit.). He had obtained the information from his father, Rhys Jones of Loughor, who was a mineral surveyor and colliery viewer at Llanelli before 1802, and some credibility must be attached to the statement. The date certainly corresponds with the known coal leases to Squire, Evans and Beynon and with the known construction of their waggon way to the Engine Fach pit. The depth of 204 feet for the pit was obtained from Mansel Lewis 2565 (1772).
19 Mansel Lewis London Collection 114, letter (5 Sep 1768) and "Report from Committee on Petition of the Owners of Collieries in South Wales 1810", evidence of Henry Smith, M.P.
20 Thomas Mainwaring, op.cit., quoting an unidentified source.
21 Ibid.
22 Ibid.
23 *The Cambrian* (24 Feb 1804).
24 Ibid, (7 Oct 1809).
25 Llanelli Public Library plan 8 (1 Jul 1812).
26 Nevill MS IV, copy letter (27 Apr 1805).
27 *The Cambrian* (16 Jan 1808).
28 Nevill MS XIX (1850-1870).
29 Numerous references to the use of horse and water-engines at Llanelli exist for the 18th century (Refer to Chapter 3).
30 The evidence of Nevill MS X, Cawdor 2/133 (1814), Cawdor 2/110 (c1814), Thomas Mainwaring, op.cit., and numerous other references to the use of horse engines, and more particularly water-engines, at Llanelli between 1800 and 1830.
31 *The Cambrian* (7 Oct 1809).
32 *The Cambrian* (3 May 1823) and Nevill MS XIX (1850-70).
33 Nevill MS IV, copy letter (17 Jun 1807).

APPENDIX C

DEVOLUTION OF THE VAUGHANS' LLANELLY ESTATE

Walter Vaughan (d.1635)
He was the fourth son of Walter Vaughan of Golden Grove (d.1597- See Appendix E). He accumulated a large estate, within and outside the Llanelli area of this study, later to be referred to as "*The Llanelly Estate*".

Francis Vaughan (d.1637),
first son of Walter Vaughan. He died unmarried.

John Vaughan (d.1669),
second son of Walter Vaughan.
He left his estate partly to his wife Margaret, and partly to his son Walter, with the proviso that Margaret's share would pass to Walter on her death.

Walter Vaughan (d.1683),
son of John Vaughan.
Walter died unmarried and pre-deceased his mother. His will specified that his inheritance should pass to his mother and then, in equal shares, to his four sisters — Jemimah, Anne, Mary and Margarett — or their heirs.

Margaret Vaughan (d.1703/04),
mother of Walter Vaughan.
Jemimah Vaughan pre-deceased her mother and her one-quarter share passed equally to her three daughters — Margarett, Dorothy and Rachell. The Llanelly Estate thus passed to six female members of the Llanelli Vaughans and, by the laws of the day, their inheritance also became the property of their husbands. The lands were divided into four equal shares, on 2 August 1705, but the Estate's coal and timber resources were kept intact, to preserve their value. The various beneficiaries and their inheritors, therefore, combined together as joint lessors in granting leases of coal and were retrospectively described as "*The Partners*".

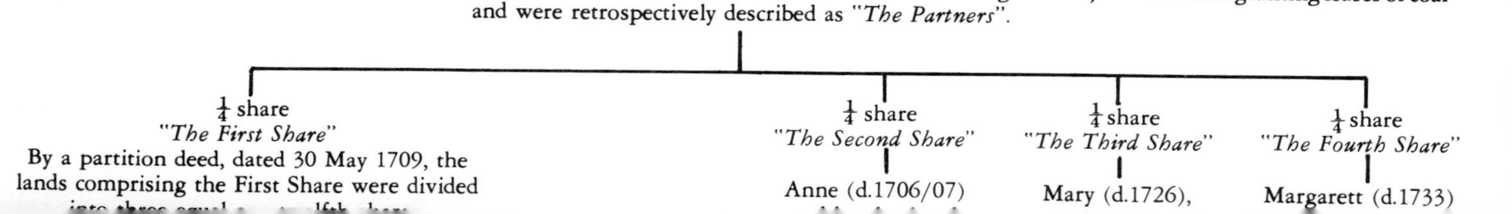

Margarett and her husband Thomas Phillips.

Charles Phillips, first son of Thomas Phillips.

Margaret Phillips, daughter of Charles Phillips.

Rachell and her husband Pendry Vaughan. She left her share to her nephew Richard Phillips.

Richard Phillips, second son of Thomas Phillips.

John Vaughan (d.1804) of Golden Grove. Son of Margaret Phillips and her husband, Richard Vaughan of Golden Grove (d.1781). He came into possession of two-twelfths of the Llanelly Estate. He left his estates to his friend, John Campbell of Stackpole Court, the 1st Lord Cawdor.

John Campbell (d.1821), 1st Lord Cawdor.

John Frederick Campbell (d.1860), first Earl of Cawdor, son of John Campbell, 1st Lord Cawdor.

Dorothy (d.1728), widow of John Parker. She left her share to her nephew Richard Phillips.

Richard Phillips, second son of Thomas Phillips.

Thomas Blackburne Phillips, son of Richard Phillips.

Sarah Blackburne. She sold her share to Admiral William Langdon in 1768.

Jane Langdon, daughter (?) of William Langdon. She left her share to her son, William Chute Hayton, and her niece, Lucinda Hayton.

William and Lucinda Hayton. They sold all, or part, of their share to Richard Pemberton, in 1809.

Richard Pemberton (d.1838). By an indenture, dated 1 May 1823, he assigned his interest in the Llanelly Estate to his three sons — Richard, Ralph Stephen and Thomas.

Richard Pemberton (d.1843).

Ralph Stephen Pemberton (d.1847).

Thomas Pemberton (d.1839).

Stepney.

John Stepney (d.1748), 6th Bart., son of Thomas Stepney, 5th Bart.

Thomas Williams, brother of Sir Nicholas. He sold the Third Share to Thomas Stepney, 7th Bart., in October 1758.

John Stepney (d.1748), 6th Bart., son of Thomas Stepney, 5th Bart.

Thomas Stepney (d.1772), 7th Bart., son of John Stepney, 6th Bart. By his purchase of the Third Share, and by inheritance, he gained possession of three-quarters of the Llanelly Estate.

John Stepney (d.1811), 8th Bart., son of Thomas Stepney, 7th Bart. He left his estate to a succession of named inheritors, with his friends given precedence over his family. By the terms of his will, Richard Henry Alexander Bennett and George James the Earl of Cholmondeley became his joint first heirs.

Richard Henry Alexander Bennett (d.1814), friend of John Stepney. On his death, full ownership of John Stepney's interests passed to George James.

George James the Earl of Cholmondeley (d.1827), friend of John Stepney. On Bennett's death, he gained full possession of John Stepney's interests.

William Chambers (d.1855), friend of John Stepney, 8th Bart., and a named inheritor in his will.

APPENDIX D

OWNERSHIP OF THE STRADEY ESTATE BEFORE 1830

John Mansell (d. c1655)
He married Mary Philips, the daughter of Sir Henry Vaughan of Derwydd, and was the first Mansell to live at Stradey, then part of the Derwydd Estate.

|

Henry Mansell (d. c1677)

|

Edward Mansell (d. 1720/21)
Stradey was granted to Edward Mansell "*for ever*", in 1672/73, by Sir Henry Vaughan of Derwydd. Edward married Dorothy Vaughan in 1684, after she had inherited the Trimsaran Estate from her father Philip Vaughan. Edward was created the 1st Baronet of Trimsaran in 1696/97.

|

Edward Mansell (d. 1754)
the 2nd Baronet of Trimsaran and Stradey.
He died childless, the "*Stradey or Mansell*" Estate, and the baronetcy, passing to Edward Vaughan Mansell, son of his late brother Rawleigh.

|

Edward Vaughan Mansell (d. 1788)
the 3rd Baronet of Stradey.

|

Edward Joseph Shewen Mansell (d. 1798)
the 4th Baronet of Stradey.
He died unmarried and the baronetcy became extinct. The Stradey Estate passed to his sister Mary Martha Ann Margaret Mansell.

|

Mary Martha Ann Margaret Mansell (d. 1808)
She married her first cousin, Edward William Richard Shewen, in 1799. He took the name and arms of Mansell in 1802 but died childless, in 1806. Mary left the Stradey Estate to her solicitor, Thomas Lewis of Llandilo.

|

Thomas Lewis (d. 1829)

|

David Lewis (d. 1872)

APPENDIX E

OWNERSHIP OF THE GOLDEN GROVE ESTATE BEFORE 1830

John Vaughan (d.1574)
he probably built the first Golden Grove Mansion and enlarged the Estate

|

Walter Vaughan (d.1597)

|

John Vaughan (d.1634) created 1st Earl of Carbery in 1628

|

Richard Vaughan (d.1686) the 2nd Earl of Carbery

|

John Vaughan (d.1713) the 3rd Earl of Carbery

|

Anne Vaughan (d.1751)
she married Charles Pawlett who inherited the title of Duke of Bolton on his father's death in 1721/22. Anne became the Duchess of Bolton at that time. She died childless and left the Golden Grove Estate to her distant cousin, John Vaughan of Shenfield.

|

John Vaughan (d.1765)

|

Richard Vaughan (d.1781)

|

John Vaughan (d.1804)
he died childless and left the Golden Grove Estate to his close friend John Campbell of Stackpole Court, the first Lord Cawdor.

|

John Campbell (d.1821) created 1st Lord Cawdor in 1796

|

John Frederick Campbell (d.1860) created 1st Earl of Cawdor in 1827

APPENDIX F

THE MEASURES USED IN LLANELLI'S COAL TRADE BEFORE 1830

Up to 1829, coal quantities in the Llanelli area were almost invariably assessed by volume. Reference to coal quantities by weight (the ton), was being made by the mid 18th century,[1] but all official documentation and legal transactions before 1830 continued to be based on volume measurement.

Coal exports from the area, before 1830, were reported in weys or chaldrons.[2] The wey was a nominal and variable quantity, never measured because of its large size. The chaldron at Llanelli appears to have corresponded to the London Chaldron, and it maintained a more or less constant value throughout the entire period, varying in weight between 1.3 and 1.8 tons.[3]

Within the Coalfield, at least 6 different measures were extensively used before 1830. In order of increasing size these were:- the gallon; Winchester bushel; bag or drag; cart; basket; and wey. With the exception of the wey, these measures corresponded to the sizes of containers used in raising, transporting or selling coal from the pits, or were small measures which allowed check measurements to be made by representatives of the coal lessors or by customs officials. The chaldron does not appear to have been used within the Coalfield and the wey, made up of quantities of the other measures, was, without question, the predominant measure.

The Wey (sometimes referred to as the Way, Weye, Weigh or Wayghe)

The measurement of coal quantities in the first reference seen to coal exports from Llanelli in 1566,[4] and in the earliest known coal leases specifying royalty payments for coal in the first decades of the 17th century,[5] were all based on the wey. It was not superseded until the late 1830s/early 1840s, when the assessment of coal quantities by weight (the ton)* came into general use in the Coalfield.[6]

The wey was a nominal quantity, consisting of a fixed number of smaller measures whose values were assessed by volume. This volume approach meant that the weights of the smaller measures varied, because coal from different seams possessed different unit weights. This was partly due to varying specific gravities, but the chief reason was the

* The Imperial ton = 1.02 tonnes.

manner in which the coal broke up when cut and transported.* Evidence also exists to show that the number of smaller measures to the wey varied, both geographically and with the passage of time, over the South Wales Coalfield[7] and, following this general pattern, Llanelli's wey represented different nominal weights at different times. To clarify this complex situation, the wey used at Llanelli will be examined in terms of its weight in relation to the ton.

The 16th and 17th centuries — no direct evidence relating to the Llanelli Coalfield has been seen to allow the wey's equivalence to be assessed, but previous writers have expressed the opinion that it was equal to 4 tons until the first half of the 17th century, increasing to 5 tons by the early years of the 18th century.[8]

1700 to 1829 — direct evidence is available for the 18th century and the early 19th century, which allows the relationships between the wey and the ton to be established for the Llanelli area. More than one relationship existed, and ranges of values have to be accepted.

Relationship 1

Leases dated 1729,[9] 1733,[10] 1754,[11] 1759,[12] 1768,[13] and 1789[14] specified a wey of 60, 61 or 62 Bags or Baggs or, in the case of the 1759 lease, Bags or Drags. The last four leases equated the Bag to 2 Winchester bushels and the leases of 1759 and 1789, together with other leases of 1765[15] and 1768,[16] equated the Bag to 24 gallons. Leases of 1797,[17] 1799,[18] 1806,[19] and 1819[20] specified a wey of 120 Winchester bushels.

This evidence allows the following relationships to be established for the years between 1729 and 1819:-

1 Wey = 120 to 124 Winchester bushels.

Taking an equivalent weight of 0.042 tons for the Winchester bushel:-[21]

1 Wey = 5.0 to 5.2 tons.

This wey was referred to as the Llanelly wey.

Relationship 2

A lease of 1765[22] specified a wey of 77 Bags of 24 gallons each, and a 1768 lease,[23] to the same person who was working in the Pwll area, referred to a wey of 77 Bags of 2 Winchester bushels of 24 gallons each.

---

* The difficulty of allocating an average equivalent weight to a measured volume of coal was highlighted by a dispute between Thomas Lewis and the Rabys, between 1824 and 1827, when it was found that a standard basket used to raise coal held, on average, 4.3 Winchester bushels of Fiery Vein coal, compared to 6.1 Winchester bushels of Bushy Vein coal, a difference of 42%.[24]

This evidence provides the following relationship for the years between 1765 and 1768, for one particular part of the Llanelli Coalfield:-

1 Wey = 154 Winchester bushels = 6.5 tons.

Relationship 3

Leases of 1749[25] and 1750[26] specified a wey of "*24 loading Carts, the same measure as used at Swansea and Neath*". Leases of 1752,[27] 1762[28] and 1785[29] gave more detail, and specified a wey of 24 carts, "*filled top full*", of dimensions 4'-0" long by 2'-1" broad by 1'-3" deep.

This evidence provides the following relationship for the years between 1749 and 1785:-

1 Wey = 24 Carts of dimensions 4'-0" x 2'-1" x 1'-3" "*filled top full*"

The volume of each cart would be 10.4 cubic feet and, allowing for the fact that they were "*filled top full,*" i.e. heaped up above the top of the cart, an approximate coal volume per cart of some 12.5 cubic feet is probable. Taking an average unit weight of 80 pounds per cubic foot (1281 Kg per cubic m) for bituminous coal,[30] this yields the relationship:-

1 Wey = 10.7 tons.

Although documents subsequently referred to this as the Swansea wey, no use of the term has been seen for the years between 1749 and 1785.

Relationship 4

Leases of 1794[31, 32] and 1802[33] specified a wey of between 216 and 220 Winchester bushels.

This gives the following relationship for the years between 1794 and 1802:-

1 Wey = 216 to 220 Winchester bushels = 9 to 9.2 tons.

This wey was generally referred to as the Swansea wey.

Relationship 5

Evidence exists to show that a wey of 108 Winchester bushels was used in the Llanelli area before 1808,[34] although no lease earlier than 1832[35] has been seen which specified this relationship. The purpose of this measure was to allow the smaller wey used in the Llanelli area to correspond to one half of the Swansea wey.

This allows the following relationship to be established for the early years of the 19th century:-

1 Wey = 108 Winchester bushels = 4.5 tons.

This wey was also referred to as the Llanelly wey.

The wey used in the Llanelli area, therefore, varied between 4 and almost 11 tons between the 16th century and 1829, a number of different equivalent weights being used at the same time. This situation was not as complicated as it seems, however.

Firstly, the wey of 6.5 tons (Relationship 2) can be dismissed because it was unique. No other reference to its use in the Coalfield has been seen, and we do not know why it was used in the Pwll area for a short period only.

The wey generally used seems to have increased from its original value of 4 tons in the 16th and 17th centuries, to 5 tons by the early 18th century. It retained this value throughout the 1700s, and was known as the "*Llanelly wey*" before 1800.[36]

The arrival of industrialists with interests in the Swansea and Neath areas, who brought the measures used in those places with them, introduced a new larger wey after 1749. Initially specified in terms of loading carts, this new wey appears to have been equivalent to 10-11 tons but, by the end of the century, a standardised "*Swansea wey*" of just over 9 tons had been introduced in its place. The Llanelly wey of 5 tons and the Swansea wey of 10-11 tons, and then 9 tons, were used in different parts of the Llanelli area at the same time.

To maintain the 1:2 relationship between the Llanelly and Swansea weys, a new Llanelly wey of 4.5 tons was introduced in the early 19th century, and remained in use until volume measurement of coal was superseded by weight measurement in the 1830s. The new Llanelly wey of 4.5 tons did not completely supersede the Llanelly wey of 5 tons and, between 1800 and 1829, the Llanelly wey of 5 tons, the Llanelly wey of 4.5 tons, and the Swansea wey of 9 tons were all in use at the same time.

The practice of over-measure

The estimates which relate the wey and other measures to the ton were based almost entirely on formal statements contained in leases. Substantial evidence exists, however, to show that coalmasters, or their agents, used the uncertainties inherent in volume measurement, and the complexities caused by the introduction of new sizes of wey, to their own advantage. The extraction and transportation of coal quantities greater than those officially declared, made it possible to evade royalty payments and also reduced transport dues and and customs duties but, perhaps more importantly, it also gave wide scope for mutually profitable negotiations with masters of vessels. The success of the export trade depended on their patronage, and the masters themselves were subjected to the pressures of giving over-measure to merchants or traders at their ports of destination.[37] There is evidence that there was open acceptance, in the Llanelli area, of the practice of giving over-measure, although landowners and customs officials often tried to alter the situation. These considerations bring into doubt the accuracy of all coal export returns which, it may be correct to assume, were greater than those recorded throughout most of the period covered by this work.

The open acceptance of the practice of over-measure is confirmed by many sources, the earliest reference seen for the Llanelli area being in 1752, when Sir Thomas Stepney, 7th Bart., was both a supplier and a shipper of coal. He found that, to develop trade with France, he would have to give over-measure to the importer, which he was unwilling to do. The importer wrote to him "*I see that you accept of my offer concerning the price of the Barrel of Coal, to be delivered up to me at four livres tournois but that you won't give the twenty one for twenty. We can't at this condition agree together nor undertake any dealings in this matter, for it is the Custom of this Country to give twenty one for twenty... no coal is bought by this Government but at this condition*".[38] Coal leases also confirm that all parties accepted the practice of over-measure. A lease granted by John Vaughan to Thomas Bowen, in 1758,[39] stipulated that Bowen would pay the wages of the "*winder*" and the "*planker or bearer*" (haulier), who were to act for Vaughan in checking quantities of coal raised and transported from pits. A similar clause was included in a lease granted by Elizabeth Lloyd to Robert Morgan, in 1759, although it was stated that the man nominated by Lloyd to check coal quantities loaded aboard ship "*is usually paid by the Master of such ship*".[40] The fact that the wages of the men safeguarding the lessor's interests were paid by the people who stood to benefit if the coal quantities raised and exported were underestimated, points to an understanding between lessor, coalmaster and shipper of the need for over-measure. The only other possible explanations for this arrangement would be the acceptance of complete honesty on the part of the nominated winders and hauliers, or extreme naivety on the part of the lessors. Both possibilities are most unlikely in the context of mid18th century Llanelli. There are other references which prove that over-measure continued as an integral part of Llanelli's coal trade, despite the passing of Acts of Parliament between 1812 and 1815 designed to curb this malpractice on a national scale.[41] It was still accepted procedure in the Llanelli area as late as 1841, when evidence given by a local coal agent summarised the situation thus "*Most of the men and boys are paid according to the quantity worked, or by the 'weigh', or 'wey', which 'wey' in the men's account at the Llanelly Collieries contains from six to eight tons, but the 'wey' sold to the shipping is accounted five tons although frequently overweight*".[42]

Although coal lessors must have accepted the practice of over-measure, it is likely that there was an unwritten understanding between lessor, coalmaster and shipper about the allowable margin. This understanding would naturally become strained if the coalmaster experienced difficulties, and could not maintain the level of royalties expected by the lessor, or the degree of over-measure required by the ship's master. This is illustrated by Alexander Raby's relationships with the owners of the Stradey Estate. Before leasing coal to Raby, in 1797, Dame Mary Mansell attempted to define the method to be used to measure the amount of coal raised. Mansell had mined her coal since 1788 and she, and her agents, understood the procedures adopted in the Llanelli Coalfield. She wrote to Raby, in February 1797, "*Each wey of coal is to contain 120 Win. Bushels of Coal, heaped measure, that is, filled only with the shovel, not placing the large coal with hands. Every waggon, cart etc., carrying the said coal to be measured and marked for the quantity they contain by a person appointed by Lady Mansel for that purpose and not to be filled higher than the wood... the coal to be delivered by weight as different parts of the vein do not correspond in weight, a wey of the heaviest coal does not exceed five tons and the general average is only 4 ½ tons per wey*".[43] When the lease was granted in the following month, however, the wey on which royalty payments were based was defined as consisting of 120 Winchester Bushels or 5 ½ tons.[44] Although a clause in the lease referred to the possibility that 120 bushels would not weigh 5 ½ tons, there can be little doubt that Raby had resisted Mansell's attempt to introduce accurate estimation of coal quantities, and had persuaded her to accept the practice of, or need for, over-measure.

Raby's subsequent financial difficulties led to low royalty payments, and relationships with the lessor deteriorated. Thomas Lewis, initially as solicitor to the Stradey Estate and later as its owner, became involved in a long drawn-out dispute with Raby, one major aspect of which was questioning the quantities of coal extracted. In 1808, Lewis accused Raby of loading ships with weys of 120 Winchester bushels whereas, for royalty purposes, the coal was booked in weys of 108 Winchester bushels. Replying, Raby claimed that over-measure was a necessary aspect of the coal export trade and stated "*We have this year sent several Cargoes to Ireland and all the complaints we receive of the Coals is that we do not give as much measure as either Neath, Newport or Swansea which L^{rs}* (letters) *you may see whenever you please*".[45]

Despite this explanation it would seem that, in addition to the normally accepted overfilling of carts and waggons, Raby was also taking full advantage of the confusion resulting from the introduction of the smaller Llanelly wey. If Raby could have paid royalties at a high enough level to satisfy the lessor, it is likely that the dispute over quantities of coal mined would not have occurred, because all parties would have been content to accept the practice of over-measure.

It can, therefore, be concluded that the measures employed for the assessment of coal quantities in the Llanelli area are known with some certainty, and their relationship to the Imperial ton can be obtained to an acceptable degree of approximation. It is considered, however, that the quantities of coal declared for royalty payments and customs duties were usually less than the actual quantities because of over-measure, the principle of which was tacitly accepted by the industry.

REFERENCES — APPENDIX F

1 Stepney Estate 1101, letter from Sir Thomas Stepney to Mr Shepard (7 Sep 1749).
2 "Welsh Port Books 1550-1603" by E. A. Lewis; "The Rise of the British Coal Industry" by J. U. Nef; "The Welsh Coal Trade during the Stuart Period 1603-1709" by B. M. Evans, M.A. Thesis, Un. of Wales 1928; "The Industrial Development of South Wales 1750-1850" by A. H. John; "Report from the Select Committee of the House of Lords appointed to take into consideration the state of the Coal Trade in the United Kingdom", 8 Feb 1830, Appendix I, pp 108-127.
3 J. U. Nef, op.cit, Vol 2 pp. 367-8; A. H. John, op.cit. pp. 187-8; "Report from Committee on Petition of the Owners of Collieries in South Wales 1810", evidence of H. Smith, p. 5; "Report on the State of the Coal Trade", 23 June 1800, evidence of A. Raby; "Report from the Select Committee of the House of Lords etc, . . ." 8 Feb 1830, evidence of W. Dickson, W. Brandling and Captain Cochrane.
4 E. A. Lewis, op.cit.
5 J. U. Nef, op.cit., Vol I, p. 324, quoting Star Chamber proceedings, James I 155/5; Derwydd 689 (29 Sep 1618).
6 Legislation was introduced, in 1831, specifying that all coal duties would have to be assessed by weight (1 and 2 William IV c16) but it was at least 10-15 years before the wey ceased to be used in the Llanelli area.
7 A. H. John, op.cit., Appendix D; J. U. Nef, op.cit., Appendix C.
8 A. H. John, op.cit., quoting Jersey Estate Archives, letter of the Steward dated 16 May 1717; J. U. Nef, op.cit., Vol 2, p. 373.
9 Stepney Estate 1100 (1 Aug 1729).
10 Cawdor (Vaughan) 13/372 (3 Sep 1733).
11 Cawdor (Vaughan) 21/623 (28 Jun 1754).
12 Carmarthenshire lease 118 (1 Oct 1759) (CCL).
13 Stepney Estate 1099, summary of lease dated 21 Nov 1768.
14 Castell Gorfod 99 (22 May 1789).
15 Stepney Estate 1094 (28 Jul 1765).
16 Cawdor (Vaughan) 114/8430 (11 Jan 1768).
17 Mansel Lewis 480 (25 Mar 1797).
18 Nevill 1738 (17 Jun 1799).
19 Mansel Lewis 1928 (c1805/06).
20 Nevill 410 (18 May 1819).
21 Alexander Raby stated in 1808 that, in the Llanelli area, 108 bushels were equal to 3 London Chaldrons. The London Chaldron was generally accepted to weigh 1 ½ tons, giving 1 bushel = 0.042 tons. (Mansel Lewis 1637, letter dated 28 Dec 1808). This evidence is confirmed by J. U. Nef, op.cit., Appendix C, p. 369 and A. H. John, op.cit., p. 188. The allocation of a weight to a volume of extracted coal can only be treated as an average value, however, and actual weights would vary from seam to seam.
22 Stepney Estate 1094 (28 Jul 1765).
23 Cawdor (Vaughan) 114/8430 (11 Jan 1768).
24 Mansel Lewis 206 (16 Jan 1824); Mansel Lewis London Collection 10 (29 Dec 1825); Mansel Lewis London Collection 13 (5 Jun 1827).
25 Stepney Estate 1099, summary of lease dated 13 Dec 1749.
26 Mansel Lewis 58 (21 Feb 1750).
27 Nevill 19 (24 Oct 1752).
28 Nevill 25 (2 Aug 1762).
29 Nevill 27 (27 Jun 1785).
30 "Report from Select Committee of the House of Lords etc . . .", 8 Feb 1830, evidence of John Buddle, p. 44, gave the unit weight of bituminous coal as being between 76 to 80 lbs. per cubic foot. Appendix 24, p. 390, in the same report, gave a unit weight of 82.3 lbs per cubic foot for *"Pemberton's Llangennech"*, which was steam coal mined in the Llanelli area.
31 Cawdor (Vaughan) 21/615 (1 Oct 1794).
32 Cawdor (Vaughan) 8112 (1 Oct 1794).
33 Carmarthenshire lease (2 Oct 1802) (CCL).
34 Mansel Lewis 1637, letter from A. Raby to T. Lewis, dated 28 Dec 1802.
35 Mansel Lewis London Collection 10, proposed lease dated 8 May 1832.
36 A lease of 1793 (Nevill 363 dated 7 Oct 1793) contained the first reference that has been seen to this terminology. It specified a coal royalty of 2s 3d per *"Llanelly or Flats wey"* or 4s 6d per *"Canal or Swansea wey."* The Flats referred to was certainly Llanelli Flats but the Canal may have been the Swansea Canal, or the Yspitty Canal which had always used the larger wey.
37 A. H. John, op.cit., p. 189, mentioned some of these points and considered that the need to attract export trade vessels by giving over-measure was the main reason for malpractice in the measurement of coal quantities.

38 Cilymaenllwyd 217, letter from an un-named person at Brest to Sir Thomas Stepney, dated 12 Apr 1752.
39 Cawdor (Vaughan) 21/624 (22 Apr 1758).
40 Carmarthenshire lease 118 (1 Oct 1759) (CCL).
41 52 Geo III c9; 55 Geo III c18; 56 Geo III c35.
42 Children's Employment Commission — Appendix to First Report of Commissioners — Mines, Part II, p. 695 (1842).
43 Mansel Lewis London Collection 97, copy letter dated 16 Feb 1797.
44 Mansel Lewis 480 (25 Mar 1797).
45 Mansel Lewis 1637, copy letter from T. Lewis to A. Raby, dated 28 Dec 1808, and letter from A. Raby to T. Lewis, of the same date.

APPENDIX G

ESTIMATED TOTAL COAL EXPORTS FROM THE LLANELLI AREA 1550 TO 1829

(i) *Explanatory Notes*

(a) Between 1551 and 1561, J. U. Nef's estimate of an average of 300 tons per annum from "*Llanelly and Burry*" has been adopted.[1] The official exports from the Llanelli area alone would have been less than this value.

(b) Between 1566 and 1603, E. A. Lewis's estimates, based on study of the Welsh Port Books, have been adopted.[2] Nef's estimates for Llanelly and Burry have been taken as a secondary source for these years.[3]

(c) Between 1604 and 1709, B. M. Evans's estimates, based on the Welsh Port Books, have been adopted.[4] Up to 1684 the Port Books recorded exports from North Burry and South Burry separately but, in that year, the two ports were merged for customs' purposes. After this, the combined foreign exports from North Burry and South Burry were recorded under North Burry alone, although coastwise exports continued to be recorded separately for North Burry and South Burry. It is considered that a reasonable approximation to the total exports from North Burry, between 1684 and 1709, can be achieved by adding North Burry's coastwise exports to one-half of the combined foreign exports for North and South Burry (throughout the period coastwise exports significantly exceeded foreign exports). Nef's estimate for Llanelly and Burry have been taken as a secondary source between 1604 and 1711.[5]

(d) For 1744 to 1746, 1772 to 1777, 1788 to 1800 and 1804 to 1815 estimation of the total export quantities has been based upon the number of ships sailing from Llanelli with coal. This estimation entails the allocation of an average tonnage to the coal ships frequenting Llanelli in different years. This can only be an approximation, but the following average tonnages are considered to be reasonable for the periods:-

1744 to 1746	—	30 tons average
1772 to 1777	—	35 tons "
1788 to 1799	—	35 tons "
1800 to 1815	—	40 tons "

The allocation of an average of 30 tons just before 1750 is based on the fact that coal ships operating around the South Wales coast up to 1700 were between 10 and 60 tons.[6] Llanelli's harbour facilities were undeveloped at this time and it is doubtful if large ships would have regularly frequented its rivers and pills. This state of affairs continued until c1750, when the first phase of Llanelli's industrialisation commenced and Sir Thomas Stepney tried to develop its shipping trade. As this industrialisation was not sustained, and harbour facilities were largely undeveloped, the class of ship frequenting Llanelli probably increased only slightly, so an average capacity of 35 tons has been allocated up to 1799. Between 1800 and 1815, the docks and the navigation of the Estuary were significantly improved as Llanelli's sustained second phase of industrialisation got under way, so it is assumed that vessels then averaged 40 tons. This assumption is verified by two references. Firstly, official returns of coal exports for 1816-20, divided by the number of ships leaving Llanelli with coal in that period, gives an average capacity of 41 tons.[7] Secondly, following improvements to docks and Estuary and the completion of the breakwater, it was said, in 1828, that "*vessels of from 40-50 tons were formerly considered large for the Flats while there are now loading there vessels of 450 tons*".[8]

(e) Between 1816 and 1828 there are official returns of shipments of coal from Llanelli.[9] No return is given for 1829, when a period of deep industrial recession started, but it is probable that exports were less than in 1828.

(ii) *Estimated coal exports from the Llanelli area*

w = weys, taken as being equivalent to 4 tons weight in the 16th and 17th centuries.

ch = chaldrons, taken as being equivalent to 1.4 tons weight in the 16th, 17th and early 18th centuries.

The figures in brackets are estimated coal export quantities given in tons. (Table IV).

TABLE IV — COAL EXPORT QUANTITIES

Years	Primary source[10]	Secondary source[11]
1551 - 1561		300 tons per annum.
1566	2½ w (10 tons)	
67	67ch (94)	
72	8w (32)	
86	26w (104)	
1591 — 1600	—	400 tons per annum.
1593	29w (116)	
94	19w (76)	30w (120)
99	57w (228)	
1602	86w (344)	
03	119w (476)	125w (500)
04	21ch (30)	153w (612)
05	38ch (53)	91w (364)
06	—	
07	45ch (63)	100w (400)
08	51ch (71)	223w (892)
09	38ch (53)	66w (264)
10	80ch (112)	
12	—	
13	—	68w (272)
18	334ch (468)	
19	—	200w (800)
20	—	
21	—	351w (1404)
22	—	400w (1600)
23	152ch (213)	
26	—	
27	—	10w (40)
28	24ch (34)	145w (580)
29	8ch (11)	
30	—	450w (1800)
31	—	14w (56)
33	—	
34	—	6w (24)
35	—	102w (408)
36	75ch (105)	351w (1404)
47	8ch (11)	

Years	Primary source	Secondary source
No Port Books for the Protectorate 1649 to 1659 inclusive.		
1660	210ch (294)	
61	433ch (606)	12ch (17)
62	—	401ch (561)
63	—	266ch (372)
64	—	692ch (969)
65	—	337ch (472)
66	—	
67	—	3433ch (4806)
68	742ch (1039)	
69	825 ch (1155)	650ch (910)
70	168ch (235)	
71	943ch (1320)	
72	1092ch (1529)	
73	727ch (1018)	742ch (1039)
74	372ch (521)	
75	1480ch (2072)	
78	923ch (1292)	
79	1273ch (1782)	
80	—	
81	—	
82	—	679ch (951)
83	—	475ch (665)
84	1479ch (2071)	
85	5290ch (7406)	
86	5512ch (7717)	
87	5579ch (7811)	
88	5525ch (7735)	7561ch (10,585)
89	2521ch (3529)	
90	3651ch (5111)	
91	2285ch (3199)	
92	1911ch (2675)	
93	291ch (407)	
94	2301ch (3221)	
96	2970ch (4158)	
97	1120ch (1568)	
98	1810ch (2534)	
1701	1788ch (2503)	
02	889ch (1245)	
03	555ch (777)	
04	938ch (1313)	
05	763ch (1068)	
06	974ch (1364)	
08	1641ch (2297)	
09	1777ch (2488)	
10	—	893ch (1250)
11	—	2332ch (3265)

Years	Primary source
1744-45[12]	144 ships @ 30 tons (4320)
45-46	144 " " " (4320)
72-73	398 ships @ 35 tons (13930)
73-74	339 " " " (11865)
74-75	428 " " " (14980)
75-76	411 " " " (14385)
76-77	388 " " " (13580)
88-89	267 " " " (9345)
89-90	272 " " " (9520)
90-91	284 " " " (9940)
1793[13]	430 " " " (15050)
94	411 " " " (14385)
95	458 " " " (16030)
96	468 " " " (16380)
97	370 " " " (12950)
98	286 " " " (10010)
99	343 " " " (12005)
1800	298 ships @ 40 tons (11920)
04[14]	458 " " " (18320)
05	409 " " " (16360)
06	417 " " " (16680)
07	352 " " " (14080)
08	590 " " " (23600)
09	835 " " " (33400)
10	889 " " " (35560)
11	927 " " " (37080)
12	1091 " " " (43640)
13	1148 " " " (45920)
14	— —
15	1290 ships @ 40 tons (51600)
16[15]	(41755)
17	(60486)
18	(49287)
19	(58101)
20	(60427)
21	(69187)
22	(71718)
23	(63703)
24	(70629)
25	(73248)
26	(68400)
27	(70822)
28	(71345)

REFERENCES — APPENDIX G

1 "The Rise of the British Coal Industry" by J. U. Nef, Vol I, p. 53.
2 "The Welsh Port Books 1550-1603" by E. A. Lewis.
3 J. U. Nef, op.cit., Vol II, Appendix D (ii).
4 "The Welsh Coal Trade during the Stuart Period 1603-1709" by B. M. Evans, Appendix A, M.A. Thesis, Un. of Wales 1928.
5 J. U. Nef, op.cit., Vol II, Appendix D (ii).
6 B. M. Evans, op.cit.
7 "*Ship News*" in *The Cambrian* newspaper 1816 to 1820 and "Report of the Commissioners appointed to inquire into the Several Matters relating to Coal in the United Kingdom" 1871, Vol III, Report of Committee E.
8 *The Cambrian* (20 Sep 1828).
9 "Report of Commissioners etc" 1871.
10 The primary source for coal export quantities has been taken as follows:- 1566 to 1603, E. A. Lewis, op.cit.; 1604 to 1709, B. M. Evans, op.cit.; 1744 to 1746, 1772 to 1777 and 1788 to 1791, Cawdor (Vaughan) 21/614; 1793 to 1800, "Hanes Llanelli" by D. Bowen; 1804 to 1815, "*Ship News*" in *The Cambrian* newspaper; 1816 to 1829, "Report of Commissioners etc" 1871.
11 J. U. Nef, op.cit., has been taken as a secondary source for coal export quantities between 1551 and 1711 because his estimates, based on study of the Welsh Port Books, cover certain years which were not reported by E. A. Lewis and B. M. Evans. Nef's returns were for both North and South Burry and therefore over-estimate the coal exports from North Burry alone.
12 The vessels listed in the keelage returns in Cawdor (Vaughan) 21/614 between 1744 and 1791 have been used to estimate coal export quantities. The percentage of ships engaged in the coal trade during this period cannot be assessed, although it is suspected that most of them would have carried coal.
13 It has been assumed that 90% of the sailings between 1793 and 1800 listed in "Hanes Llanelli" by D. Bowen were coal sailings. Llanelli was undergoing its real industrialisation at this time and many ships may have sailed in with required commodities and left in ballast.
14 Ships listed in *The Cambrian* between 1804 and 1815 which sailed out loaded with culm (anthracite) have been omitted from the estimates. These were the only years (before 1840) when significant quantities of anthracite were shipped from Llanelli because the Carmarthenshire Railway, the only link with the anthracite coalfield, soon became moribund.
15 The 1871"Report of the Commissioners etc" listed coal and anthracite exports together for the years 1816 to 1818 and separately from 1819 onwards. The coal quantities for 1816 to 1818 have been derived by assuming that 7% of the listed exports were anthracite. This assumption is based on the fact that anthracite exports averaged 7% of the total between 1819 and 1821.

APPENDIX H

ACCIDENT RECORDS BEFORE 1830

The number of deaths and injuries in Llanelli's coal mines before 1830 will never be known because records were seldom kept. Up to 1829, colliery owners were not required to submit official returns of fatal accidents* and their correspondence seldom mentioned the existence of a work-force, let alone the hardships it faced. Even the emergence, in 1804, of a newspaper (*The Cambrian*) reporting local happenings produced little in the way of accident reports. Consequently, reference to published sources has yielded proof of only 22 fatalities in Llanelli's coal mines between 1754 and 1829. A summary of these known fatal accidents, together with a number of other accidents which may have caused deaths, is given in Table V.

There is no doubt that the summary bears no relationship to the actual loss of life which occurred, and this is confirmed by evidence, given to the Children's Employment Commissioner, in 1841, by a Llanelli colliery agent, who said "*Within the last 25 years I recollect about 24 fatal accidents at one works*".** The accident reports in the published sources between 1816 and 1841 give no indication that so many fatalities could have occurred in a single colliery. When it is considered that a large number of collieries, many with extensive workings after 1749, operated in the Llanelli Coalfield before 1830, then the probable loss of life must have been high. Even allowing for the fact that Llanelli's coal industry underwent an unprecedented expansion between c1795 and 1830, the total number of deaths due to colliery accidents must have run into many hundreds.***

Non-fatal accidents were seldom mentioned before 1830 and yet they often resulted in crippling disabilities and a shortened life. The extent of these accidents in Llanelli's coal mines will never be known but one particularly distressing aspect was the fact that children, sometimes as young as 5 years of age, were exposed to this form of danger, before and after

---

* The Coal Mines Act of 1850 first compelled colliery owners to notify the Home Secretary of all mining fatalities.

** Children's Employment Commission, Appendix to First Report, Mines Part II (1842), report by R. W. Jones, p. 695.

*** This conclusion is supported by the official returns made by colliery owners after the passing of the Coal Mines Act of 1850. 100 deaths by accident occurred at the collieries within the area covered by this work in the 13 years between 6 January 1851 and 21 April 1864.

1830. The tragedy of this situation was summarised by two surgeons at Llanelli (D. A. Davies and T. L. Howell) in their evidence to the Children's Employment Commissioner (Rhys Jones of Loughor) in 1841 when they spoke of the hazards of accident and loss of health due to conditions in the mines: "*In this neighbourhood it is customary for children under 10 years of age to be employed in the coal-mines, and their physical condition appears to us to be somewhat deteriorated by their being put to work at so early an age, as well as by the frightful accidents to which their inexperience necessarily exposes them; and they are not unfrequently, from the nature and extent of injuries received at this age, maimed for life. Their form and health we believe to be influenced by the nature of their occupation and mode of life; as regards the former, an evident smallness of stature being the consequence, the frame being frequently ill-formed and cramped. The health is generally delicate, and a sickly appearance marks the inhabitants of the mine*".*

* Children's Employment Commission, Appendix to First Report, Mines Part II (1842). Report upon the physical condition of the children of the district of Llanelly, Carmarthenshire by D. A. Davies and T. L. Howell dated 9 Aug 1841.

TABLE V — ACCIDENTS IN LLANELLI'S COAL MINES BEFORE 1830

Date	Colliery	Person(s) killed or injured*	Cause of death or injury and remarks	Reference Source
1754	Un-named colliery worked by Chauncey Townsend	William Bowen	Not known	"Llanelly Parish Church" by A. Mee, p. 97
1783	Un-named pit located on marshlands	Thomas Morris and David Bowen	"smothered in a coal-pit"	A. Mee, op.cit., p. 97 and *The Cambrian* (12 Jul 1805)
3 Jul 1805	,,	Jenkin Stephen	"suffocated by the damp in a coal-pit only 4 fathoms deep" (the same location as the 1783 accident)	*The Cambrian* (12 Jul 1805)
19 May 1808	Bres	Francis Richards	fell down pit	*The Cambrian* (21 May 1808)
Dec 1810	Hogspit, Cefen near Bynea	David Rees	fall of roof	Coleman 332 (1810) (NLW)
29 Mar 1813	Un-named colliery	2 un-named colliers	"lost their lives by unfortunately getting enveloped in fire-damp"	*The Cambrian* (3 Apr 1813)
1816	Plas Issa	A number of un-named colliers (not known if any deaths resulted)	"several men dreadfully burned" by an explosion of firedamp	"Annals of coalmining and the coal trade" by R. L. Galloway, First Series, p. 406.
c1816	Un-named colliery	Un-named person	"fell down ladders at pit"	Appendix to First Report of Children's Employment Commission, Part II, p. 695
c1816	,,	,,	"lost his way in the dark and died before he was found"	,,
Dec 1816	,,	3 or 4 un-named persons (not known if any deaths resulted)	entombed by pit falling in	*The Cambrian* (7 Dec 1816)

28 Aug 1820	Box	1 un-named person killed 1 un-named person injured	fall of roof	*The Cambrian* (2 Sep 1820)
Sep 1821	Un-named colliery	William Williams	fell down pit; the rope by which the workmen were lowered down the pit having been partly cut by some unknown person	*The Cambrian* (6 Oct 1821)
May 1822	Un-named colliery worked by General Warde	Un-named person	"two men quarrelling a coal of above two hundred weight fell upon the head of one of them and literally dashed his brains out."	*The Cambrian* (1 Jun 1822)
18 Sep 1826	Caemain	1 lad killed 7 un-named persons injured	explosion of firedamp	*The Cambrian* (30 Sep 1826)
26 Mar 1827	Box	Un-named person	fall of roof	*The Cambrian* (7 Apr 1827)
c1829	Un-named colliery	2 un-named persons	coal-carrying basket fell on them	Appendix to First Report of Children's Employment Commission, Part II, p. 695
c1829	Un-named colliery	2 un-named persons	fall of roof	,,
c1829	Un-named colliery	3 un-named persons	fall of a basket during the sinking of a new pit	,,

* All listed accidents fatal unless a statement to the contrary given.

APPENDIX I

KNOWN COAL MINING ACTIVITY BEFORE 1830

The plan is a reduced scale composite formed from 1:10560 Ordnance Survey plans SN40SE, SN50SW, SN50SE, SS49NE, SS59NW and SS59NE. The following features are superimposed over the Ordnance Survey detail:-

1. 231 pits and adits known to have been worked in the Llanelli area before 1830. Descriptions of entries and the reference authorities for their locations are given in the accompanying schedule. Adits have been distinguished from pits wherever possible, but early plans seldom differentiated between them, and many pits shown at seam outcrops may well have been adits. It is stressed that some of the locations given are approximate, because old plans often contained cartographic errors, particularly of orientation, making positional translation on to modern plans difficult. The final locations shown can, therefore, vary from very accurate to somewhere within 80 feet (25m) of the exact position.

It is known that the total number of pits and adits worked before 1830 was significantly in excess of the 231 shown. The locations of the earliest entries will never be known and many old plans, which would have allowed pits to be located, have not survived to the present day. Additionally, numerous old, undated and un-named entries, shown on post-1830 plans, have not been included in this work, although many of them were probably worked before 1830. (Over 600 separate entries have been located for the area studied up to 1947, but the date of sinking of the majority is not known with any certainty. Many examples have survived at Llanelli of 18th and early 19th century entries being incorporated into later colliery complexes, with the result that they remained in use until the early 1900s).

2. Coal seam outcrops and geological faults which governed the pit and adit sitings.

3. Canals, railways, shipping places and metalworks constructed before 1830.

APPENDIX I

(i) SCHEDULE OF PITS AND ADITS SHOWN ON PLAN

The following abbreviations are used:-

AP — abandoned mine plan held by the National Coal Board.
LPL — Llanelli Public Library plan.

Entry No.	Grid Location	Authority for location	Description
1 to 3	SN475009	AP 1175 (c1880)	These pits could have been worked by Edward Rees and Martyn John Roberts. They were the main pits to the Pwll Colliery on abandonment c1880.
4	SN487013	Mansel Lewis 2538 (1805)	Un-named "Col pit" adjacent to Dulais Mill. Possibly worked by Alexander Raby and Richard Williams.
5	SN500039	Mansel Lewis 2590 (1822)	"Pit on Trevenna, 6 fthms deep".
6 to 9	SN507039 SN509039	"	"Trial pits on the Trevenna Vein".
10 to 13	SN505037 SN508037 SN510037	"	"Trial pits on the Pantylliedyfawr Seam".
14 to 17	SN508035 SN507035	"A Map of Hengoed 1744" (NLW).	"Old Collierys" in "Ka Pant Cynan".
18	SN516036	Mansel Lewis 2590 (1822)	"Opening made into a small seam of coal".
19	SN517032	"	"Coal level now at work for private purposes".
20	SN503025	"	"Coal level in Cwm Trebeddod".
21	SN516024	"	"Trial pit sunk by Messrs. Driver to coal" and "Coal pit 3 fthm".
22	SN504020	"	"Brickyard Level".
23	SN505012	"Old Llanelly" by J. Innes and Llanerch Building plan (1919) at Llanelli Town Hall.	"Slip" on map of 1809 in "Old Llanelly" and "Adit Level from Bushy Vein" on 1919 plan. Considered to be Alexander Raby's Old Slip Colliery.

Entry No.	Grid Location	Authority for location	Description
24	SN509014	LPL 9 (1814) and Llanerch Building plan (1919).	Branch line from the Carmarthenshire Railway to this point on the 1814 plan. Called "Pentre-porth" Colliery on 1919 plan. Considered to be Alexander Raby's New Slip Colliery.
25	SN515014	LPL 10 (c1822)	Un-named pit near Penyfinglawdd Farm.
26	SN519014	"	Un-named pit.
27	SN520015	"	Un-named pit.
28	SN522015	Mansel Lewis London Coll. 60 (1800).	Un-named "Coal pit".
29	SN518011	Nevill MS XVIII (1836).	General Warde's Penygare Colliery.
30	SN513009	LPL 10 (c1822)	Un-named pit.
31	SN513009	"	Un-named pit.
32	SN508008	LPL 9 (1814)	Un-named pit (in Caerhalen).
33	SN505009	"	Un-named pit (at Cilfig).
34	SN502008	Mansel Lewis 1966 (1825).	"Horseway" on the outcrop of the Golden Vein. Probably used to take horses into the Caemain Colliery workings.
35	SN501007	LPL 10 (c1822); Mansel Lewis 1966 (1825).	"Engine Fach" on LPL 10. "Engine House" on Mansel Lewis 1966. Possibly the site of Alexander Raby's Trevithick steam engine.
36	SN499008	Mansel Lewis 1966 (1825); LPL 498 (c1828); first edition of O.S. 1:10560 plan Carm. LVIII NW (1891).	"Old pit" on Mansel Lewis 1966. "Furnacė" on LPL 498 "Caerelms pit" on Ordnance Survey plan. This information suggests that the Caerelms pit was an upcast furnace ventilation pit to Raby's colliery complex.
37	SN494007	Mansel Lewis 1841 (1825).	Un-named "old pit".
38 and 39	SN499006	Mansel Lewis 1966 (1825).	Caebad Colliery pits.
40	SN498005	Mansel Lewis London Collection 42 (1821)	Un-named pit.

Entry No.	Grid Location	Authority for location	Description
41 to 45	SN 502004/5 SN503005	LPL 9 (1814); AP 2341 (1889)	Pits to Raby's Caemain Colliery. On final abandonment of the colliery in 1889 only pits 44 and 45 were still in use.
46	SN502005	Cawdor (Vaughan) 8660 (1787)	"Landing pit" worked by a water engine. This is considered to be Charles Gwyn's water-engine pit. It is possible that Entry 46 was not a separate pit but was, in fact, either Entry 43 or 44, orientation errors on the 1787 plan giving a false location.
47	SN505006	LPL 10 (c1822)	Un-named pit (in Weinelly).
48	SN505007	LPL 9 (1814)	Un-named pit (in Caeperson).
49	SN508002	LPL 5 (1806); AP 2341 (1889)	"Bresfawr pit" on LPL 5. "Bres pit" on AP 2341.
50	SN507001	LPL 10 (c1822)	Un-named pit. Possibly a winding pit worked by the Pembertons in conjunction with the Bres pumping pit.
51	SS499996	Mansel Lewis London Collection 69 and 87 (1846)	Un-named "old coal pit".
52	SS500997	"	Un-named "old water wheel" pit.
53	SS500997	LPL 10 (c1822)	Un-named pit.
54	SS500998	Mansel Lewis London Collection 69 and 87 (1846)	Un-named "Old Coal pit" from which John Allen drove a westward heading in the Golden Vein.
55	SS500998	"	Un-named "old coal pit".
56 to 58	SS501998	Mansel Lewis 2578 (c1813)	Un-named pits.
59	SS501997	Mansel Lewis London Collection 69 and 87 (1846)	Un-named "Old Coal pit" known to have been worked by Thomas Bowen.
60 and 61	SS502998	Mansel Lewis 2578 (c1813)	Un-named pits.
62	SS503998	Mansel Lewis London Collection 41 (c1820)	Un-named pit.
63 to 65	SS503999	LPL 9 (1814); Stepney Estate Office plan 46 (1814)	3 pits constituting General Warde's original "Old Castle Colliery".

Entry No.	Grid Location	Authority for location	Description
66	SS504999	Mansel Lewis 2578 (c1813)	Un-named pit.
67 to 69	SS504997 /8 SS505998	Mansel Lewis London Collection 41 (c1820)	Un-named pits.
70	SS505999	Mansel Lewis 2578 (c1813)	Un-named pit.
71	SS505999	Mansel Lewis 2578 (c1813)	Un-named pit, but considered to be Pwllycornel worked by Roderick & Co.
72	SS507998	Mansel Lewis London Collection 57 (1800)	Un-named pit in "Cae'r Pulpit". Considered to be Pwll Cae Pulpud worked by Roderick & Co.
73	SN508000	LPL 5 (1806); LPL 15 (c1822)	"Engine pit" on LPL 5. "Wern Colliery" on LPL 15.
74	SS508999	Mansel Lewis 2510 (1808)	Un-named pit.
75	SS508999	LPL 5 (1806)	Pwllmelin (pit).
76	SS508998	"	"An old pit on the Golden Vein, 11 fathoms deep".
77	SS508998	"	Un-named pit.
78	SS508998	"	Un-named pit.
79	SS508997	"	"Old pit on the Cae Eithen Vein, 17 fathoms" also "water wheel" pit.
80	SS510998	LPL 5 (1806); AP 2341 (1889); O.S. 1:10560 plan Carm. LVIII NE (1891)	"pit to Golden Vein" on LPL 5 "Bigyn pit" on AP 2341 "Tregob Colliery" on O.S. plan.
81	SS510999	LPL 10 (c1822)	Un-named pit.
82	SN510000	LPL 5 (1806)	Hill pit "204ft to Fiery Vein".
83 to 85	SS511999	LPL 10 (c1822)	Un-named pits.
86	SN512000	"	Un-named pit.
87	SN512000	LPL 6 (1806)	Un-named pit.
88	SS515999	LPL 5 (1806)	"An old pit on the Fiery Vein".
89	SS506982	Nevill 800 (1848)	Un-named pit but considered to be the unsuccessful sinking by the Llanelly Copperworks Company in 1813.
90	SS519999	LPL 10 (c1822)	Un-named pit.

Entry No.	Grid Location	Authority for location	Description
91	SN520000	LPL 10 (c1822)	Un-named pit.
92	SN522000	LPL 1 (1751)	Information superimposed post-1751 shows a pit described as: "Mr Bowen sunk this pit, coal small being on the crop".
93	SN518002	Local Collection 252 (1828); AP 5244 (1908)	"Llwyncyfarthwch or Park pit" on Local Collection 252. Retrospectively termed "Pemberton pit" on AP 5244.
94	SN513008	Stepney Estate Office plan 46 (1814); AP 2454 (1890)	"Messrs. Pembertons' New pit" on the 1814 plan. An air pit to the Talsarnau Colliery on the 1890 plan.
95	SN513007	,,	Un-named pit on the 1814 plan. "Talsarnau Colliery" on the 1890 plan.
96	SN513007	LPL 10 (c1822); Stepney Estate Office plan 46 (1814)	Un-named pit on both plans.
97	SN516006	LPL 9 (1814); LPL 10 (c1822)	Alexander Raby's Box Colliery winding pit.
98	SN516006	LPL 9 (1814); LPL 10 (c1822); LPL 387 (c1856)	Alexander Raby's Box Colliery pumping pit. Later General Warde's pumping pit.
99	SN516006	LPL 10 (c1822)	Box Colliery pit (probably sunk by General Warde between 1819 and 1823).
100	SN517007	LPL 10 (c1822); LPL 386 (c1854)	General Warde's Box Colliery winding pit. Called "Fiery Vein Pit" on LPL 386.
101 to 106	SN516008 /9 SN517008	LPL 10 (c1822)	Un-named pits.
107	SN519008	Mansel Lewis London Collection 63 (1820)	Un-named "coal-pit".
108	SN519008	LPL 10 (c1822)	Un-named pit.
109	SN520008	Mansel Lewis London Collection 63 (1820)	Un-named "coal pit".
110	SN522008	,,	Un-named pit (at Penyvigwm).

Entry No.	Grid Location	Authority for location	Description
111	SN524008	Mansel Lewis London Collection 63 (1820)	"Horseway" on the outcrop of the Swansea Four Feet Vein. Probably used to take horses into the Box Colliery workings.
112 to 116	SN523006 /7 SN524005	"	Un-named pits.
117	SN524005	"	"Old Level".
118	SN525003	Stepney Estate 17 (1828)	Un-named pit. Probably a winding pit to the Daven Colliery.
119	SN525003	LPL 1 (1751); MS 3.299 (CCL) (c1810); LPL 386 (c1854)	"Pit begun by John Smith and finished by Gen. Warde" superimposed on 1751 plan. "Davon Engine" on MS 3.299. "Daven Pit" on LPL 386.
120	SN526001	LPL 1 (1751)	Information superimposed at a later date: "pit on the crop of the Fiery Vein".
121	SS525999	LPL 10 (c1822)	Un-named pit (at northern end of Trostre Canal).
122	SS525999	LPL 10 (c1822); Cawdor (Vaughan) 8660 (1787)	Un-named pit on LPL 10 (at northern end of Trostre Canal). "Coal pit" on 1787 plan.
123	SN531000	AP R.11640 (1925) AP 9265 (1927)	Old Cefncaeau Slant. Possibly worked by General Warde before 1830.
124	SN529003	LPL 412 (1841); AP R11641	Llandafen pit (Colliery). Possibly sunk by General Warde before 1830.
125	SN528007	LPL 10 (c1822); LPL 386 (c1854)	Un-named pit.
126	SN530008	LPL 412 (1841); Geol. Survey 1:10560 plan Carm 58 NE (1907)	"Old Horseway to Llandafen" on 1907 plan. Workings, dated 1827, shown at this location on 1841 plan. Probably used to take horses into the workings in the Swansea Four Feet Vein at the Llandafen Colliery.
127 to 132	SN535025 SN534023 SN538023 /4	LPL 14 (c1825)	Un-named pits.
133	SN539022	"	Clyngwernen pit.
134	SN546027	"	Un-named pit.

Entry No.	Grid Location	Authority for location	Description
135	SN550023	LPL 14 (c1825)	Un-named pit.
136 to 140	SN551023 SN551022 SN552023 SN553022 SN554022	Cawdor (Vaughan) 8660 (1787)	"Coal pits" at the Allt. The 1787 plan contained the statement: "The Collieries on this Common belong to John Vaughan Esq."
141	SN555021	LPL 14 (c1825)	Un-named pit.
142	SN551020	AP 12823 (1885)	"Brynshaffre Colliery". Considered to be the original "Brynshaffrey Colliery", mentioned in Edward Martin's 1803 report on the Llangennech Collieries and may have been worked by Thomas Jones as early as the 1770s.
143	SN555017	LPL 14 (c1825)	Un-named pit (at Glanmwrwg). Considered to be the Glanmwrwg Colliery mentioned in Edward Martin's 1803 report on the Llangennech Collieries.
144	SN556017	MS 3.299 (CCL) (c1800); O.S. 1:10560 plan Carm 59 NW (1891)	"Glanmwrwg Colliery", with more than one pit or level shown at this location on MS 3.299. "Glanmwrwg Slant" on 1891 plan.
145 to 148	SN544019 SN541019	LPL 14 (c1825)	Un-named pits.
149	SN540018	AP R78 (1845)	Penprys pit. Possibly inherited or sunk by the Llangennech Coal Company before 1830.
150 to 151	SN539016	LPL 14 (c1825)	Un-named pits (locations approximate).
152 to 153	SN536013	LPL 10 (c1822); Penller'gaer B.29.21 (c1828)	"Gellygille Pits" on c1828 plan.
154	SN539013	Penller'gaer B.29.21 (c1828)	"St David Pit". In the process of being sunk by the Llangennech Coal Company in 1829.
155 to 156	SN539011 /12	LPL 10 (c1822); Penller'gaer B.29.21 (c1828)	"Bryn Pits" on c1828 plan. Probably worked by Martyn John Roberts and perhaps by others before him.
157	SN540010	Penller'gaer B.29.21 (c1828); AP 3564 (1896)	"Bryn Pit" on c1828 plan "Roberts Pit" on 1896 plan. Considered to have been sunk by Martyn John Roberts in the early 1820s.

Entry No.	Grid Location	Authority for location	Description
158	SN541008	LPL 10 (c1822); Penller'gaer B.29.21 (c1828)	Un-named pit.
159 to 160	SN541007	LPL 10 (c1822); Penller'gaer B.29.21 (c1828); Local Collection 402 (c1880)	Un-named pits on c1822 and c1828 plans. Entry 159 called the "Gellywhiad Pit" on Local Collection 402.
161 to 164	SN543004 /5 SN544004	LPL 10 (c1822)	Un-named pits.
165 to 167	SN558009 SN560009	LPL 14 (c1825); AP 12843 (1902)	Un-named pits on c1825 plan. "Plasissa Colliery" on 1902 plan.
168	SS535997	LPL 1 (1751)	Information superimposed after 1751 states: "Pit made by Chauncy Townsend 9½ fthms deep upon the Fiery Vein".
169	SS536997	LPL 348 (1878); AP 1395 (1882)	Llwynhendy Slant or Level. Considered to have been worked by General Warde and perhaps by Townsend and Smith before him.
170	SS537994	LPL 1 (1751)	Information superimposed after 1751 states: "Coal pit, 14 fthms". Possibly worked by Townsend and Smith.
171	SS537994	"	Information superimposed after 1751 states: "Water Engine Pit". Possibly worked by Townsend and Smith.
172	SS537994	LPL 1 (1751); LPL 10 (c1822)	Information superimposed after 1751 states: "Old Pit". Possibly worked by Townsend and Smith.
173 to 176	SS538997 SS539997 SS540998 SS541996	LPL 10 (c1822)	Un-named pits.
177	SS543995	LPL 1 (1751)	Information superimposed after 1751 states: "2nd Pit". Worked by John Smith and perhaps by Townsend before him.

Entry No.	Grid Location	Authority for location	Description
178	SS543994	LPL 11 (1821)	"Level" on 1821 plan. Worked by John Smith and perhaps by General Warde. Oral tradition at Llwynhendy has perpetuated the name "Cae Level" for this entry.
179	SS540992	LPL 10 (c1822)	Un-named pit.
180	SS544996	"	"
181	SS545995	LPL 3 (1772); AP 5245 (1908)	Coal raising pit to Chauncey Townsend's colliery at Llwynhendy on the 1772 plan. Upcast pit to the Genwen Colliery on the 1908 plan.
182	SS545995	"	"Llwynhendy Fire Engine Pit" on 1772 plan. Downcast and pumping pit to the Genwen Colliery on the 1908 plan.
183	SS545995	LPL 1 (1751); LPL 3 (1772)	Information superimposed on LPL 1 after 1751 states: "Townsend's Coal Pit". "27 fthm Pit" on LPL 3.
184 to 185	SS545994	LPL 3 (1772)	Un-named pits.
186 to 190	SS546994 SS547994	"	"Old Coal pits".
191	SS547994	LPL 3 (1772); Mansel Lewis 2565 (c1772)	"Old Coal pit" on LPL 3. "12 fthms" on Mansel Lewis 2565.
192	SS547994	LPL 3 (1772)	"Old Coal pit".
193 to 194	SS548993	Mansel Lewis 2565 (c1772); Mansel Lewis 2578 (c1813)	Pits forming part of Squire, Evans and Beynon's Water Level on Mansel Lewis 2565. Un-named pits on Mansel Lewis 2578.
195	SS548994	LPL 3 (1772); Mansel Lewis 2565 (c1772)	Un-named pit on LPL 3. A pit forming part of Squire, Evans and Beynon's Water Level on Mansel Lewis 2565.
196	SS548994	"	"Old Water Engine Pit" on LPL 3. A pit forming part of Squire, Evans and Beynon's Water Level on Mansel Lewis 2565.
197	SS548994	"	A pit forming part of Squire, Evans and Beynon's Water Level.

Entry No.	Grid Location	Authority for location	Description
198	SS547995	"	Un-named pit on LPL 3. "32 fthms" on Mansel Lewis 2565.
199	SS547995	"	"Dr Evans's Fire Engine Pit" on LPL 3. "34 fthms" on Mansel Lewis 2565. This is Squire, Evans and Beynon's Engine Fach pit and the site of the first steam engine used in the Llanelli Coalfield.
200	SS548995	"	A pit forming part of Squire, Evans and Beynon's Water Level.
201	SS548996	LPL 10 (c1822)	Un-named pit.
202	SS548996	LPL 3 (1772); Mansel Lewis 2565 (c1772)	The coal-raising pit de-watered by Squire, Evans and Beynon's Water Level.
203	SS548996	Mansel Lewis 2565 (c1772)	Un-named pit.
204	SS551996	LPL 3 (1772)	"Coal pit worked 80 yards to the Deep of the Level upon the Fiery Vein".
205	SS551995	"	Un-named pit.
206	SS551995	Mansel Lewis 2565 (c1772)	"
207	SS550995	LPL 3 (1772)	"
208	SS550994	Mansel Lewis 2578 (c1813); AP 1545 (1883)	Apple pit.
209	SS552994	LPL 3 (1772)	Ox pit.
210	SS551993	"	"21 fthm" pit.
211 to 212	SS550993	"	Un-named "Old pits".
213	SS551993	"	Un-named pit with the wording: "Struck the Old Stock from this pit".
214	SS552993	LPL 10 (c1822)	Un-named pit.
215	SS552992	"	"
216	SS553993	LPL 3 (1772)	"
217	SS553995	LPL 3 (1772); Mansel Lewis 2565 (c1772)	"Pit upon the Fiery Vein worked 120 yards to the Deep" on LPL 3. "12 fthms" on Mansel Lewis 2565.
218	SS555992	LPL 3 (1772)	Water engine pit from which Thomas Bowen worked "all the Coal" to the southern outcrop of the Swansea Five Feet Vein at Ffosfach.

Entry No.	Grid Location	Authority for location	Description
219	SS555993	Mansel Lewis 2578 (c1813)	Un-named pit.
220	SS556995	LPL 3 (1772); Mansel Lewis 2565 (c1772)	Un-named pit on LPL 3. "14 fthms" on Mansel Lewis 2565.
221	SS558995	LPL 3 (1772)	"Old pit".
222	SS558995	"	"Old Water Engine pit".
223	SS559995	Mansel Lewis 2565 (c1772)	Un-named pit.
224 to 227	SS558993 SS557992	LPL 3 (1772)	"Sundry Trial Pits upon the Crop" also "Pits sunk by Mr Bowen in order to make a new winning but was prevented by Water".
228 to 229	SS549985 SS548984	Local Collection 37 (1824); AP 8967 (1927)	"Carnarvon Colliery" pits on 1824 plan (also referred to as "Baccas Colliery").
230 to 231	SS547981 SS548980	LPL 10 (c1822)	Un-named pits.

INDEX

C

H

I

M

N

O

P

Q

R

T

U

V

SOUTH WALES
MINERS LIBRARY